Aristotle's *Physics*

Aristotle's study of the natural world plays a tremendously important part in his philosophical thought. He was very interested in the phenomena of motion, causation, place and time, and teleology, and his theoretical materials in this area are collected in his *Physics*, a treatise of eight books which has been very influential on later thinkers. This volume of new essays provides cutting-edge research on Aristotle's *Physics*, taking into account recent changes in the field of Aristotle in terms of its understanding of key concepts and preferred methodology. The contributions reassess the key concepts of the treatise (including nature, chance, teleology, art, and motion), reconstruct Aristotle's methods for the study of nature, and determine the boundaries of his natural philosophy. Because of the foundational nature of Aristotle's *Physics* itself, the volume will be a must-read for all scholars working on Aristotle.

MARISKA LEUNISSEN is Assistant Professor of Philosophy at the University of North Carolina, Chapel Hill. Her recent publications include *Explanation and Teleology in Aristotle's Science of Nature* (2010).

CAMBRIDGE CRITICAL GUIDES

Titles published in this series

Hegel's *Phenomenology of Spirit*
EDITED BY DEAN MOYAR AND MICHAEL QUANTE

Mill's *On Liberty*
EDITED BY C. L. TEN

Kant's *Idea for a Universal History with a Cosmopolitan Aim*
EDITED BY AMÉLIE OKSENBERG RORTY AND JAMES SCHMIDT

Kant's *Groundwork of the Metaphysics of Morals*
EDITED BY JENS TIMMERMANN

Kant's *Critique of Practical Reason*
EDITED BY ANDREWS REATH AND JENS TIMMERMANN

Wittgenstein's *Philosophical Investigations*
EDITED BY ARIF AHMED

Kierkegaard's *Concluding Unscientific Postscript*
EDITED BY RICK ANTHONY FURTAK

Plato's *Republic*
EDITED BY MARK L. MCPHERRAN

Plato's *Laws*
EDITED BY CHRISTOPHER BOBONICH

Spinoza's *Theological-Political Treatise*
EDITED BY YITZHAK Y. MELAMED AND MICHAEL A. ROSENTHAL

Aristotle's *Nicomachean Ethics*
EDITED BY JON MILLER

Kant's *Metaphysics of Morals*
EDITED BY LARA DENIS

Nietzsche's *On the Genealogy of Morality*
EDITED BY SIMON MAY

Kant's *Observations and Remarks*
EDITED BY RICHARD VELKLEY AND SUSAN SHELL

Augustine's *City of God*
EDITED BY JAMES WETZEL

Descartes' *Meditations*
EDITED BY KAREN DETLEFSEN

Kant's *Religion within the Boundaries of Mere Reason*
EDITED BY GORDON MICHALSON

Kant's *Lectures on Anthropology*
EDITED BY ALIX COHEN

Kierkegaard's *Fear and Trembling*
EDITED BY DANIEL CONWAY

Kant's *Lectures on Ethics*
EDITED BY LARA DENIS AND OLIVER SENSEN

Aristotle's *Physics*
EDITED BY MARISKA LEUNISSEN

Aristotle's *Politics*
EDITED BY THORNTON LOCKWOOD AND THANASSIS SAMARAS

ARISTOTLE'S
Physics
A Critical Guide

EDITED BY

MARISKA LEUNISSEN

CAMBRIDGE
UNIVERSITY PRESS

CAMBRIDGE
UNIVERSITY PRESS

University Printing House, Cambridge CB2 8BS, United Kingdom

Cambridge University Press is part of the University of Cambridge.

It furthers the University's mission by disseminating knowledge in the pursuit of education, learning and research at the highest international levels of excellence.

www.cambridge.org
Information on this title: www.cambridge.org/9781107031463

First published 2015

A catalogue record for this publication is available from the British Library

Library of Congress Cataloguing in Publication data
Aristotle's physics : a critical guide / edited by Mariska Leunissen.
pages cm. – (Cambridge critical guides)
ISBN 978-1-107-03146-3
Includes bibliographical references.
1. Aristotle. Physics. 2. Physics – Early works to 1800. 3. Philosophy, Ancient. 4. Science – Philosophy. I. Leunissen, Mariska, 1979– editor.
Q151.A8A76 2015
530–dc23
2015014691

978-1-107-03146-3 Hardback

Contents

Contributors

JAMES ALLEN is Professor of Philosophy at the University of Toronto, Canada. His principal interests are in ancient Greek and Roman philosophy. He is the author of articles concerning ancient conceptions of expertise, ancient scepticism, ancient medicine, Aristotelian logic, Epicureanism, Stoicism, and Cicero, as well as of the monograph *Inference from Signs: Ancient Debates about the Nature of Evidence* (2001).

ROBERT BOLTON is Professor of Philosophy at Rutgers University, USA. He is the author of *Science, dialectique et éthique chez Aristote* (2010), and author or editor of numerous books and articles on Aristotle's methodology, epistemology, and psychology, and on other aspects of ancient philosophy. He is a former Rhodes Scholar and Fellow of the Centre National de la Recherche Scientifique, Paris.

DAVID CHARLES is Professor of Philosophy at Yale University, USA. He is the author of *Aristotle's Philosophy of Action* (1984) and *Aristotle on Meaning and Essence* (2000), and editor of *Definition in Greek Philosophy* (2010). He has also written many articles on Aristotle's discussion of action and teleology.

URSULA COOPE is Professor of Ancient Philosophy and Fellow of Corpus Christi College, Oxford University, UK. She is the author of *Time for Aristotle:* Physics IV.*11–14* (2005), as well as of articles on Aristotle's physics and philosophy of action.

ANDREA FALCON is Associate Professor in the Department of Philosophy at Concordia University, Montreal, Canada. He is a specialist on Aristotle and the Aristotelian tradition in antiquity. He was a member of the Institute for Advanced Study in Princeton in 2008. He is the author of *Corpi e movimenti: Il* De caelo *di Aristotele e la sua tradizione nel mondo antico* (2001), *Aristotle and the Science of Nature: Unity without Uniformity* (2005), and *Aristotelianism in the First Century* BCE: *Xenarchus of Seleucia* (2011).

DEVIN HENRY is an Associate Professor of Philosophy at the University of Western Ontario, Canada. He received his PhD from King's College London, where he wrote a dissertation on the metaphysical foundations of Aristotle's account of biological generation. He is the author of several articles on topics in Aristotle's philosophy of science (including classification, teleology, inheritance, and sexism), as well as Plato's late epistemology.

SEAN KELSEY is Associate Professor of Philosophy at the University of Notre Dame, USA. He is a specialist on Aristotle's natural science and metaphysics, and his papers have appeared in *Ancient Philosophy, Phronesis, Pacific Philosophical Quarterly,* and *Oxford Studies in Ancient Philosophy.*

JAMES G. LENNOX is Professor of History and Philosophy of Science at the University of Pittsburgh, USA. He is the author of a translation, with introduction and commentary, of *Aristotle:* On the Parts of Animals (2001) and of *Aristotle's Philosophy of Biology: Studies in the Origins of Life Science* (2001), as well as of several edited volumes and numerous articles on Aristotle's science and biology. He has held fellowships at the Center for Hellenic Studies (1983–4); Clare Hall, University of Cambridge (1986–7); and the Istituto di Studi Avanzati, University of Bologna (2006).

MARISKA LEUNISSEN is Associate Professor of Philosophy at the University of North Carolina at Chapel Hill, USA. She is the author of *Explanation and Teleology in Aristotle's Science of Nature* (2010), co-editor of *Interpreting Aristotle's* Posterior Analytics *in Late Antiquity and Beyond* (2010), and has written several articles on Aristotle's theory of demonstration, on topics in his natural science, and on the relation between natural character and virtue. She has held fellowships at the Center for Hellenic Studies (2010) and at Leiden University (2013).

DIANA QUARANTOTTO is Professore Aggregato at the University of Rome Sapienza, Italy. She is the author of *Causa finale, sostanza, essenza in Aristotele* (2005), as well as of articles on Aristotle's science of nature, biology, metaphysics, philosophy of language and ethics.

JACOB ROSEN is Assistant Professor of Philosophy at Harvard University. Previously he was the Academic Coordinator in the Graduate Program in Philosophy and Researcher in the TOPOI Excellence cluster 'Place, Space, and Motion' at the Humboldt University, Berlin, Germany. His papers on Aristotle's logic and physics have appeared in *Phronesis, Oxford Studies in Ancient Philosophy*, and *Mind.*

MARGARET SCHARLE is Associate Professor of Philosophy at Reed College, USA. She works in ancient philosophy, with special interest in Aristotelian natural philosophy. Her papers have appeared in *Oxford Studies in Ancient Philosophy, Phronesis*, and *Apeiron*, as well as in edited volumes.

STASINOS STAVRIANEAS is Lecturer at the University of Patras, Greece. He has published papers on Aristotle's natural science and biology, and a modern Greek translation and commentary of Aristotle's *Parts of Animals*.

CHARLOTTE WITT is Professor of Philosophy and Humanities at the University of New Hampshire, USA. She is the author of two books on Aristotle's metaphysics, *Ways of Being: Potentiality and Actuality in Aristotle's Metaphysics* (2003) and *Substance and Essence in Aristotle* (1989). She has also written many articles on Aristotle's philosophy, and on feminist theory and feminist history of philosophy.

Introduction

Mariska Leunissen

Throughout his life, Aristotle was deeply committed to the study of natural phenomena: About one-third of the surviving *Corpus Aristotelicum* investigates and explains the motions and attributes of things that have a nature, that is, of things that have an internal principle of change and rest. The *Physics* – an intellectual masterpiece in itself and one of the most widely read Aristotelian treatises – forms Aristotle's most fundamental treatise in his studies of natural philosophy.

In this treatise, Aristotle investigates the principles and causes of all natural things in general, and, in the course of doing so, defines a large number of key concepts of his natural philosophy, such as motion and change, space and time, matter and form, causal explanation, luck and spontaneity, teleology, and necessity. In addition, Aristotle specifies in the *Physics* the methodological guidelines for how one should study natural entities and their properties if one wants to gain scientific knowledge of them, which includes the famous – but still ill-understood – recommendation to start from things that are "more known and clearer to us" and to work from there to what is "more known and clearer by nature." In this way, the *Physics* lays out Aristotle's conceptual apparatus and methodological framework for *all* of his natural philosophy, including his psychology, biology, and other inquiries into the more specific and more complex segments of the natural world preserved in Aristotle's remaining natural treatises.

The *Physics* is relevant not just for Aristotle's natural philosophy, however. For, since the objects of metaphysics do not, for the most part, exist independently of the objects of physics and can thus only be studied through those, the science of metaphysics often has to rely on the same concepts, definitions, and approaches as are presented in the *Physics*. And the same might be said for certain aspects of Aristotle's political science: For instance, the very notion of the perfection of human nature builds on Aristotle's on natural teleology and involves a type of change ("perfection") that is introduced in the *Physics*. In this way, the *Physics* forms the conceptual entry-way into much of the Aristotelian Corpus.

Despite what its ancient title might suggest, Aristotle's "Lecture concerning Nature" (Φυσικὴ ἀκρόασις) barely counts as a unified treatise. The eight books naturally divide into five separate sections or investigations: Book I identifies the principles of change at a very general level; Books II–IV inquire into nature and natural motion and its preconditions more properly; Books V–VI provide a technical discussion of how motion works; and Book VIII establishes rather independently of the rest of the *Physics* the existence of an unmoved mover. The place of Book VII in the treatise is disputed: Parts of the book – dealing mostly with the relation between the mover and what is moved – are handed over in two quite different versions, and when Eudemus produced his paraphrase of Aristotle's *Physics* he did not include it, suggesting that it was either not part of the edition of the *Physics* he possessed or that he did not consider it part of the same investigation.[1]

The current volume is not overly concerned with the history, unity, or structure of the edition of the *Physics* as we have it today, nor does it try to provide a comprehensive treatment of the rich materials presented in it. The existing scholarship on Aristotle's *Physics* is wide-ranging and voluminous, and it would be impossible to engage with it all. Instead, as is the case with the other volumes in the *Cambridge Critical Guide* series, the fourteen chapters collected in this volume all attempt to make optimal use of the recent changes in the field of Aristotle studies – changes both in terms of its understanding of key concepts in Aristotle's philosophy and in terms of its preferred methods for gaining such understanding – and thereby to push forward the scholarship on Aristotle's *Physics*. Each of the chapters engages with these changed perspectives on Aristotle in at least one of the following three ways:

(1) **Reassessing the key concepts of Aristotle's natural philosophy.** First, most of the chapters provide a challenge to existing interpretations of some of the key concepts of Aristotle's *Physics* and argue for alternative understandings: The notions of nature, chance, teleology, and art are discussed in Chapters 2 to 7, while traditional understandings of the notion of *kinêsis* (translated as change, process, or motion) are revised in Chapters 8 to 13. These chapters not only draw from the latest research in the field, but also exhibit a greater sensitivity towards the richness and

[1] The best discussion of the structure, unity, and transmission history of Aristotle's *Physics* is Brunschwig 1991.

complexity of Aristotelian concepts, as well as towards the extent to which Aristotle builds on and reshapes his concepts in different explanatory contexts. Instead of just studying the main concepts of Aristotle's natural philosophy in the immediate context in which they are first introduced in the *Physics* and trying to provide a unified account of their definitions and roles, the chapters presented in this volume also pay attention to some of the (methodologically) later uses to which the concepts are put in other (natural) treatises, provide explanations of why these other uses require conceptual changes, and answer the meta-question about why Aristotle needs the specific understandings of, for instance, his concepts of nature and cause for his natural philosophy as a whole.

(2) **Reconstructing Aristotle's methods for the study of nature.** Second, several of the chapters provide a reconstruction of the methods Aristotle uses and/or describes for the study of nature, and do so either directly, as in Chapter 1, or as part of a reassessment of one of Aristotle's key notions in his *Physics*, as for instance in Chapters 3, 8, 9, and 14. Scholars working on Aristotle have become increasingly interested in the relationship between Aristotle's "geometric-style" theory of scientific demonstration and investigation as presented in the *Posterior Analytics* and his practice in the natural treatises, and although much work has been done on the methodological connections between the *Posterior Analytics* and Aristotle's biology (i.e. his study of living nature), the connections with his *Physics* remain largely unexplored territory. The chapters in this volume aim to work towards closing this gap in the existing scholarship by offering interpretations of (a) what it means according to Aristotle to investigate things *physikôs* – i.e., in the manner of a natural scientist; (b) how this method relates to other methods available to a philosopher (such as conceptual analysis or dialectic), as well as to the scientific method outlined in the *Posterior Analytics*; and (c) how methodological concerns stemming from the *Posterior Analytics* drive the investigations in the *Physics*.

(3) **Determining the boundaries of Aristotle's natural philosophy.** Third, some chapters – most notably Chapter 14, but also Chapters 1, 9, 11, 12, and 13 – concern themselves with the boundaries (and the extent to which these boundaries are crossed) between Aristotle's natural philosophy and his metaphysics or even his ethics. In recent years, there has been a growing awareness among scholars of Aristotle that Aristotle's full views can rarely be plucked out of single passages, or even single treatises. This has led to modifications in how we believe one should conduct conceptual analysis in Aristotle (e.g. as taking into account the different uses to which concepts are put in different parts of a science, as described under item

number one above), but it has also highlighted problems concerning Aristotle's division of the sciences and his requirement that principles that belong properly to one science cannot be used in the generation of knowledge in another science (unless the two sciences are subordinated to each other). Several of the chapters in this volume examine the ways in which Aristotle demarcates the science of nature and sets it apart from other theoretical sciences, as well as how his accounts in the *Physics* relate to or are fundamental for Aristotelian views that belong properly to other sciences, such as his metaphysics, but also his ethics.

The opening chapter by James Lennox addresses the three methodological questions concerning the science of nature head on, while also providing a reconsideration of several of Aristotle's key notions in the *Physics*. Lennox argues that even though the order of and method for the investigation of nature Aristotle presents in the *Physics* reflect in important ways the recommendations for scientific investigation he had already laid out in the *Posterior Analytics*, ultimately Aristotle believes that every scientific domain is governed by norms for inquiry that are quite specific to that domain. This means that an important task of Aristotle in the *Physics* is to specify how *natural* entities – which undergo change and are always enmattered – ought to be studied, and thereby to establish the norms for scientific inquiry at a general level for the whole domain of natural science. These norms will then have to be specified even further in the other natural treatises which deal with particular kinds of natures. In his chapter, Lennox identifies these "local" norms for natural science and also shows how Aristotle's concern for establishing these norms for natural inquiry drives much of the conceptual analysis that can be found in Books II and III regarding especially the notions of change, nature, and motion.

Chapters 2 to 7 offer reassessments of some of the most fundamental notions Aristotle introduces in *Physics* II, such as nature, chance, teleology, and art, and often do so by drawing from other Aristotelian texts or from the pre-Socratic and Platonic traditions.

Sean Kelsey, in Chapter 2, turns to Aristotle's definition of nature in *Physics* II.1 with the idea of tracing its role in the remainder of Book II, where explicit appeals to the definition are surprisingly scant. According to Kelsey, the definition ultimately functions as a kind of instruction for how to interpret the phenomena of nature. Aristotle defines nature as a principle of motion and rest; taken together with the idea that nature is above all form, this implies that *form* is a principle of motion. Kelsey argues that this is a result that Aristotle expects us to find difficult: How can form, which is immobile, be a principle of movement? This difficulty, Kelsey suggests, sets a tacit agenda for the rest of *Physics* II, and the key to unraveling it lies in realizing that the *way* form is a "principle" (*archê*) is by being an "end" (*telos*). Put another way, the suggestion is that the definition of nature, together with the identification of nature with form, instructs us (in effect) to put a certain interpretation on natural phenomena; it tells us to read the "behaviour" of things due to nature – their characteristic ways of moving and staying at rest – as for the sake of perpetuating their distinctive forms of being. The moral is that if nature as form stands at the beginning of natural phenomena, it must do so by being its end.

Chapter 3 also engages with Aristotle's definition of nature in *Physics* II.1, but, where Kelsey focused on its role, Stasinos Stavrianeas provides a reassessment of the content and meaning of this definition. However, rather than arguing that the definition yields just one criterion for judging what counts as a nature for Aristotle (such as the standard criterion that is offered in the scholarship on this topic, namely the possession of an internal principle of motion and rest), Stavrianeas believes that the definition provided in the *Physics* is left intentionally vague and general (in much the same way as Aristotle leaves his definition of soul vague and general in *On the Soul*), thereby allowing for further specifications and filling-in within the individual natural treatises. By treating Aristotle's definition of nature in this way, Stavrianeas is able to handle difficult cases in other natural treatises, such as the motion of the elements, and even hypothetical, miraculous *automata*. The aim of *Physics* II.1, then, is not to identify the one thing that characterizes all natural things, but rather to demarcate the realm – and thereby the science – of nature at the most general level possible.

In Chapter 4, James Allen reinterprets Aristotle's notions of luck and spontaneity (which are both a type of chance) in *Physics* II.4–6 by drawing not just on these chapters, but also on Aristotle's treatment of these notions in the ethical treatises. Allen shows that chance for Aristotle is not an alternative to teleological explanation, which he is determined to reject, as most standard readings hold, but rather that chance is an inevitable byproduct of final causation and unintelligible apart from it. Accordingly, the account of chance in *Physics* II.4–6 should be read as a complement to Aristotle's defense of natural teleology in *Physics* II.8. Thus, under the interpretation of Aristotle's own account of chance as a "cause by accident" as Allen defends it, chance events are a marginal exception to the teleological rule that prevails in nature, and thereby in fact constitute themselves proofs of the existence of natural teleology.

Margaret Scharle, in Chapter 5, offers a new interpretation of Aristotle's rainfall-example in *Physics* II.8. By building on Aristotle's account of the teleology of the elements in especially the *Meteorology, On Generation and Corruption*, and *On the Heavens*, Scharle shows that Aristotle understands winter rain as a natural phenomenon that is teleological in its own right, and that this teleological view of winter rain is indeed required by the dialectic Aristotle engages in with the Empedoclean opponent. As a result of this, Scharle is also able to offer a unified interpretation of the dialectic in Book II of the *Physics* as a whole: The underlying agenda that drives Aristotle's discussions in this book is his attempt to carefully carve out his own position that the natural world owes its order and regularity to teleology, and he does this against the pre-Socratic tradition that fails to properly distinguish nature from chance and against the Platonic tradition that fails to properly distinguish nature from art.

In Chapter 6, Charlotte Witt continues the focus on Aristotle's defense of natural teleology in *Physics* II.8, this time by turning to his analogy between art and nature. Against standard readings which have downplayed the importance of Aristotle's analogy between art and nature for his argument in favor of natural teleology, and which have assumed a radical ontological difference between the two (which sometimes lead them to conclude that artifacts are not substances), Witt argues that artifacts have intrinsic ends and proper functions just like natural beings (and are thus substances), and that this similarity is of crucial importance to Aristotle's appeal to the craft analogy in order to argue for natural teleology. Moreover, she shows that Aristotle is very careful to bracket off the question of the origin of change causing a creation or generation whenever he appeals to the analogy between art and nature as part of his argument for natural teleology. She thereby counters the often-heard objection that Aristotle's analogy between art and nature is mistaken or misleading, and re-establishes its importance for Aristotle's defense of natural teleology in the *Physics*.

While Chapters 5 and 6 focused on Aristotle's defense of natural teleology, Robert Bolton in Chapter 7 offers an account of its origins, and thereby corrects some longtime misunderstandings of the nature of his theory. Crucial to his innovative account is that we should appreciate the differences between Aristotle and his pre-Socratic predecessors such as Empedocles, as well as the differences between him and Plato, especially as these differences are emphasized by Aristotle himself. Aristotelian final causality, as Bolton understands it, is not – as in Plato – reducible to any other kind of causality such as efficient causality, but is instead based on the notions of fitness and natural regularity. And in its turn, this notion of fitness is not something that holds of the natural world by chance – as Empedocles thinks – but rather something that holds of it *by nature*. In this way, Bolton introduces a thorough reassessment of Aristotle's natural teleology.

Chapters 8 to 13 offer reinterpretations of Aristotle's notion of *kinêsis* – variously translated as change, process, or motion – while also exploring issues related to Aristotle's natural scientific method broadly construed and offering suggestions about the demarcation of (the science of) nature.

In Chapter 8, Devin Henry provides a critique of the traditional view that according to *Physics* I.5–7 Aristotle holds that every change – including substantial change – requires a persistent subject of change. Instead, Henry argues that, even though the evidence in *Physics* I.7 is silent on the matter of substantial change, evidence from other natural treatises (most prominently from *On Generation and Corruption* and *Generation of Animals*) suggests that Aristotle only believes that substantial change requires a subject from which the change proceeds, not one that persists as a constituent of the substance that comes into being. Henry thereby reshapes our understanding of change, while also showing the importance of reading the *Physics* in the context of Aristotle's broader natural scientific views.

In Chapter 9, Diana Quarantotto also analyzes Aristotle's concept of change and examines the relationship between change and substantial being (*ousia*). She observes that Aristotle's treatment of this issue exhibits a development from what is more familiar to us to what is "closer to nature" and claims that the result of this inquiry is a major innovation by Aristotle within the Greek tradition of natural philosophy. She identifies this development as mostly taking place within the *Physics*, especially in Books I–III and VIII. In Books I–III, where Aristotle presents the view that is more familiar to us, there is a clear-cut distinction between substantial being and change. However, in Book VIII (as well as in passages from other natural treatises), Aristotle presents views that are "closer to nature," and this is where Aristotle presents his innovative idea that being is itself dynamic. Critical to this inquiry is the relationship between the definition of change given in III.1–3 and the treatment of eternal change in Book VIII.

David Charles, in Chapter 10, analyzes three aspects of Aristotle's account of processes (*kinêseis*): namely, (1) Aristotle's definition of process in *Physics* III.1–3 and what exactly this definition entails when read in the broader context of the *Physics* and *Metaphysics Theta*; (2) Aristotle's account of the individuation of processes in *Physics* III.3; and (3) the nature and role of the concept of actuality in Aristotle's definition of process, especially as it is used in *Metaphysics Theta*. One major upshot of Charles' account is that it is a mistake to think of *kinêseis* as events rather than processes; if he is right, this has major repercussions for our understanding of Aristotle's notions of action, time, and causation.

In the next chapter, Chapter 11, Jacob Rosen explores the relationship between Aristotle's account of the basic structure of motions and other continuants in *Physics* v–vi and his arguments in *Physics* viii.8 that there can be no eternal motion along a straight line, and that therefore eternal motion (which he thinks must exist) must be circular. Rather than trying to read these various accounts as presenting one unified theory of motion, Rosen zooms in on the tensions between these accounts, especially regarding Aristotle's thesis about continua as formulated in *Physics* viii.8, according to which continua do not possess any actually existing proper parts or middle-points, the defense of which conflicts with an important theorem presented in *Physics* vi.5. From this he concludes that Books v, vi, and viii – even though not completely unrelated – are, most likely, not part of one single lecture that was composed in one sitting, but also, and more importantly, that Aristotle's physics and cosmology face some difficulties establishing the priority of circular motion over rectilinear motion if some of his strongest arguments from *Physics* viii.8 in favor of this thesis are already pre-emptively defeated by claims made in *Physics* v–vi.

In my own chapter, Chapter 12, I turn to Aristotle's account in *Physics* vii.3 of the changes one undergoes when acquiring virtues of character. In this chapter, Aristotle argues, somewhat surprisingly, that "conditions" – such as the virtues of the body or soul – do not belong to the category of "quality" and do not come to be as a result of qualitative change, but that they are rather "some kind of perfections" that "exist in virtue of a particular relation," thereby seemingly introducing a fifth type of change in addition to the traditional four (substantial change, qualitative change, quantitative change, and locomotion). This chapter analyzes Aristotle's account of the type of change involved in the acquisition of virtues in the *Physics* and shows how it offers a physiological and naturalistic explanation for his account of habituation or "the perfection of human nature" in the *Nicomachean Ethics*. In a way, the *Physics* thus provides a physical grounding for Aristotle's political science, as only in the *Physics* do we learn what kind of change is involved in moral development.

In Chapter 13, Ursula Coope turns to Aristotle's account of self-motion in *Physics* viii.5. In this chapter, Aristotle characterizes self-movers as involving two components – a part that produces the movement while being itself unmoved, and a part that is moved. Although Coope also offers an analysis of Aristotle's notion of self-motion in her chapter – in particular of how it forms a response to Plato's suggestion that all motion must ultimately be grounded in self-motion – and solves some problems that seem particular to Aristotle's alternative, she is mostly interested in determining the role of Aristotle's appeal to self-motion in the overall project of establishing the existence of an unmoved mover (an appeal that seems surprising, given Aristotle's arguments in *Metaphysics L*, which establish its existence without such an appeal to self-motion). According to Coope, Aristotle's overall aim is to show how motion must ultimately be caused by something that is not itself moving. She argues that his account of self-movers in Physics VIII.5 gives necessary but not sufficient conditions for being a self-mover properly speaking. As such, it is able to apply both to genuine self-movers, such as animals, and also to the conjunction of the first unmoved mover together with the thing it causes to be in eternal motion. Providing a single account that covers both these cases helps Aristotle to show how physics, in so far as it is the study of motion, is a single unified science.

For this volume, the issue concerning the relation between the different sciences arises most prominently with regard to Aristotle's concept of the unmoved mover – the origin of all motion in the universe – which he introduces in *Physics* VIII. Since this unmoved mover is a divine being that is pure form and is without matter, it is technically speaking not part of the physical world, but rather belongs to Aristotle's "first philosophy" or metaphysics. Aristotle's treatment of the unmoved mover in the argument of *Physics* VIII and the questions it raises for the boundaries between Aristotle's physics and metaphysics are the topics of the fourteenth and final chapter, by Andrea Falcon. Falcon argues that, against traditional readings of *Physics* VIII, Aristotle's treatment of the unmoved mover there does not, and need not, go beyond the boundaries of natural philosophy, but that instead he offers a single extended natural scientific argument concerned with eternal motion. This argument proceeds in two stages that follow the two stages of inquiry as presented in the *Posterior Analytics*: Aristotle first sets out to determine *whether* there is eternal motion and then tries to identify *what* eternal motion is. The unmoved mover comes into play as the efficient causal factor that ultimately needs to be picked out by the definition of eternal motion. In this way, Falcon's chapter also contributes to our understanding of Aristotle's scientific methods in the *Physics*.

CHAPTER I

How to study natural bodies: Aristotle's μέθοδος

James G. Lennox

Introduction

If, as the opening sentences of the *Physics* strongly suggest, the basic framework for inquiry in the pursuit of scientific knowledge presented in *Posterior Analytics* II is shaping Aristotle's views about *natural* inquiry, then there are a number of features one would expect to find. In light of *Posterior Analytics* II.1–2, there ought to be two intimately related *goals* of natural inquiry: knowledge of *what* natural beings are, ideally to be formulated in definitions; and knowledge of causes used to demonstrate *why* natural beings have the necessary but non-essential attributes that they do; *and* fact-establishing stages *on the way* to those goals.[1] The intimate relationship between these two lines of inquiry stems from Aristotle's conviction that the essences of things are, in various ways, causally responsible for their non-essential features – if one has scientific knowledge of *what* something is, one will thereby be able to explain *why* it has the other non-accidental features it has.

Nevertheless, the conviction motivating this chapter is that, notwithstanding the generality of this framework, Aristotle sees different subject matters or domains as governed by norms that are quite specific to them.[2] This specificity derives from (at least) three sources: (1) differences in the objects being investigated; (2) differences in our epistemic access to those objects; and (3) differences in the perspective we take on those objects.[3]

This chapter has benefitted greatly from comments on an earlier draft by Marko Malink, Tom Ainsworth, and our editor, Mariska Leunissen; and from discussions with Christopher Shields, David Charles, Michael Peramatzis, and Alan Code about this chapter's concerns. I am reasonably confident none of them will fully agree with the conclusions I have reached.

[1] See *Posterior Analytics* II.1–2. The view is, of course, much more complicated than this brief summation suggests. For detailed discussion of the complications, see Charles 2000, 2010b; Lennox 2004.

[2] This is the primary thrust of my forthcoming *Seeking and Knowing: Aristotle on Norms of Inquiry*.

[3] For concrete examples of how each of these sources might affect the way in which inquiry is carried out, consider Aristotle's remarks in *Parts of Animals* I.1 on the differences between studying natural

Thus, while the *Posterior Analytics* provides what I will call an *erotetic framework* for scientific inquiry, it does not provide the domain-specific norms that are needed to guide actual research. This chapter aims to articulate those domain-specific norms that Aristotle thinks guide *natural* inquiry. A key concept in Aristotle's discussion of this issue is μέθοδος. Tellingly, it is a concept completely absent from the *Posterior Analytics*, and yet extremely prominent whenever Aristotle is discussing how to proceed successfully in a specific inquiry – including, as we will see, many of the opening sentences of his treatises in various fields of inquiry. It is with this concept that I begin.

I A discourse on μέθοδος

Physics III opens by noting that there has been a serious lacuna in the discussion up to that point:

> Since the nature [of a natural thing] is a source of motion and change, and our μέθοδος is concerned with nature, [the question] what is motion must not escape our notice; for necessarily when we are ignorant of this we are also ignorant of nature. (*Phys.* III.1, 200b12–15)

Our μέθοδος is concerned with nature – but what is a μέθοδος, and what is distinctive about one concerned with nature? I begin by attending to the concept of μέθοδος in Aristotle generally, and return to the question of what is distinctive about a natural μέθοδος in the next section.

Based on how often Aristotle uses this concept in the introductions to his treatises,[4] this is a concept worthy of more attention than it has received.

The opening lines of the *Physics* can be used to illustrate the problems people have in deciding how to translate μέθοδος. The relevant Greek phrase at 184a10 is περὶ πάσας τὰς μεθόδους, which Charlton translates "in all disciplines," Pellegrin "toutes les recherches," Waterfield "in any

objects that undergo generation and perishing, and those that do not for (1); for (2), consider Aristotle's remarks about how limited our access is to phenomena about the objects in the heavens compared with "the animals and plants around us" in *Parts of Animals* I.5; for (3), notice that Aristotle's study of the rainbow in *Meteorology* III involves two different perspectives – that of natural science and that of mathematical optics. Different norms are engaged for these different perspectives on the same natural phenomenon.

[4] E.g. *Phys.* I.1, 184a10–11, *PA* I.1, 639a1–2, *NE* I.1, 1094a1–3, *Meteor.* I.1, 338a25–6, *Top.* I.1, 100a18. One could argue that this opening passage of *Phys.* III should be included, on grounds that the lists of Aristotle's works in Diogenes Laertius and Hesychius include two references to a book entitled Περὶ κινήσεως (V. 23). For a judicious discussion of these references and what they might imply about the structure of our *Physics*, see Ross 1936: 2–7 and, more recently, Menn (in press).

subject," Hardie and Gaye "in any department," and Irwin and Fine "in every line of inquiry." There appears to be general agreement that the term refers to some sort of human endeavor, but there agreement ends.

To make matters more difficult, Aristotle also deploys the concept in a different (though, I will argue, closely related) way, to refer to a *method*, a *way of proceeding*. Its use in the first chapter of *On the Soul* provides a rich illustration of this usage.

Its opening lines tell us that inquiry into the soul (τὴν περὶ τῆς ψυχῆς ἱστορίαν) will contribute greatly to all truth, but *especially truth related to nature*. But we are warned of the extreme difficulty of this investigation, because we are inquiring not merely into the soul, but into its "substantial being and essence," which raises questions about how to proceed (402a10–14). Is the way forward, Aristotle asks, "demonstration, division or some other method" (τις ἄλλη μέθοδος)? The μέθοδοι on offer bring to mind the dialectical development of *Posterior Analytics* II.1–10,[5] which explores the relationship between definition and demonstration. Given that we suppose we have scientific knowledge of something when we have a demonstration of it, he considers whether we can come to know definitions, i.e. statements that identify what something is, through *demonstration*. This possibility is discussed, and apparently rejected, in chapters 3 and 7. The possibility that we can come to know definitions through *division* is considered and rejected in *Posterior Analytics* II.5.[6] Chapters 8–10 articulate another option, which he summarizes by relating three kinds of definition to demonstration: one states the *conclusion* of a demonstration (and thus can be known through demonstration), one can be *reconfigured* as a demonstration, and one is the *starting-point* of a demonstration.[7]

Here, then, μέθοδος refers to different possible *ways of pursuing* a specific goal of inquiry, in this case knowledge of what the soul is; the transliteration "method" seems to be a perfectly reasonable translation.

The term is deployed in the opening lines of *Parts of Animals* in a manner similar to the opening of the *Physics*.

> Regarding every study [*theoria*] and every *methodos*, the more humble and more valuable alike, there appear to be two sorts of state, one of which may

[5] The background is Academic: Dialogues such as the *Phaedrus, Sophist*, and *Statesman* view division as the primary tool for inquiry into being and essence.

[6] Division can be helpful in *hunting* for the essence: see *Posterior Analytics* II.13.

[7] There is controversy about whether *Posterior Analytics* II.10 discusses three or four sorts of definition: see Barnes 1993: 222–223; Bolton 1976; Charles 2000: 23–56; and Ross 1949: 634–636.

> properly be called scientific knowledge of the subject matter, the other a certain sort of educatedness. (*PA* I.1, 639a1–5)[8]

There are two cognitive states associated with every *methodos*, and *Parts of Animals* I is a guide to acquiring the *second*, a special form of *paideia*, which provides one with the ability to make critical judgments about what is well or poorly expressed.[9] Aristotle goes on to distinguish one who has a very general form of this skill, and one who has it about a specific discipline (say, the art of medicine), and transitions to the business at hand with the following words: "So it is clear, for natural *inquiry* too (καὶ τῆς περὶ φύσιν ἱστορίας), that there is need of some such standards (ὅροι)" (639a12–13). The ability to make such judgments requires certain standards, and if one's *paideia* is about a specific field, then it will be standards appropriate to that field that one needs to acquire. This will be the topic of discussion for the remainder of *Parts of Animals* I.

It is notable that θεωρίαν and μέθοδον are conjoined not simply by "and" (καὶ), but by "both and" (τε καὶ), which makes it quite clear that these terms are conveying *different* ideas, and are not merely synonyms. It could be that Aristotle has two different categories of cognitive endeavor in mind. However, there is an attractive alternative – that he is referring to two different *aspects* of a cognitive endeavor, aspects which are immediately identified as two different states associated with them – scientific knowledge (ἐπιστήμη), on the one hand, and that general critical judgment he identifies as a certain sort of *paideia*, on the other. Support for this reading comes from the end of chapter 4, where Aristotle summarizes what has been accomplished up to that point:

> We have said, then, how the *methodos* of nature (τὴν περὶ φύσεως μέθοδον) should be appraised, and in what way (καὶ τίνα τρόπον) the study (ἡ θεωρία) of these things might come about methodically and with greatest ease (ὁδῷ καί ῥᾷστα). (*PA* I.4, 644b15–17)

This summation echoes the use of μέθοδος and θεωρία in the opening lines, and here one can see that the two words emphasize different aspects of a single study: μέθοδος emphasizes the way in which the investigation is to be

[8] Cf. also *NE* I.1, 1094a1–3 and *Phys.* I.1, 184a10–11.

[9] Cf. *Pol.* III.6, 1282a1–8. Although medicine is the subject being considered, just as in the *Parts of Animals* I.1 passage, there is a contrast between those with ability to judge and those with knowledge of the subject matter.

carried out; θεωρία emphasizes the object towards which the investigation is directed.[10]

The other use of μέθοδος is on display in the concluding lines of *Parts of Animals* I, looking forward to Book II:

> Enough said about our manner of proceeding (περὶ μὲν τοῦ τρόπου τῆς μεθόδου); we must now attempt to discuss the causes concerning both the common and distinctive [parts], beginning first, as we have determined, from the primaries. (646a2–4)

The combination of τρόπος and μεθόδος here, which is fairly unusual, is difficult to capture in translation, since the idea of proceeding in a certain manner seems already *implicit* in μέθοδος itself.[11] The combination stresses, however, that what has been defended in *Parts of Animals* I is one among a number of possible ways of proceeding, one distinctive (in this case) to the study of animals. The discussion in *Parts of Animals* I, then, supports a hypothesis about the connection between the two uses of the term μέθοδος: A μέθοδος is an inquiry that is carried out according to a distinctive set of *norms or standards* that are appropriate to a specific domain of investigation. That is, to refer to an inquiry as a μέθοδος is to emphasize that the inquiry is characterized by a specific μέθοδος, i.e. by *specific norm-governed procedures that are appropriate for that subject matter.*

Note, however, that *Parts of Animals* I does not explicitly restrict its scope to zoology; the norms outlined are said to be appropriate to *natural* inquiry (639a12, 644b16).[12] Having thus explored the semantic landscape of the concept of μέθοδος in that discussion, let us now return to our original question – what sort of μέθοδος is the μέθοδος concerning nature?

[10] It is the use of ὁδῷ adverbially to characterize the way in which the θεωρία stresses the connection to which I am drawing attention. The noun ἡ ὁδός refers to a road, path, or track, and is used metaphorically about inquiry in much the same way as those English expressions are. The dative form used here often has adverbial force, conveying the idea of staying on the road to your destination, thus my "on course." Μέθοδος is formed from that noun and a prepositional prefix, μετά, which when used as a prefix carries the sense of "going after" and "in quest of." The basic idea, then, is a path taken in quest or in pursuit of something. It is already used in Plato's *Sophist* (218d, 235c, 243d) to refer to an inquiry, and *Republic* VII 533c refers to "the dialectical method" (ἡ διαλεκτικὴ μέθοδος) as the only way to advance to first principles.

[11] Indeed at *PA* I.4, 644b17–21 Aristotle uses τίνα τρόπον twice as a sort of substitute for μέθοδος in this second sense – perhaps because of the conceptual confusion he would have caused by using the term in both its senses in the same passage!

[12] For a discussion of why this might be so, see Lennox 2006.

II "Our *methodos* concerns nature"

The *Physics* opens, recall, in a manner strikingly similar to *Parts of Animals*. Both start with a general claim about *every* μέθοδος and then apply what is said generally to the same specific inquiry, the inquiry into nature:

> Since, concerning every *methodos* of which there are starting points, causes or elements, knowing, i.e. scientific knowing (τὸ εἰδέναι καὶ τὸ ἐπίστασθαι) comes about from knowing (γνωρίζειν) these things . . . it is clear that, concerning the scientific knowledge (ἐπιστήμη) of nature too we must first attempt to determine the facts about the starting-points. (184a10–16)

Aristotle then reminds us that we are here focused on the proper way to proceed:

> And the path is by nature (ἡ ὁδὸς . . . πέφυκε) from things more knowable (γνωριμωτέρον) and clear to us to the things that are by nature (φύσει) clearer and more knowable; for what is clear to us and what is clear without qualification are not the same. (184a16–18)

It is a path towards knowledge of the starting-points, causes, and elements regarding nature that Aristotle is characterizing – it is by achieving these that we acquire scientific knowledge of nature. And the natural way to proceed is to start with what is better known and clearer to us and proceed towards what is better known and clearer "by nature" or without qualification, that is, without epistemic relativization to the state of the knower. Up to this point, however, we have been told nothing specific about *natural* inquiry, since Aristotle considers this the proper and "natural" way to proceed in *any* discipline whatsoever.[13]

So what is the proper μέθοδος for the μέθοδος concerning nature? For Aristotle's answer to that question I will explore a variety of texts, all concerned with φυσική, but not all of them from the *Physics*. I begin with a discussion in *Metaphysics* E.1.

Metaphysics *E.1*

In *Metaphysics* E.1, Aristotle aims to clarify the special sort of wisdom that has been the object of his inquiry from the beginning by differentiating it from other kinds of inquiry. Yet the chapter has a great deal to say about what is distinctive about scientific knowledge of nature (1025b18–19:

[13] Cf. *APo.* 1.2, 71b33–72a5; *EN* 1.4, 1095b2–4.

ἡ φυσικὴ ἐπιστήμη), and the reason why becomes clear near the end of the discussion:

> If then there is no other substantial being apart from those constituted naturally, the science of nature would be primary; but if there is some changeless substantial being, this would be prior and <the science of it> would be primary philosophy, and universal in this way, because primary. And it would then be for this science to study being *qua* being, both what it is and the things that belong to it *qua* being. (1026a27–32)

That is, one of Aristotle's chief concerns is to determine whether the science of nature is primary, or whether there is a study prior to it.

Yet this is a puzzling concern.[14] The chapter opens by reiterating the distinction, introduced at the beginning of *Metaphysics* Γ.1, between the investigation of being *qua* being and investigations focused on some one part of being (1003a21–25); and the science of nature is introduced in *Metaphysics* E.1 within that very framework as the science of the kind of being in which there is an inherent source of change and rest (1025b18–21). This would appear to disqualify natural science immediately as a candidate for first philosophy, since, in the language of this chapter, it is one of those disciplines that "circumscribe a certain being and a certain kind, and treat of it, and not of being without qualification nor *qua* being" (1025b8–10). After looking carefully at the ways in which Aristotle demarcates natural investigation, we will return to this puzzle.

One clear mark of φυσικὴ ἐπιστήμη has already been identified in the process of distinguishing it as a theoretical rather than a practical or productive activity: The particular kind of being which the natural investigator takes as his object consists of objects which have inherent sources of change and rest. This was already alluded to in the passage from *Physics* III.1 with which we began – we need to investigate what change is because natures are sources of change in natural things. He next argues that if all scientific knowledge is theoretical, or productive, or practical, scientific knowledge of nature would be a particular kind of *theoretical* knowledge. Which kind?: "about the sort of being that is capable of undergoing change, and for the most part about substantial being according to account (οὐσίαν τὴν κατὰ τὸν λόγον), only not separable."[15]

[14] See Shields 2012b: 343–371 on this puzzle, and Frede 1996: 81–95 (following Patzig) for one sort of solution.

[15] Cf. Ross 1924: 354 at 1025b27n. Ross prints ὡς οὐ χωριστὴν μόνον following mss. E and T, and omits the preceding comma of the OCT. I don't think the addition of ὡς is necessary. I'm following Alexander's reading, taking μόνον adverbially (LSJ, 1145 cites this passage for this use of the neuter μόνον). On its own it is difficult to understand, but the next few lines make Aristotle's point clear.

Commentators typically assume that οὐσίαν τὴν κατὰ τὸν λὸγον is a way of referring to substantial being as *form*.[16] But caution is called for: In light of what follows, it is likely that Aristotle is leaving the question of what to substitute for οὐσία here open. For immediately after this very puzzling sentence, Aristotle issues a caution:

> It is necessary not to lose sight of how the-being-what-it-is and the account is (τὸ τί ἦν εἶναι καὶ τὸν λόγον), because without this the inquiry is destined to be unproductive. (1025b28-30)

What is significant and suggestive about this caution is that it introduces Aristotle's characterization of the objects to be investigated by the natural scientist. It strongly suggests that what one identifies as "the-being-what-it-is" of something and what an account of that thing should include will differ between mathematics and natural science.

> Among things defined and among the "what-it-is" of things, some are in the manner of the snub and others are in the manner of the concave. These differ in that the snub is comprehended with the matter (for the snub is concave nose) while the concave is <comprehended> without perceptible matter. And if all natural things are spoken of in like manner to the snub, for example, nose, eye, face, flesh, bone – generally, animal – leaf, root, bark – generally, plant (for the account of none of these things is without change, but always has matter), then it is clear in what way one ought to inquire into and define the what-it-is in natural things, and the reason why it is also for the natural inquirer to study certain things about the soul, as many as are not without matter. (1025b30–1026a5)

Thus the reference to "substantial being in accordance with the account" in the initial specification of the objects to be investigated by the natural scientist may well have been intentionally left open – and, after being cautioned about attending to the ways that different inquiries approach "the account" and "the what-it-is-to-be," Aristotle introduces us, conditionally, to the thought that *all* natural things are spoken of like "the snub." These are things the accounts of which are "not without change and always include matter." Unfortunately, although he claims that on the assumption that natural things are spoken of like the snub, "it is clear in what way one ought to inquire into and define the what-it-is in natural things," he doesn't actually spell out the way of inquiry and definition at all. Why would he – it's clear!

[16] Ross 1924: 354 at 1025b27n; Kirwan 1971: 185; Peramatzis 2011: 100.

A number of different options have been suggested, since Aristotle's views about what the definition of a natural composite should look like, and even whether there are such definitions, are anything but clear. Indeed, scholarship on *Metaphysics* Z–H has focused on a tension, especially apparent in Z.10–11, which arises when one asks whether, in Aristotle's view, the material aspect of a natural substance should be part of its definition or not. Determining the answer to this question is directly relevant to the question of what a μέθοδος of nature will look like, since arriving at definitions that identify the what-it-is of natural substances must surely be a central part of the goal of such an inquiry.

What then, are the options?

Option 1. Natural substances are matter/form composites, but a proper *account* of such composites will only identify the form of the composite, not the matter.

Option 2. Natural substances are matter/form composites, but a proper *account* of such composites will refer to matter only as a part of the form (or conversely to form as matter-involving).

Option 3. Natural substances are matter/form composites, and a proper *account* of such composites will be an account of a universal *composite*, i.e. will refer to the material and formal aspects of the natural substance.[17]

Finally, although I won't label this as a distinct option, it is possible to combine options 1 and 3, by arguing that there are primary and derivative definitions of natural substances, definitions that only refer to their form and definitions that refer to both form and matter.

In this chapter I will assume the truth of a version of option 3, one which I have defended elsewhere[18] and for which I present additional evidence in its favor by indicating how this option coheres best with Aristotle's methodological recommendations for natural inquiry. But before turning to those recommendations, I will highlight two features of the above passage that are rarely mentioned: First, all the examples of snub-like natural things are biological, both non-uniform[19] and uniform parts of animals and plants, as well as animals and plants themselves; second, the fact that all natural things are snub-like not only has implications for how one ought to

[17] Option (1) Devereux 2011, Frede 1990; (2) Charles 2010b, Peramatzis 2011; (3) Balme 1987, Gill 1989, Whiting 1986.

[18] Lennox 2008, 2010.

[19] The fact that he mentions noses first in his list of parts that are "snub-like" may be an instance of Aristotelian humor.

define them, but (Aristotle says) also for *how one ought to inquire* into what they are.[20] The last line quoted above also has significant implications for how one interprets the opening chapter of *DA* and for how one understands the relationship of *DA* to the *Smaller Natural Treatises* (*Parva Naturalia* [*PN*]). For that line makes it clear that it is by no means a foregone conclusion that the natural philosopher should study the soul, and it leaves open, as does *DA* I.1, the possibility that there is some part of the soul that is without matter. The mere fact that all animals, plants, and their parts are spoken of like the snub and are not without matter does not imply that all soul is to be thought of in the same way.[21]

What then does this passage tell us about the μέθοδος of ἡ μέθοδος ἡμῖν περὶ φύσεως? It quite straightforwardly says that the account we are aiming for in natural inquiry must include reference to matter and change; and that therefore the natural scientist, while acknowledging the priority of form, must make the material make-up of natural things and the changes they undergo central to his investigations. Thus, the sort of abstraction that is appropriate in mathematics, where the proper account of (say) a parallelogram does not mention perceptible material or change at all, would be inappropriate in natural science. However, little or nothing is said in this chapter about the definition referring as well to form, nor to the nature of the relationship *between* matter and form in natural objects, and how that relationship should be studied or represented in a proper definition or causal demonstration. Similarly, nothing is said about the implications of this characterization of the objects of natural science for how one ought to *inquire* into them. However, that is the central focus of *Physics* II.2, and the "snub-like" character of the objects of natural investigation is Aristotle's starting-point.

Physics II.2 *and snub-like inquiries*

Physics II.2[22] begins with the assumption of one major result of the previous chapter, namely the number of ways in which nature is spoken,[23] and turns to a question he apparently thinks follows more or less directly

[20] . . . δῆλον πῶς δεῖ ἐν τοῖς φυσικοῖς τὸ τί ἐστι ζητεῖν καὶ ὁρίζεσθαι . . .

[21] As discussed in Lennox 2010, *Parts of Animals* I.1 states unequivocally that *nous* and *dianoia* are *not* proper objects of inquiry for the natural scientist.

[22] Some results of Lennox 2008 are assumed here.

[23] Reading λέγεται at 193b22. Later in the chapter, at 194a12 we are told that nature is διχῶς τό τε εἶδος καὶ ἡ ὕλη, a result Aristotle argues for from 193a9–b5, after which he makes the case for the form being a composite's nature "more than" the matter.

from that result – in what way does the mathematician differ from the person who investigates nature? Aristotle generates a concern about this by highlighting two related but distinct points: First, the objects investigated by the mathematician, such as points, lines, planes, and solids are features of natural bodies; and second, it is apparently a goal of the natural scientist to grasp the nature of the sun and moon, and to determine whether the earth or the cosmos is spherical or not. But these would appear to be the concerns of the astronomer, who is a type of mathematician (193b22–30).

From his responses to these problems it becomes clear that at this point in the *Physics* Aristotle is concerned to make a case for the *unity* of the objects being investigated by the natural scientist, as a precondition for there being a single science of nature. This becomes critical because chapter 1 has argued that the nature of a natural thing is, and is spoken of, in two ways, as matter and as form. Chapter 2 points to the errors of "those who speak of Ideas" (193b36), indicating his concern clearly to be about what it means to say that the natural investigator should be pursuing knowledge of the *forms* of natural things. He precisely criticizes them for attempting to *separate* the natural things, which are "less separable than the mathematicals" (193b36–37). He explains:

> This would become clear if one were to attempt to state the definitions of each – both of the objects and of their attributes. For the even and odd and the straight and curved, as well as number, line, and figure will be [defined] without change, while flesh, bone, and human being will not – these are spoken of as is snub nose (ῥίς σιμή), not as the concave. (194a1–7)[24]

As in *Metaphysics* E.1, here the contrast with mathematical definitions is intended to highlight the fact that the definitions of natural things refer to change – and, once again, they are said to be referred to in the way one refers to a nose as snub, not as concave. But what exactly are we referring to when we refer to a natural object in that way?[25]

It is of the very essence of a living thing, and more generally of any natural object, to move, behave, act, and change in specific ways; in fact, in specifying what it is to be an eye or a leaf – or even air – one must mention capacities to function or change in specific ways.[26] But it is also the case that those capacities are the capacities of specific materials, and thus any such definition must refer to a material structure constituted in precisely the way it must be in order to have the capacities to move and change as an eye or a leaf.

[24] Cf. *Metaph.* Z.10, 1035a4–5, Z.11, 1036b21–31. [25] Cf. *Metaph.* Z.11, 1036b24–31.
[26] Cf. *Meteor.* IV.12, 390a10–12.

And this has implications for how one carries out a natural inquiry, as Aristotle indicates in *Metaphysics* Z.11, while explaining why first philosophy bothers to discuss natural composites:

> the study of perceptible substances is in a way the task of the study of nature, or second philosophy. For the natural investigator must have knowledge not only of the matter of things but also, and more especially, of the substance given by the formula. (1037a15–17)

To stress this point in *Physics* II.2, Aristotle once again invokes the analogy of natural substances and the snub, but now as an implication of the fact that natural things have a twofold *nature*:

> Since the nature [of natural substances] is in two ways (διχῶς), both the form and the matter, we should study as if we were investigating, about snubness, what it is; that is, we should study such things neither without matter nor in accordance with matter. (194a12–15)

The message appears to be this: Natural things have a nature in two ways – a material nature and a formal nature.[27] For that reason, we need to investigate the "what is it" of natural things while keeping their twofold nature in focus; we cannot ignore their matter, but we cannot focus only on their matter, for they have a formal nature as well – indeed, *Physics* II.1 gives a series of (more or less compelling) arguments for giving more attention to the formal nature. It is somewhat surprising, then, that Aristotle immediately raises an *aporia* about which of the two natures the natural scientist should study! Have we not just been told that he should study both? The full statement of the *aporia* is worth reviewing:

> For indeed one might also raise a problem about the following: since the natures of things are two, which of the two is it for the natural scientist to study – or is it rather for him to study that which derives from both natures? But if [it is for him to study] that which is derived from both, [then it is] also [for him to study] each of them. (194a15–17)

A provisional solution to the *aporia*, which flows from the idea that the natural scientist should study the hylomorphic *composite*, is immediately challenged, on grounds that one would still be studying the two natures, though not treating form as separate from matter. The question at issue is:

[27] Aristotle identifies the two natures in relation to one another in various ways: at *Phys.* II.1, 193a28–31, the contrast is of the primary underlying matter with the shape and form in respect of the account (cf. *Phys.* II.1, 193b3–5); at *PA* I.1, 640b28–29 it is of the nature in respect of form with the material nature; at *PA* III.2, 663b22–24 the nature in respect of the account is contrasted with the necessary nature; and at *GA* IV.4, 770b9–18 the nature in respect of the form is contrasted with the nature in respect of the matter.

Even if not *ontologically* separate, can these two very different natures be studied by a single science (ἐπιστημή)? That question has purchase because, within an Aristotelian framework, although the objects of mathematics are not ontologically separate from natural objects, they are "separated in thought" and studied by distinct sciences, sciences which use very different methods guided by very different norms from those of the natural scientist.

To this point he has claimed that natural substances have two natures, that they and their attributes are spoken of like snub nose and are not defined without reference to change, and that we are to study them "not without matter yet not according to matter." He is now prepared to present his solution to the above *aporia*, albeit in conditional form:

> *If* art imitates nature, and it is for the same science (ἐπιστημή)[28] to know the form and the matter to a certain extent (e.g. for the doctor to know health and bile and phlegm, in which health is present, and similarly also for the house builder to know both the form of the house and the matter, that it is bricks and wood, and likewise in other cases), *then* it would also be *for natural science to know both natures. And again: it is for the same science to know that for the sake of which and the goal and as many things as are for the sake of these.* (194a21–27; emphasis added)

The knowledge the natural scientist seeks is of *both* the matter *and* the form; and they can be the subject of one science, on the model of the crafts, if the materials are both present for the sake of, and defined by reference to, the formal nature. That is the reason why, although natural things have two natures, they are properly the subject matter of a single science.

Aristotle picks up on the qualification "to a certain extent" in the passage I have just quoted, near the end of the chapter:

> To what extent is it necessary for the natural scientist to know the form and the what-it-is? Is it just as [it is necessary] for the doctor [to know] sinew or the sculptor bronze, to the extent of knowing what each thing is for the sake of, and concerning those things which are separable in form, yet in matter? (194b9–13)

It appears that the issue of the extent to which the two natures are to be studied by the natural scientist is only explicitly addressed with respect to form, a point I will return to shortly. First we should recall the wider context of this question, specifically issues concerning the demarcation of natural philosophy from first philosophy. Charlton translates the chapter's

[28] Here ἐπιστημή is being used in a "big-tent" sense to include the crafts.

closing sentence thus: "What it is which is separable, and how things are with it (πῶς δ' ἔχει τὸ χωριστὸν καὶ τί ἐστι), it is the work of first philosophy to determine" (194b14–15). Whatever Aristotle exactly means by the Greek here, one thing is certain: He is making this point because he has just made a claim about the limited nature of the separability of the forms to be investigated by the natural scientist.[29] And this passage surely looks back to the end of Book I, where Aristotle says that the discussion to follow (in Book II) will investigate the forms of natural and perishable things,[30] while it will be "the work of first philosophy to discuss with precision the principle in respect of form, whether it is one or many and what it is or they are" (192a34–36). Together, these passages are saying that the natural philosopher's investigation of form is distinct from that of the metaphysician. Specifically the natural philosopher is to study the formal nature *in so far as it is that for the sake of which the materials that make up the material nature are present*. Thus, the natural scientist should study matter to the extent that it is for the sake of the form. This chapter, then, looks forward to chapter 8 and its defense of the *teleological dependence* of material nature on formal nature, succinctly summed up in the following sentence:

> And since the nature [of a thing] is twofold, on the one hand as matter and on the other as form, and the nature as form is an end, while the others are for the sake of the end, this [nature as form] would be the cause for-the-sake-of-which. (199a30–32)[31]

This passage, moreover, looks forward to the discussion of the relationship of conditional necessity between materials and what they are for in chapter 9.

We now begin to make contact with norms that govern Aristotle's actual practice as a natural investigator, such as the following, during a critical review of his predecessors' attempts to explain respiration:

> The most significant reason for [previous thinkers] not discussing these [questions about respiration] well is a combination of (τε) their being

[29] Cf. Waterfield 1996: "Questions remain – in what sense is anything separable? What is it that is separable? – but it is the job of first philosophy to answer them"; Hardie and Gaye 1930: "The mode of existence and essence of the separable it is the business of the primary type of philosophy to define"; Pellegrin 2000: "Ce qu'il en est du separable et ce qu'il est, c'est l'affaire de la philosophie première de le determiner"; Angioni 2010: "Mas delimiter como se comporta e o que é o separável é tarefa da filosofia primeira." Certainly the nature of the separability of the objects of mathematics, Platonic Forms, the forms of natural substances, and a first unmoved mover is a theme that runs through the entirety of the *Metaphysics*.

[30] I defend this reading (which follows the Λ mss.) in Lennox 2008.

[31] καὶ ἐπεὶ ἡ φύσις διττή, ἡ μὲν ὡς ὕλη ἡ δ' ὡς μορφή, τέλος αὕτη, τοῦ τέλους δὲ ἕνεκα τἆλλα, αὕτη ἂν εἴη ἡ αἰτία, ἡ οὗ ἕνεκα.

> inexperienced with the internal parts and (καὶ) not grasping that nature makes them all for the sake of something; for seeking what respiration belongs to animals for, and examining this question in the presence of the parts, e.g. in the presence of gills and lungs, they would have discovered the cause more easily. (*Resp.* 3. 471b23–29)

That is, one must carry out a natural investigation of a living activity like respiration on the assumption that it and the structures that perform it are present in animals for the sake of something; and one must carry out such an investigation making use of comparative dissection so that one knows exactly which internal parts are involved in performing the activity of respiration. Recalling an oft-cited passage from the beginning of Aristotle's investigation of animal locomotion, we should not be surprised by what Aristotle says about the study of respiration, for he there claims that this is an assumed starting-point of natural inquiry generally:

> The starting-point of our investigation is achieved by supposing principles we are accustomed to use often in natural inquiry (πρὸς τὴν μέθοδον τὴν φυσικήν) – assuming this is the way things stand in all the works of nature. One of these principles is that nature does nothing in vain, but always, given the possibilities, does what is best for the substantial being of each kind of animal (τῇ οὐσίᾳ περὶ ἕκαστον γένος ζῴου τὸ ἄριστον). (704b11–17)

The assumed starting-points Aristotle introduces here are starting-points of *investigation* (not of proof) and are said to be used often in the μέθοδος of nature. Yet the first one that is stated, the nature-does-in-vain principle, is stated in a way that is specific to animals. There is, of course, absolutely nothing inconsistent about that – a principle that is often used in the investigation of animals will, by that very fact, be used often in natural inquiry. But he also says that it is appropriate to assume that these principles (he lists two others) are applicable to "all the works of nature." He could mean that they collectively apply to all works of nature, or he could simply be allowing that, at the beginning of an investigation of nature, we have no grounds for restricting these principles to a limited part of nature.[32]

In *Physics* II.2, the primary focus was on the question of whether, given that natural substances have both material and formal natures, they could be investigated by a single science – where the goal of such an investigation tended to be stated in terms of defining and knowing the essence of such things. However, the possibility of a unified science of natural, "snub-like"

[32] Cf. Lennox 2009a: 187–214, and Leunissen 2010: 215–237.

beings turned on there being a *teleological dependence of their material nature on their formal nature.* Indeed, the natural scientist's knowledge of the matter and form of a natural substance was restricted to knowing what the material constituents of a natural being are *for*, which means knowing the form in so far as it is that for the sake of which the matter is as it is. Chapter 2 thus concludes by implicitly grounding the "what is it?" inquiry in a *causal* inquiry, the sort that is in focus in chapter 9, namely an inquiry into the causal relationship between ends, on the one hand, and materials and their movements, on the other. But that is precisely what the erotetic framework of *Posterior Analytics* II would have led us to expect – that the search for an answer to the "what is it?" question will be intertwined with causal inquiry in precisely this way.

That is how *Physics* II.2 ends – but recall how chapter 3 *begins*, echoing the opening sentence of the *Physics* (which in turn echoes *Posterior Analytics* I.2 and II.11).

> Having determined these things, we must investigate concerning the causes, both of what sort and how many in number they are. For since our study is for the sake of knowing, and we do not think we know each thing until we grasp that-on-account-of-which (τὸ διὰ τί) about each thing (and this is to grasp the primary cause), it is clear that we must do this concerning generation, destruction, and all natural change, so that, knowing their starting-points, we may try to bring each of the objects of inquiry back to them. (194b16–22)

However, though introduced as important for knowledge of nature, the discussion of the four types of cause in this chapter (as well as in the next three chapters on chance) says little that is specific to the inquiry into nature. In fact, chapter 3 is, as Ross puts it, "a doublet which corresponds in all except small details with *Met.* Δ.2."[33] It is not until chapter 7 that Aristotle explicitly discusses the idea that there are four distinct kinds of cause in the context of *natural* inquiry. It begins by reminding us of the four kinds of answer to the "on account of what?" question that were discussed in chapter 3, still without particular specificity to nature (198a14–21). He then proceeds to make the connection:

> So then, that the causes are these and this many is apparent; and since the causes are four, *it is for the natural scientist to know about them all*, and by referring to all of them – the matter, the form, the mover, and that for the

[33] Ross 1936: 511.

> sake of which – he will display that-on-account-of-which (τὸ διὰ τί) in a way appropriate to natural science (φυσικῶς). (198a21–24; emphasis added)[34]

Notoriously, he goes on to say the last three often come to one – and then, substituting τὸ τί ἐστι for τὸ εἶδος, he explains what he has in mind. In cases of substantial generation, the what-it-is and that-for-the-sake-of-which are one, while the primary source of motion is the same as these in form (198a24–26). The goal of generation is the form, so in these cases there is referential (and numerical) identity between them; while the parent and the off-spring are not numerically identical, they are *the same in form.* The remainder of the chapter is, though puzzling in a number of respects, equally important, since its focus is, as with chapter 2, on distinguishing natural inquiry from other sorts of inquiry. The mention of the primary source of motion, he realizes, might well be misunderstood – thus, he immediately restricts the set of such movers to those that "move while being moved" (a25) and goes on:

> for as many as are not like this are not natural; for they move <other things> without having a change or a source of change in themselves, but unmoved; for which reason there are three subjects of study – that concerned with the immoveable, that concerned with what moves but is imperishable, and that concerned with perishable things. (198a28–31)

This way of dividing up theoretical investigations is quite different from what we found in *Metaphysics* E.1; assuming that the last two subjects mentioned correspond to a study of the heavens (which, while imperishable, partake of eternal circular motion) and a study of sublunary natural objects, these are both appropriate objects of natural inquiry.[35] And the third subject area is described in such general terms that it could refer to a number of things in Aristotle's universe, including souls, objects of desire, and the heavenly unmoved movers, all of which move other things while being themselves unmoved.[36] It is best, I think, to read this as having the

[34] Cf. *Metaph.* H.4, 1044a32–b20, where, however, after citing the four causes and saying that in the study of generated natural substances it is necessary to discuss all the causes, Aristotle states that we should proceed by "another *logos*" when discussing the natural and eternal substances (1044b2–6), and that certain natural processes, such as eclipses, perhaps don't have final causes (1044b9–13); no final cause is ever provided in the *Meteorology* for any of the cases discussed here.

[35] Cf. *PA* 1.5, 644b22–25: "Among the substantial beings constituted by nature, some are ungenerated and imperishable for all eternity, while others participate in generation and perishing." See, too, *Cael.* 1.1, 268a1–6 and *Metaph.* H.4, 1044b3–6.

[36] The discussion in *Metaphysics* E.1 depends on comparing the objects of mathematics, natural science, and a still hypothetical first philosophy along two axes, separability and changeability, whereas here separability is not discussed. Recall that in *Physics* 11.2 he refers that topic to first philosophy.

limited purpose of clarifying what sort of first mover the natural scientist can investigate, for he goes on to state that "the sources of things moving naturally (φυσικῶς) are of two sorts, one of which is not natural; for it does not have a source of motion in itself" (198a35–b1).

Things that are by nature have a source of change or rest within themselves (192b13–15, b20–23, b32–34). He postulates two sorts of things that fall into the category of things that are sources of natural motion without being natural in that sense: One of these is not to be investigated by the natural scientist, the prime unmoved mover; the other is the form viewed as final cause, "the what-it-is and form, for it is an end and for the sake of which" (198b2–3). He goes on:

> So since nature is for the sake of something, and it is necessary to know this, it is also necessary to display that-on-account-of-which in all ways, e.g. that this comes from this necessarily (either without qualification or for the most part); <what must be> if this is to be (as the conclusion from premises); that this is what it is to be; and for what reason it is better in this way, not better without qualification but <better> in relation to the being of each thing. (198b1–9)[37]

These concluding remarks build on the idea that there are two different kinds of thing that are sources of natural change while not themselves changing, and that one of those, while not itself a nature, is an appropriate subject of natural inquiry, namely that which nature, understood as an inherent source of change, acts for the sake of. Note that both here, and in the opening sentence of chapter 8, Aristotle says that nature is "among the causes for the sake of something" (198b4, b10), not that nature is "that for the sake of which." The concept of nature is here being used strictly to refer to the inherent source of change delineated in *Physics* II.1–2. And here we see the importance of the distinction between form understood as that for the sake of which and form understood as source of change.[38] For the defense of natural science as a unitary science in *Physics* II.2 depended on the idea that the two natures of a natural substance are teleologically unified – and that unity depended on the formal nature of a natural substance being *that for the sake of which* its material nature comes to be and persists. This conception of natural science depends on nature not only being among the causes that act for the sake of something – it must also be "the cause for the sake of which." This claim is asserted in *Physics* II.2, but

[37] On the many puzzling aspects of this remark, see Ross 1936: 527–528; Philoponus *In Ar. Phys.* 303.8–306.12; and Simplicius *In Ar. Phys.* 366.30–369.13.

[38] Cf. *DA* II.4, 415b8–28.

only comes to the fore in chapters 8 and 9, beginning with the following (already quoted) passage.

> And since the nature [of a thing] is twofold, on the one hand as matter and on the other as form, and the nature as form is an end, while the others are for the sake of the end, this [nature as form] would be the cause for-the-sake-of-which. (*Phys.* II.8, 199a30–32)

This teleological way of conceiving the relationship between the materials that constitute a natural being and the form of the thing so constituted provides the ontological underpinnings for a new, conditionalized, way of conceiving of necessity, the subject of *Physics* II.9: Given that a certain goal is to be realized, certain materials, with appropriate dispositional properties, must come to be in a certain temporal and spatial order.[39] And this, in turn, provides yet another norm that is specific to *natural* inquiry, to the μέθοδος of nature.

> It is apparent, then, that the necessary in natural things is that which is spoken of as matter and its motions. And it is for the natural scientist to discuss both causes, but more the cause for the sake of which; for this is a cause of the matter, rather than the matter being the cause of the end; and the end is that for the sake of which, and the starting-point is from the definition and the account. (200a30–35; emphasis added)[40]

Again, questions of causation and definition are integrated tightly. As so often when teleological causation is under consideration, Aristotle stresses that the goal, that for the sake of which, has priority because in these cases the account that specifies the identity of the goal must be understood in order to properly understand the materials and processes that are needed to achieve that goal. That "that for the sake of which" will be prioritized in accounts of natural substances is clear; however, *Physics* II ends by considering whether "the necessary" will as well:

> Perhaps the necessary is also in the account. For in defining the work of sawing, that it is a certain sort of dividing, surely this will not exist unless the saw has teeth of a certain sort; and such teeth will not be, unless they are made of iron. Thus, even some parts of the account are in the account as matter. (200b4–8)[41]

39 Cf. Charles 2012: 230–232.

40 Cf. *PA* I.1, 639b12–16 and 642a14–18.

41 This example should be compared to *PA* I.1, 642a10–13, where Aristotle makes the analogy of the axe to the living body explicit; whereas in *Physics* II.9 the analogy to living bodies is mentioned (rather, the act of *sawing* is being defined, perhaps analogous to Aristotle's study of respiration or sleep); the *Parts of Animals* example also makes clear that it is the *dispositional property* that is conditionally necessary, not a specific uniform material.

Once again, then, form, understood as the end for the sake of which materials are as they are, is prioritized, but matter is not ignored. Indeed, it is argued that the necessary materials should be brought into the account by way of conditional necessity.

As with the teeth of saws, so too with the teeth of animals. In discussing the uniform materials out of which hard parts are made in *Parts of Animals* II, Aristotle considers the material out of which teeth are made:

> The nature of teeth is . . . in some cases present for a single function, the preparation of nourishment, while in other cases it is present both for this and defense, e.g. in all those with saw-like teeth . . . . Of necessity all of these parts have an earthen and hard nature; for this is the defensive potential (ὅπλου δύναμις). (*PA* II.9, 655b8–13)

When he moves on to the non-uniform parts to discuss the teeth he assumes the material nature of teeth is of this character, and focuses on differences in their shape and organization in relationship to the different lives that animals lead:

> Some animals have teeth . . . for the sake of nutrition alone. But those which have them for protection as well as for strength in some cases have tusks . . . and in other cases have sharp, interlocking teeth, for which reason they are called saw-toothed. For since their strength lies in their teeth, and this comes about because of their sharpness, teeth which are useful for strength fit together in an alternating pattern, so as not to be worn down by being rubbed against one another. (*PA* III.1, 661b17–22)

Here we see in his actual biological practice exactly the sort of teleological account of teeth that are intended both for nutrition and defense, from which it follows that they must have a certain structure and organization, and be made of a suitably hard material.

III Conclusion: One natural μέθοδος or many?

We were provoked to explore the question of what a μέθοδος of nature might be by Aristotle's comment, at the beginning of *Physics* III, that "[s]ince the nature of a natural thing is a source of motion and change, and our μέθοδος is concerned with nature, the question 'what is motion?' must not escape our notice."[42] And indeed, for the next three chapters of Book III he

[42] The concept μέθοδος appears four more times in the *Physics* – once in III.5, 204a34–b11, and the remainder in Book VIII (251a5–8, 253a32–b10, 261a27–31). As one would expect if the position I have been defending is correct, all are concerned with delimiting the approach appropriate to natural inquiry, as opposed to mathematics or first philosophy.

pursues that question and comes up with a notoriously puzzling answer. But before doing so, he notes that, since motion seems to be continuous, there are a number of *other* questions – about the unlimited, place, void, and time – that must be taken up after the question of what motion is. Before turning to these topics, which will occupy him for the next two books, he makes a comment that sheds light on how this treatise and its method differ from the rest of Aristotle's natural investigations. The lines I am especially interested in might be translated as follows:

> It is clear, then, that, both on account of these things and because the being of all these things is common and universal (κοινὰ καὶ καθόλου), we must investigate by getting hold of each of them (for the study of the proper things (τῶν ἰδίων) follows the study of the common things (τῶν κοινῶν)). (200b21–25)

In any particular investigation of nature, whether of meteors, rainbows, minerals, animal parts or activities, or the motions of the heavens, one needs to have a clear conception of motion and change in general, as well as of the unlimited, place, void, and time in order to proceed. These concepts are, to borrow Waterfield's translation of καθόλου here, *pervasive* to the study of nature. The "Physics"[43], then, is a work in the conceptual *foundations* of natural science; not in the sense that its methodology is any more "conceptual" than any other, but in the sense that the concepts about which it aims to get clear – nature, cause, chance, motion, place, time, and so on – are presupposed by any of the more specialized natural inquiries in which Aristotle was engaged, each of which may have distinctive norms of inquiry. In *Physics* I.2–3 Aristotle makes it clear that he doesn't think it is appropriate for the natural scientist, *qua* natural scientist, to question *whether* change exists – obviously it does. At the beginning of *Physics* III, however, he makes it equally clear that *what* change is is anything but obvious, and to fail to answer that question is paramount to being ignorant of nature itself. Moreover, in the case of the unlimited, void, time, and place, it is even legitimate to consider the question of *whether* they exist, and part of the exploration of what they are involves exploring *in what sense they exist.*

[43] I remain neutral on how the various books of our *Physics* came to be related to one another (but see Menn [in press]), but note that this opening of Book III both looks back to the earlier books, in the sense that it notes a lacuna in that discussion which he now intends to fill, and also previews all of the topics covered in Books III and IV. Book V can be viewed as a natural appendix to the general discussion of change; and since all of the topics in Books III and IV arose out of the undefended assumption that motion "seems" (δοκεῖ, 200b16) to be continuous, the discussion of continuity in Book VI fills an obvious gap in the earlier discussion as well.

CHAPTER 2

Aristotle on interpreting nature

Sean Kelsey

Physics II begins (more or less) with a definition of nature. I want to consider how this definition is used in the rest of Book II. I don't think this has been done before, and in any case a first look turns up a surprise: It seems as though it is hardly used at all.

A closer look suggests that this initial appearance, though resilient, is ultimately misleading. Although the definition seldom functions as an explicit premise in an important argument, it does influence the course of the subsequent discussion. A little more fully, it serves as both source and signal of an important theme that sets a tacit agenda for the book as a whole.

It is not hard to see that the case for such a suggestion must rely on what we might call "the argument from salient omission" – always perilous. But although I do not think such arguments establish the suggestions they are invoked to defend, I do think it quick to suppose they simply refute them. A case in point: Aristotle does not say, in introducing his discussion of chance and luck, that its purpose is to supply the backbone of an argument that nature is a final cause (*Physics* II.4–6, 8). On the contrary, he introduces that discussion as almost an afterthought – a kind of appendix to the discussion of causes in II.3 (195b31–36, 196b6–9). Yet it is impossible to believe that the opening of II.4 gives a fair representation of the real purpose of this material. Despite the absence of any explicit statement to this effect, it is clear that the function of these chapters – their place in the design of the book as a whole – is to support Aristotle's first and principal argument that nature is a final cause. I mention the case to exhibit a principle – in this text, at least, silence is not probative against.[1]

I would like to thank David Ebrey, Joseph Karbowski, Mariska Leunissen, Yannig Luthra, Gretchen Reydams-Schils, Wiebke Marie Stock, and especially David O'Connor for conversation and comments on earlier drafts of this chapter.

[1] Another example is *Physics* I, where it emerges in retrospect that the problem raised at the end of I.6 was staged: Its solution requires certain distinctions made in I.7, just the ones (lucky chance!) necessary to solve "the" puzzle discussed in I.8. Or again in *On Generation and Corruption* I.3,

I

It will be useful to have the passages in which the definition is used collected in one place, if only to show how little appears to be gotten from it in the body of *Physics* II.

So first, in II.1, having said what nature is (an internal source of movement and rest), and what sorts of things have natures (substances), and what sorts are "by" or "in accordance with" nature (the substances that have natures, and whatever belongs to them per se), and that it would be ridiculous to try to show that there are natures, Aristotle turns to identify the natures of those beings that have them – to say what "in" those things *is* their nature (193a9ff.). Here is a place we might expect him to use the definition. Having just established (as he thinks) that nature is an internal source of movement and rest, we might expect him to assess something's claim to *be* a nature by considering whether it *is* such a source. But in fact he refers to the definition but twice in the ensuing discussion, both times in stating a result already arrived at, and even then only to specify that nature, whether it be "the first underlying matter," or "the shape and form," is that, not just of any old thing, but of things "having in themselves a principle of motion" (that is, nature is the "matter" or "form" of things that *have* natures) (193a29–30, b3–4).

So much for II.1; the definition is not referred to again until II.6, in a passage discussing the difference between chance and luck (197b32–37). Having said that this difference is most visible in the case of things that happen "contrary to nature" – we ascribe such things to chance, not to luck – Aristotle enters a qualification: This is not quite right, inasmuch as the cause of things due to chance is "external," whereas the cause of things that happen contrary to nature is "internal" (namely, to the extent that even such things are caused in a way by nature) (197b36–37). This remark does make an appeal to the definition, but by way of establishing a point that (at least on its face) could hardly be less important.

The definition is next appealed to in II.7, where it is used twice to make the point that certain causes are *not* "natural" (φυσική) or "within the purview of natural science" (φυσικῆς). The idea is that these causes do not themselves have natures, because, being immobile, they do not themselves have an internal principle of motion (198a28–31, 198a35–b1). Last come two passages towards the end of II.8, about which it suffices for now just to say

where the point of the indirection is made explicit: Aristotle introduces a new puzzle, about why generation is everlasting, in order to solve an old puzzle, about the very possibility of unqualified generation and corruption (317b33–35, 318a10–13).

that they occur long after the chapter's main business has ostensibly been completed.

A total of seven passages, in none of which is the definition obviously the crux of a matter of any consequence.

II

I now want to revisit these passages, starting with II.1. This chapter highlights two results in particular: that nature is a principle of motion (and rest), and that form is nature ("more so" or "rather than" matter). Although these results imply that form is a principle of motion and rest, this is nowhere stated in II.1. Indeed, the very idea of a principle of "motion and rest" – that is, of a thing's characteristic *behavior* – is more or less idle in the latter half of II.1; as we will see, Aristotle's focus in identifying the natures of things is not on their behavior, but on their "substance" or "being" or "essence" (οὐσία) – not on what they "do," but on what they most fundamentally "are." Being and behavior are importantly related, of course, but this only makes Aristotle's disregard of behavior in this context all the more striking. The effect, I submit, is to make the (unstated) implication of the results he does highlight – the implication that *form* is a principle of *motion* – something of an elephant in the living-room.

The point that wants showing is that in *Physics* II.1, in identifying the nature(s) of things due to nature, Aristotle doesn't consider their behavior – their characteristic forms of motion and rest. Consider first his handling of the idea that matter is nature (193a9–28). He begins with the remark that "the nature and substance of things due to nature" is thought by some to be "the first [thing] present in it, which is in and of itself unformed" (193a9–12). Significant here is the tacit identification of "nature" with "substance"; although it was said earlier that things that *have* natures are substances (192b33–34), this is the first indication that a thing's nature *is* its substance. Doubtless one reason this identification is made here is that it is useful for understanding the view being considered; this is borne out by the remarks that follow, in which that view is represented as resting on the thought that "the first thing present in" each thing is, not the source of its characteristic motion and rest, but its real substance. It is true that the thought experiment attributed to Antiphon might suggest a focus on behavior: "if you planted a bed and the rotting wood acquired the power of sending up a shoot, it would not be a bed that would come up, but *wood*" (193a12–14). But this suggestion is canceled by Aristotle's statement of the experiment's moral: "the one, the conventional arrangement and the art, belongs but incidentally; the substance is that other, which persists continuously while undergoing these things" (193a14–17). As this statement makes clear, Antiphon, in asking after a thing's "nature," is focused on its *substance*, which he conceives of not as the source of a thing's movement and rest – of its behavior while it exists – but as what survives its demise. Indeed, this approach betrays an attitude in which the nature of things is positively concealed by their behavior, being manifested instead in their destruction.

This reading is strengthened by the lines that follow, in which Aristotle extends Antiphon's reasoning and, in so doing, widens his net to take in other thinkers:

> And if each of these experiences the same thing in relation to something else (e.g. bronze and gold in relation to water, bones and wood in relation to earth, and so on), that (they say) would be their nature and substance. That's why some say fire, and others earth, and some air, and others water, and some [say] some of these and others all of them are the nature of the things that are. For whatever any one of them supposed to have this character – whether one thing or more than one thing – this or these he declared to be the whole of substance, all else being its affections, states, or dispositions of these. Every such thing they held to be eternal (for it could not pass into anything else), but other things [they held] to come into being and cease to be again and again interminably. (193a17–27)

Aristotle represents these thinkers as conceiving of the nature of things as something that is above all permanent, un-generated, and indestructible. But on this way of thinking, it would be a mistake to look for the nature of something in the source of its characteristic behavior, which is as ephemeral as the thing itself is; on the contrary, the place to look would be in its remains – or, equivalently, in its antecedents or origins.[2]

In giving reasons for thinking that nature is matter, then, Aristotle makes no appeal to the influence of a thing's matter on its characteristic behavior. This is at least partly explained by his conducting the discussion in a historical mode; the reasons he gives are represented as reasons that actuated

[2] To be clear, Aristotle does not think these theorists are wrong to identify a thing's nature with its substance; this is an identification he himself would endorse. The mistake must lie elsewhere, and apparently (if *Parts of Animals* I.1 is any guide) in different places: e.g. in simply not considering the substance of things (640a10–12), or in considering the substance of the wrong things (640b4–6, 17–22), or of the right things, but in the wrong way (640b22–25, 640b29–31, 641a14–17) – in any case and in a word, in *not* treating the substance of things by reference to their characteristic behavior or *work* (640b35–641a3). (*Parts of Animals* I.1 *is* a guide to *Physics* II, and *vice versa*, as is shown by the careful and illuminating discussion of the relationship between these two texts in a recent unpublished conference paper by Stephen Menn [in press].)

his predecessors. But the case is different in his discussion of the idea that nature is form; as we will see, although there too he makes no appeal to behavior, the reason he gives in support of the view is his own (193a30–b5).

The case for the view that nature is form is made to rest on two points: First, that things that have natures are all "natural" or "due to nature" or "in accordance with nature," and, second, that nothing is any of these things before it lays hold of the form whereby its type is defined.[3] These two points are recommended by the fact that analogous points hold for works of art, and taken together they imply that nothing "has" its nature before it "has" this form; this in turn implies (or anyhow suggests) that the nature of things that have natures *just is* this form. Notice that this line of reasoning makes no reference at all to what things that have natures "do" – to their characteristic ways of moving and resting (let alone to the sources thereof). It focuses instead on what they are: or rather, on *when* they first properly are called that (a moment here specified just as coeval with the advent of form).[4]

Things are much the same for the reasons Aristotle gives for thinking that form is "more" nature than matter (193b6–21). The first simply repeats the line of argument we were just considering (193b7–8). The second re-appropriates "Antiphon's bed" (193b8–12) and thus should probably be read in line with Aristotle's earlier use of that example (193a12–14).[5] The third is based on a use of the word "nature" as the name for a kind of process or "road" (a road "towards nature"); since this "road" towards nature is also a road towards form, the fact that it is a road towards nature is taken to show that nature is form (193b12–13, 18). Whatever we make of this

[3] Aristotle takes pains to make clear that the "form" he has in mind is associated with definition; though twice referring to it as "the shape" (ἡ μορφή), he offers several accompanying clarifications: it is not just "the shape," but "the shape i.e. the form" (ἡ μορφὴ καὶ τὸ εἶδος), "the one in accordance with the definition" (τὸ κατὰ τὸν λόγον), "by which in defining we say what a thing is," "not being separable except in account" (193a30–31, b1–2, 4–5). This specification excludes certain perceptible forms, of the sort whereby finished works of nature and art are recognized (one might think) for the kinds of things they are. (Cf. *PA* 1.1, 640b32–33, where Democritus is referred to as holding that "it is evident to everyone what form (μορφή) it is that makes the man, seeing that he is recognizable by his shape (σχῆμα) and color.") It also excludes the forms posited by "those who talk of ideas," who are said to "separate the objects of natural science, though they are less separable than those of mathematics" (*Phys.* 11.2, 193b36–194a1).

[4] This needn't imply that a thing's matter is not its nature at all. After all, Aristotle does call the nature a thing doesn't yet have, namely when the thing is [what it is] only potentially, *its* nature (193a36–b1: οὔτ' ἔχει πω τὴν ἑαυτοῦ φύσιν). This suggests that he thinks of such things as somehow already being what as yet they are not, in which case they might already somehow "have" what as yet they have not, i.e. their own nature (which perhaps they would not, were matter not nature).

[5] The principle underlying Aristotle's reasoning here is admittedly difficult to make out. Most conservatively, it is that a thing's nature is revealed by "what comes to be from it." But it is left unclear why *that* should be: whether because what comes to be from a thing reveals its substance or essence, or because it reveals the principle of its characteristic behavior.

reasoning, it hardly appeals to the idea of nature as a kind of "principle" or "beginning" (ἀρχή) of behavior; on the contrary, the key point is that nature is *not* "that from which" (ἀφ' οὗ or ἐξ οὗ), but "that towards which" (εἰς ὅ) – not the beginning, but the end.[6]

In this chapter, then, in identifying the nature(s) of things, Aristotle makes no appeal to the sources of their behavior; this is contrary to the expectation created by his definition of nature, especially in the case of his treatment of form, where he argues *in propria persona*. The effect (if not the design) is to draw our attention to the idea that, if his discussion is in good order, form must be a principle of motion too.

III

I have suggested that, in II.1, the idea that form is a principle of motion is conspicuous by its absence. But it is not just its absence that makes the idea conspicuous; there is the further point that, as Aristotle is very much aware, the idea presents difficulties of a kind that are salient in this context.

The definition of nature, in conjunction with the idea that form is nature, implies that form is a principle of motion. The reason Aristotle will expect this result to be difficult is that his predecessors (he thinks) were unanimous in the view that *nothing* can be a source of movement that is not itself in motion. (This is clear from his account of earlier theories of soul in *On the Soul* I.) Surely Aristotle realizes that the obstacles in the way of thinking correctly about how *soul* is a principle of motion will likewise stand in the way of thinking that *form* is a principle of motion. This is not merely because, in Aristotle's view, soul *is* form. There is the further point that the definition, in making nature a principle of motion, is evidently meant to correct the Platonic idea that *soul* is a principle of motion.[7] Given this, Aristotle will almost certainly have had the topic of soul on his mind in writing *Physics* II.1; not only that, but he will have expected his definition to put that topic in the mind of his audience. But in that case it is very unlikely that the difficulties involved in making form a principle of

[6] This point depends on a *contrast* between nature and art: Unlike "doctoring," which is not named after what it is "to" – it is a road to health, not to medicine – the word "nature" or "growth" (φύσις) is different. (It is true that this contrast is rooted in an underlying similarity: just as "doctoring" [ἡ ἰάτρευσις] takes its name from doctors [or more precisely their art, "medicine" (ἰατρική)], because [after all] it is an undertaking of doctors, so too "growth" [φύσις] takes its name from "growing things" [τὰ φυόμενα], or more precisely from "nature" [φύσις], and that for just the same reason, namely that *it* is an undertaking of *them* [193b17–18].)

[7] On the Platonic background to Aristotle's definition of nature, see e.g. Mansion 1946: 83ff.; Solmsen 1960: 95ff.; and Wieland 1970: 234, 240ff.

movement would simply not have occurred to him, or would have occurred to him but struck him as not salient in this context.

Here it might be objected that surely these difficulties are easily sidestepped by reflection on Aristotle's well-known comparison of nature to art (which is also very salient in this context). This comparison invites a picture of nature-as-form as being (like art) a kind of capacity or power (δύναμις); on such a picture, it seems straightforward enough just to say that form is a "principle" of movement by being a capacity for movement (perhaps better, a whole suite of capacities, of which a whole ensemble of movements and rests are the manifestation or exercise (ἐνέργεια)). Simply put, why should the idea that form is a principle of motion be any more problematic than the idea that art is?

Plausible as this may seem, it is not the way Aristotle actually goes in *Physics* II; he does not highlight the idea that form is a capacity, and that perhaps with good reason. Certainly it is no general solution to his problem – how can something altogether immobile be a principle of movement? – to reply that the issue presents no difficulties when the something in question is a capacity. There will remain particular questions about the prime mover, not to mention more general questions about whether there must not be something else, a prior principle, whereby this capacity is "energized." Note too that Aristotle's principal ground for saying that form is nature is that form is "actuality" (ἐντελέχεια); although this doesn't rule out that form is also capacity, it hardly puts the point in relief. What is more, as we will see, where Aristotle does emphasize form's causality, what he emphasizes is its role as an end.[8]

I return to the idea that form is an "end" (τέλος) below. My point now is just that the idea that form is a "principle" (ἀρχή), and that precisely of *motion*, stands out in II.1, not just from Aristotle's failure to mention it, but also from the fact that it presents difficulties he will have expected his audience to be very much aware of.

IV

I turn now to the next text on our agenda, *Physics* II.7.[9] This chapter's official business is to conclude the discussion of causes begun in II.3,

[8] This is forecast already at 193b12–18 (discussed briefly above, where it was noted that in this respect – in its role as an end – nature is precisely contrasted with art).

[9] The definition is used in passing in II.6 (197b32–37), but the issues there are too slight to warrant discussion in the main text. (The distinction between chance and luck is of some interest, considering that Aristotle often speaks of nature as a cause to be ranged alongside of "reason" or "thought" [νοῦς,

drawing its lesson for the study of nature, which is that *it* is concerned with all four (198a22–24, b5–9). This is *the* conclusion of II.7, and it is in the course of arguing for it that Aristotle appeals to the definition of nature (twice, at 198a28–29 and 198a36–b1). My ultimate focus here is on the definition's role in this reasoning, but it will take a little time to get there; the basic movement of thought in this relatively neglected chapter is not transparent.

I start with some summary, taking the chapter in two pieces (198a14–31, a31–b9). First, after a brief restatement of the main results of II.3 (198a14–21), Aristotle begins by announcing his conclusion: The student of nature will know all four causes (198a22–24). For reasons that are not immediately apparent, he next remarks that three of them (form, mover, and end – τὸ εἶδος, τὸ κινῆσαν, τὸ οὗ ἕνεκα) "often come to one"; he supports this by claiming that form (τί ἐστι) and end (οὗ ἕνεκα) flat-out are one and the same, while the first mover (ὅθεν ἡ κίνησις πρῶτον) is the same as these "in kind" (τῷ εἴδει), defending this latter point by appeal to an example familiar from II.1 ("man begets man"), which example he says generalizes to all moved movers (198a24–27). This generalization prompts an apparent aside about *un*moved movers, to the effect that they lie outside the study of nature; it is here that Aristotle makes his first appeal to the definition "such as are not of this kind are no longer inside the province of physics, for they cause motion (κινεῖ), not possessing motion or a principle of motion in themselves, but being themselves incapable of motion" (198a28–29).

Second, though making as if drawing a consequence from all this (ὥστε κτλ.), Aristotle continues by picking up a different thread entirely, dropping the three causes he has just said are one and picking up another three – matter, form (τί ἐστιν), and first mover (τὸ πρῶτον κινῆσαν). He starts by saying that it is by referring to *these* three that (as it is) people explain things (198a31–33); this is his own interpretation of what people do, which he defends by pointing to the questions that actually preoccupy them (198a33–35).[10] With this description of the *status quo* in hand, he then proceeds to enlarge it so as to encompass his preferred ideal. His first step is to "divide"

διάνοια] and contrasted with "chance" or "luck" [196a30; cf. 196b22, 198a4, 6, 10, 12]. The contrast is marked by the presence or absence of order, as is clearest in a passage from II.4, in which Aristotle writes as if it were accepted by all parties that "nature, reason, and the like" differ from chance and luck in that the proceedings ascribed to them unfold in a definite order [196a28–35; cf. *PA* I.1, 641b10–28]. Such passages make clear that Aristotle regards the association of nature with order, not as an innovation, but as a piece of common ground [on this point, see also Scharle, Chapter 5 in this volume].)

10 "For in respect of coming to be it is mostly in this last way that causes are investigated – 'what comes to be after what? what was the primary agent or patient?' and so at each step of the series." Aristotle

the moving cause – what he here calls "natural kinetic principles" (αἱ ἀρχαὶ αἱ κινοῦσαι φυσικῶς) – into two kinds (198a35–36). Noting that one of these, although it causes motion "naturally" (φυσικῶς), is not itself "natural" (φυσική), because it lacks an internal principle of motion (this is his second appeal to the definition), he then claims that this non-natural sort of cause encompasses all unmoved movers, giving as examples the prime mover and form, and defending the latter (or is it both?) on the grounds that it is an end (198b1–4). Only then does he finally draw the conclusion announced at the outset, that the student of nature should explain in terms of all four causes, and this on the grounds that such a one (of all people) should know nature, which is for something (198b4–5ff.).[11]

Although these two pieces of text are apparently meant to comprise a single discussion, aimed at a single conclusion, it is not obvious how to put them together; certainly their several contentions are not simply so many premises of a single deductive argument. I will work by first just identifying some overlapping threads, and then trying, with the help of a little friction and spit, to twist these together into a single strand.

So, first, both pieces at least tacitly identify formal and final cause (the first identifies them outright, while the second does so implicitly, in claiming that form is a "kinetic principle" (ἀρχὴ κινοῦσα) *because* it is an end) (198b1–4). Second, both pieces associate moving cause with formal/final cause (again, the first does this outright, while the second does it implicitly, offering form as an example of an (unmoved) kinetic principle) (198b1–3). Third, both pieces at least tacitly divide moving causes into "moved" and "unmoved," and, fourth, both appeal to the definition of nature to mark the latter as not "natural" (φυσική) or "belonging to the study of nature" (φυσικῆς).

These common elements identified, I propose to read the second piece of text as conducting the argument proper, drawing on materials supplied by the first and refining them as necessary along the way. Read this way, the

means to align these three questions with his three causes, namely (respectively) form, mover, and matter. Note the concluding phrase, "and so at each step of the series," i.e. until we come to the first.

11 "Hence since nature is for the sake of something, and this we must know, we must explain the 'why' in every sense of the term (ὥστε ἐπεὶ ἡ φύσις ἕνεκά του, καὶ ταύτην εἰδέναι δεῖ, καὶ πάντως ἀποδοτέον τὸ διὰ τί)." I take the words "and this we must know" (καὶ ταύτην εἰδέναι δεῖ) as part of the dependent clause introduced by "since" (ἐπεί). This is because I think that (1) the antecedent of "this" (ταύτην) must be "nature" (φύσις), in which case I think that (2) the words "and this we must know" must give a premise, not a conclusion. So too (on both points) Philoponus 305.11–13 (Vitelli) and (on (1)) Simplicius 368.12 (Diels). (Simplicius considers (2), although he does not prefer it.) Apropos (1), many translate as though the antecedent of "this" (ταύτην) were "cause" (αἰτία, which occurs several lines back at 198a34) (so e.g. Charlton 1970, Hardie and Gaye 1930, Irwin and Fine 1995), but I think this is difficult.

basic movement of thought is from a statement of established scientific practice to a recommendation of Aristotle's own ideal – from explaining things in terms of three causes to explaining them in terms of four (198a31–33, b5). The argument's basic strategy, encapsulated in the idea that "often the three come to one," is to bring out how a commitment to explanation in terms of the missing fourth is already implicit in a commitment to explanation in terms of the other three; in particular, the argument brings out how a commitment to asking "for what end?" is already implicit in a commitment to asking "what is it?," which in turn is already implicit in a commitment to asking after "the first moving cause" (τὸ πρῶτον κινῆσαν). In appreciating this series of commitments, it is a help to realize that final, formal, and moving cause "often coincide," at least in kind. This is a help, but it is not enough: In addition, one must also appreciate not only that kinetic principles are of two sorts, mobile *and* immobile, but also that the latter, though not themselves "natural" (φυσικαί), are indeed among the principles of natural science, and this because they do indeed cause motion, and do so "naturally" (κινεῖν φυσικῶς) (198a36) – not by "having" natures (as Aristotle concedes), but by *being* natures (as he leaves us to infer).[12] Once we appreciate this – that these principles *are* natures – we will likewise appreciate their claim on the professional attention of the natural scientist.

This anyhow is how I propose to read "the argument of II.7" – the "official" argument for the chapter's stated conclusion. I turn now to the role of the definition of nature in this argument and begin with the following observations. First, although the definition's ostensible function is to exclude certain principles – to show they are not "natural" – it is clear that Aristotle also holds, and indeed must hold, if his reasoning is to support his conclusion, that these principles are also included in the study of nature: not as a department of its subject matter, but as among its explanatory principles. Second, although Aristotle's stated conclusion is that the student of nature will be concerned with all four causes, his "argument" for this is very indirect. Forget about the present account of it and just consider some of its raw materials: the association of formal, final, and motive causes, the division of the last into moved and unmoved, or the division of the latter into God and form. Is all this necessary? Is it helpful, or at least more helpful than obfuscatory? Is this really the best way of arguing that the student of nature will be concerned with all four

[12] The distinction between "having" a nature and "being" a nature is an important theme in Scharle 2008.

causes – i.e. final causes too, and not merely the other three? Surely not. Third, not only does the chapter's argument proceed circuitously, but Aristotle appears to take up the controversial element in its conclusion – that the student of nature must seek final causes too – again in the very next chapter, in a way that is vastly more direct and more satisfying.

These observations invite the thought that if there is a method to this madness, it is not to be read off the chapter's explicit signposts. With that thought in mind, I propose that we read II.7 in light of the problem thrown up by II.1, namely, the problem of how *form* can be a principle of *movement*. We know that Aristotle thinks this a difficult problem, and that he thinks the answer to it lies in the point that "ends" (τέλη) are genuine "beginnings" (ἀρχαί) – just the point left uncovered by the meandering argument of II.7. Read this way, against the backdrop of II.1 – which II.7's use of the definition will remind us of anyway – perhaps there is some sense in a proceeding that seems otherwise rather pointless.

V

So far I have suggested that one purpose of the definition is to highlight the idea that form is a principle of movement, that this idea will be felt to create difficulties, and that resolving these difficulties requires an appreciation of the fact, mentioned in II.7, that ends are kinetic principles too. Putting these points together, the result is that, for Aristotle, nature is indeed a principle of movement, not only by being a capacity for movement, but also and primarily by being an end. Before turning to the last set of passages on our agenda, I want to pause briefly to develop this result.

The chief interest of this idea – that form is a principle of movement, primarily by being an end – lies in the shape it gives to the definition's implications for how to interpret the phenomena of nature. The definition makes nature a principle of "movement and rest"; in so doing, it places nature at the origin of *behavior* (as distinct from *genesis*) – that is, it tells us that the substance of things due to nature is implicated in, and thus revealed by, what they *do*. But to add to this, first that the nature and substance of things is their form, and then that the way form is implicated in behavior is primarily by being its end, is now to say something really striking about things due to nature. It is to say that the point of their characteristic behavior is to perpetuate their distinctive form of being.

I do not expect that this result will be surprising, considered just as a thesis Aristotle subscribes to; on the contrary, I hope it will be recognized as confirmed from elsewhere – perhaps most strikingly in *On the Soul* II.4, where Aristotle puts precisely this interpretation on reproduction, going on to claim that for just this reason reproduction is the most "natural" (φυσικώτατον) (415a26) of all the works of living things. The novelty (such as it is) lies in seeing not that this is something Aristotle holds, but that it is a quiet, controlling theme of *Physics* II as a whole. The lesson of Aristotle's definition of nature is that the key to deciphering natural phenomena – to seeing them for what they are – is to interpret the movement and rest of things due to nature as for the sake of immortalizing their being and form.[13]

VI

I have observed that II.1's definition of nature, in conjunction with the identification of nature with form, implies that form is a principle of movement, and I have suggested, first, that we are expected to find this difficult – how can something immobile be a principle of movement? – and, second, that we are to recognize in II.7 the elements of a solution, inasmuch as that chapter goes out of its way to say that forms are ends and that ends are immobile principles of movement. Notice that this solution is no good unless the particular forms we call *natures* are ends, and that just this is the burden of II.8: that nature (in the sense of form) is a final cause.[14]

So far as that goes, then, II.8 as a whole may be read as advancing the line of thought we've been following. But it remains to consider the two passages in II.8 in which Aristotle actually uses the definition of nature (199b9–26, 28–32). Both passages occur towards the very end of the chapter; I take them in turn.

In the first, Aristotle is arguing against an implication of a view he finds in Empedocles, namely that "coming to be among seeds too would have to have been as luck would have it" (199b13–14).[15] Aristotle's complaint is not

[13] This sits well with the recurring example (in II.1 and II.7): "man begets man."

[14] It is true that, in formulating the chapter's conclusion, Aristotle sometimes uses the phrase "for something" (ἕνεκά του) (rather than "that for the sake of which" [οὗ ἕνεκα]) (notably *ad init.* and *ad fin.*). But it is plain that in these passages the phrase "for something" functions to specify *which kind of cause* Aristotle is here claiming nature to be. If that is right, then when Aristotle says that nature is "among the for-something causes" (τῶν ἕνεκά του αἰτίων), or that it is a "cause in the for-something way" (αἰτία οὕτως ὡς ἕνεκά του), what he means is that it is an *end* (as opposed to *for* an end). If doubts remain, they are removed by another, fuller formulation, in which it is clarified that the "nature" this conclusion holds of is *form*, seeing as form *is* an end:

> And since "nature" means two things, the matter and the form, of which the latter is the end, all the rest being for the sake of the end, the form must be the cause in the sense of "that for the sake of which" (καὶ ἐπεὶ ἡ φύσις διττή, ἡ μὲν ὡς ὕλη ἡ δ' ὡς μορφή, τέλος δ' αὕτη, τοῦ τέλους δὲ ἕνεκα τἆλλα, αὕτη ἂν εἴη ἡ αἰτία, ἡ οὗ ἕνεκα). (199a30–32)

[15] Trans. Charlton 1970.

simply that this is false, or even impossible, but (stronger) that "a person who says that does away with nature and things due to it altogether (ὅλως δ' ἀναιρεῖ ὁ οὕτως λέγων τὰ φύσει τε καὶ φύσιν; 198b14–15)."[16] His reasoning draws explicitly on the definition of nature and goes something like this: If seeds grow by nature, then (unless something interferes) they grow into the same thing every time,[17] in which case it can hardly be a coincidence that they grow into that – which it would be if they grew "as luck would have it," which they would if their growth were not for an end (199b14–26). Of interest in this reasoning is the broader moral: that nature will be a principle (ἀρχή) only if it is an end (τέλος). For in condensed form the argument is simply this: If nature really does stand at the beginning of the proceedings whereby seeds grow into plants, then the result (if nothing interferes), namely nature as form, will be precisely what those proceedings are *for*.[18] If that is right, then Aristotle's use of the definition here complements the use he makes of it in II.7. There the moral was that nature (as form) can be a motive principle *if* it is an end; here the moral is that nature will be such a principle *only if* it is an end. That is why to deny that nature is an end is "[to] do away with nature and the things due to it altogether" (198b14–15).

The second passage I want to consider occurs at the very end of II.8:

> If the ship-building art were in the wood, it would produce the same results by nature (ὁμοίως ἂν τῇ φύσει ἐποίει). If, therefore, purpose (τὸ ἕνεκά του) is present in art, it is present also in nature. The best illustration is a doctor doctoring himself: Nature is like that (μάλιστα δὲ δῆλον, ὅταν τις ἰατρεύῃ αὐτὸς ἑαυτόν· τούτῳ γὰρ ἔοικεν ἡ φύσις). (199b28–32)

This illustration evokes a passage from II.1:

> [A] man who is a doctor might cure himself. Nevertheless it is not in so far as he is a patient that he possesses the art of medicine: It merely has happened that the same man is doctor and patient – and that is why these attributes are not always found together. So it is with all other artificial products. None of them has in itself the source of its own production. But while in some cases (for instance, houses and everything else wrought by hand) that principle is

[16] Trans. Charlton 1970.

[17] It is here that Aristotle appeals to the definition: "for 'by nature' is everything that, moved continuously from some principle in itself, arrives at some end" (φύσει γάρ, ὅσα ἀπό τινος ἐν αὑτοῖς ἀρχῆς συνεχῶς κινούμενα ἀφικνεῖται εἴς τι τέλος; 199b15–17). (As he clarifies immediately, he does not mean that all arrive at the same end, nor just that all arrive at some end or other, but that each arrives at the same end every time, unless something interferes: ἀφ' ἑκάστης δὲ οὐ τὸ αὐτὸ ἑκάστοις οὐδὲ τὸ τυχόν, ἀεὶ μέντοι ἐπὶ τὸ αὐτό, ἂν μή τι ἐμποδίσῃ [199b17–18].)

[18] Aristotle's reasoning in this passage may be usefully compared with that in *Phys.* II.8, 199a8–12 and *PA* I.1, 641b23–642a1.

> in something else external to the thing, in others those which may cause a change in themselves in virtue of a concomitant attribute – it lies in the things themselves, but not in virtue of what they are. (192b23–32)

Significant in both of these passages is the idea that nature's work is *reflexive*: As Aristotle puts it in II.1, unlike artifacts, which (being wrought by hand) have "the principle of their making outside them," things due to nature have this principle "in themselves." The claim Aristotle makes here is arresting; how could anything contain within itself "the principle of its own making" – this seems to require that it pre-exist its own coming into being. The solution, I submit, lies in seeing that, for things due to nature, the "principle of making" and "principle of movement" are one and the same: that is, although their "making" and "moving" – that is to say, their genesis and behavior – are distinct proceedings, they have the same principle, and that because they have the same end.[19]

Earlier I suggested that one of the definition's important morals is that the phenomena of nature are for the sake of perpetuating being or form. This moral is reinforced by Aristotle's use of the definition at the end of II.8, and it makes Aristotle's closing remark in that chapter highly apposite: "the best illustration is a doctor doctoring himself: nature is like that." In a word, nature mends itself.

VII

My suggestion, then, is that the definition of nature holds the "key" to interpreting both the phenomena of nature (as Aristotle understands them) and *Physics* II itself. The "key" is the idea that form is a principle of motion and rest. The definition "holds" this key by setting off the following sequence of ideas: that forms are natures, which means that they must also be principles of movement, which they can be if and only if some principles of movement – indeed, the truest ones, i.e. the ones that are most of all "first," and thus most of all "principles" (ἀρχαί) – are immobile, causing motion naturally by being its end. This sequence of ideas shows us how to interpret the phenomena of nature, because its meaning is that movement and rest are a striving for being (and thus an imitating of the "one ruler" (εἷς κοίρανος), the true "principle" (ἀρχή), "which is completely unchangeable and the first of all things (τὸ παντελῶς ἀκίνητον καὶ

[19] As Aristotle puts it in II.7, things behave as they do "because it is better thus, not without qualification, but in relation to their substance or essence" (διότι βέλτιον οὕτως, οὐχ ἁπλῶς, ἀλλὰ τὸ πρὸς τὴν ἑκάστου οὐσίαν; 198b8–9).

πάντων πρῶτον)" (*Metaph.* Λ.10, 1076a4, *Phys.* II.7, 198b2–3)). It shows us how to interpret *Physics* II by revealing that one part of that book's agenda is to clear away the obstacles to the idea that form is a principle of movement. We notice this by keeping track of and interrogating Aristotle's use of the definition in II.1, II.7, and II.8.

This last suggestion, about how to interpret *Physics* II itself, is of course very tentative. Even if it were accepted, the question would remain: Why does Aristotle proceed so obliquely? It is maybe a step towards answering this question to consider the alternative, which would have been to argue for the idea directly: Since form is nature, and nature a principle of motion, form must be a principle of motion too. What is striking about this argument is that, even allowing that its premises are in proper order, each resting on a decent set of more or less independent considerations, it is powerless to persuade; the obstacles to the idea that form is a principle of motion, not arising from the lack of an argument that would demonstrate it, are not removed by the supply of one. We might also ask exactly what Aristotle is trying to accomplish in *Physics* II: Is it to demonstrate theorems, for the books and for the ages, or rather something different, e.g. to help make a particular audience "capable of judgment" (δυνάμενος κρίνειν) (193a4–5). The latter is arguably the more difficult undertaking, in which indirect methods and even hidden themes are perhaps sometimes the better course.

CHAPTER 3

Nature as a principle of change

Stasinos Stavrianeas

Introduction

The study of nature, says Aristotle, aims at the understanding of the principles, causes, and elements of the natural world. This is the fruit of an investigation that starts from things more familiar to us, yet mixed with much confusion as to their true nature, and proceeds towards things less familiar to us but more intelligible by nature, i.e. to the discovery of this true nature (*Phys.* 1.1, 184a14–23). The things that are familiar to us are compounded, comprehensive, and closer to perception, and we understand them in a universal and undifferentiated manner; further investigation is needed in order to grasp our subject matter in a more detailed manner. Our first conception does not have the status of scientific understanding, but it has the advantage of being based on some sort of inductive reasoning that lends to it the status of an obvious and fundamental assumption.[1] One such assumption is that the natural world of changing things exists (*Phys.* 1.2, 185a12–13). Starting from this, *Physics* 1 investigates the principles necessary for change, natural or otherwise. This investigation meets the programmatic thesis that our study should move from what is more intelligible to us towards what is more intelligible by nature, but merely identifying the principles of change does not mean that the analysis of change is complete. Other principles and causes need to be added as well, such as most prominently the efficient cause. In other words, nothing in the *Physics* 1 analysis of change provides in itself sufficient material for understanding the complexity of – say – the changes involved in animal generation.

I am grateful to Mariska Leunissen for stimulating and extensive comments that helped me to elaborate and clarify many of the ideas included here, to Pavlos Kontos and Lampros Spiliopoulos for generously commenting on earlier drafts, and to Andreas Anagnostopoulos, Panos Dimas, Vassilis Kalfas, and Spyros Rangos for comments and discussion.

[1] See Bolton 1991: 27–29.

Physics II turns to the investigation of the principles of *natural* change, and first to what nature is, i.e. to a definition that demarcates the entities that possess a nature. Still, further investigation will be needed in order to determine in detail that which is characteristic of each kind of natural being. In that sense, the definition, or so it may be argued, no more gives us an articulated picture of what nature is than the analysis of change in Book I, in terms of the principles of form, subject, and privation, gives us a complete account of what change is. So although *Physics* II.1 establishes more clearly what nature is, this is by no means the complete account. It is understandable, then, that the proposed definition of nature as an inner principle of change and rest is often found wanting. Both this definition and the ideas associated with it require further elaboration before they can yield a *complete* understanding of nature.

My aim in this chapter is to examine the conditions that Aristotle's preliminary definition of nature in *Physics* II.1 introduces. I argue that the claim that nature is a principle of change need not be read as offering a complete and final articulation of what nature is, but rather should be interpreted as a general account or preliminary sketch that will be qualified in order to accommodate the nature of various natural kinds. For, ultimately, it is the detailed study of natural kinds that Aristotle presents in other works that provides a much richer network of distinctions and a more precise determination of what is characteristic and distinctive of natural beings. In section I, I specify the conditions included in Aristotle's definition of nature. In sections II and III, I explain the way in which natural generation satisfies these general conditions, while artificial production does not. Section IV shows how even the locomotion of the elements counts as natural given these conditions, despite the fact that the elements are not self-movers.

I Aristotle's definition of nature in *Physics* II.1

The list of entities that exist by nature includes animals and plants, parts thereof, the four simple bodies, and, most likely, their mixtures. The reason these entities exist by nature is, according to Aristotle, that "each of them has within itself a principle of change and rest, some in respect of place, some in respect of growth and decline, some in respect of alteration" (*Phys.* II.1, 192b13–15).[2]

[2] Animal parts and elements are excluded in other works, because they lack the kind of unity necessary for being substances, let alone natural ones (*Metaph.* Z.16, 1040b5–10). Animal parts are, I think, included here only for the sake of completeness. The case of elements is dubious, but at least according to the *Physics* they possess a principle of change in themselves, though in a special way

Two things deserve comment here. First, it is uncertain how we should interpret the expression "principle (*archē*) of change." A natural way to understand it, and the way in which it has been traditionally read, is as denoting the "efficient cause" of change.[3] This reading, however, is questionable for several reasons: (1) Nothing in what precedes in *Physics* I indicates that "principle" refers merely to the efficient cause;[4] (2) this reading seems to exclude several types of *natural* changes of entities such as the elements, their mixtures, and possibly some of the animate beings, because they are not produced by an internal efficient cause; (3) the arguments offered in *Physics* II.1 supporting the identification of nature with either the matter or the form of entities that are considered to exist by nature do not turn exclusively on the activity of matter or form as efficient causes (Kelsey 2003: 61). Rather, the ways in which something can be the nature of an entity includes (in many cases) all four Aristotelian causes.

Second, it is worth asking whether the definition of nature Aristotle offers here is supposed to count as a scientific definition in the strict *Posterior Analytics'* sense. I do not believe it is; it is rather a first approximation or nominal definition that states a common, general characteristic of natural things. More precise definitions are to be provided elsewhere, and they will track the different ways in which various kinds of entity possess a nature. Starting an investigation with a general definition is not unprecedented in Aristotle. In *On the Soul*, he argues that it would be absurd to try to provide a *common* definition of the soul without also working towards the definitions of *particular* kinds of soul that correspond to the proper indivisible species of ensouled beings,[5] as a common definition of the soul will not be able to fully track the latter. Even though Aristotle is not explicit about this, I believe that the same methodological considerations that pertain to soul also pertain to nature, especially given that nature, in animate beings, is in fact their soul (*PA* I.1, 641b9; *GA* II.4, 740b37–38).

(see section IV). This way may not, in the end, be sufficient for attributing to them a nature, thus excluding them from the list.

[3] See, for instance, Simplicius *In Ar. Phys.* 264.9–10 (cf. *Phys.* II.3, 194b29–30). However, all four causes are principles (*Metaph.* Δ.1, 1013a17).

[4] The word is used extensively in *Physics* I (forty-one occurrences), but only in one of them does it refer to efficient causes (*Phys.* I.6, 189a24–26, referring to the Empedoclean powers of Love and Strife). The word usually refers to the principles of change, such as the subject of change, its form, and its privation (this is also the case when it is used in conjunction with the word *aitia* [cause]: see e.g. *Phys.* I.7, 190b17–20 and I.9 192a13).

[5] According to *De Anima* II.3, there is a single *logos* (account) of soul only in the sense in which there is a single account of the genus "rectilinear figure." Since this account is common to all rectilinear figures but not unique to any one of them, Aristotle says it is the definition of none (414b20–28, cf. *DA* I.1, 402b6–9).

Given Aristotle's remarks about the need to provide more specific definitions of the soul to fill in the more general one already provided in *On the Soul*, we should expect the same to be the case regarding Aristotle's definition of nature in the *Physics*: It will be satisfied by different natural beings in different, yet related, ways.

The fullest formulation of the definition introduces two further conditions on nature being a principle of change and rest: namely, that nature is such a principle in that in which it inheres (a) primarily and (b) *per se*, i.e. not *per accidens* (192b21–23). Condition (a) exploits the distinction between something belonging to a subject primarily or immediately and it belonging to a subject non-primarily or non-immediately. An attribute, *F*, belongs to a subject, *s*, *non-primarily* if it belongs to *s*, *because* there is a further subject, *s'*, such that *F* belongs to *s'* and *s'* belongs to *s*. By contrast, an attribute, *F*, belongs to a subject, *s*, *primarily* if it belongs only to *s* and to no other subject (or, if it belongs to other subjects as well, it belongs to them *because s* is attributed to them). Condition (b), on the other hand, exploits the familiar distinction between belonging to a subject *per se*, i.e. necessarily, and belonging to a subject accidentally, i.e. contingently.[6]

The two distinctions are independent. First, an attribute may belong to a subject *per se*, but not primarily, in those cases where it belongs to it in virtue of belonging to some other property which in its turn belongs to the subject *per se*: Having the sum of its three angles equal to two right angles belongs *per se*, though not *primarily*, to an isosceles triangle, because it belongs primarily and *per se* to triangle and triangularity in its turn belongs *per se* to isosceles. Second, an attribute may belong primarily to a subject, though not *per se*, in those cases where the relation is not mediated by a third term and yet the attribute is not essential to the subject. This is the way white, for instance, belongs to a surface. Aristotle provides illustrations for both conditions in his text, namely essentiality and immediacy.

The condition of essentiality is used in order to block one way in which things that do not exist by nature may be thought of as possessing an inner principle of change. Aristotle's example is a physician who happens to be ill, and who can thus cure *himself*. The restoration of health is the product of his craft, i.e. medicine, so in this case the principle directing the change belongs to the person who suffers the change. They are one in number, and this may be taken as suggesting that products of craft *may* possess an inner principle of change. Hence, Aristotle needs to explain why the definition does not apply in such cases. He does that by noting that here the agent and

[6] See Simplicius *In Ar. Phys.* 267.21–268.3, followed by Ross 1936: 501.

the patient, i.e. the source and the subject of the change, may belong to one and the same entity but they belong to it under different causally relevant descriptions. Hence, not only *can* source and subject be separate in number (no physician needs to be a patient and *vice versa*), but they also *must* be separate in definition or essence (*GA* I.18, 724a43–45; II.1, 735a2; II.4, 745b28–29). So, in craft, the principle *may* be internal, accidentally, but *needs* to be external, essentially.

The condition, of immediacy, is plausibly at work in the following passage:

> On the other hand, a bed and a coat and anything else of that sort do not in so far as they happen to receive these predicates – i.e. in so far as they are products of craft – have an innate impulse to change. But in so far as they happen (*sumbebēken*)[7] to be composed of stone or of earth or of a mixture of the two, they do have such an impulse, and [they have it] just to that extent. (*Phys.* II.1, 192b16–20)

The examples here illustrate a different way in which artifacts may be thought of as possessing a principle of change, compared with the self-doctoring doctor. The latter exemplifies the coincidence of active and passive powers in one entity. The former focuses on a principle of change belonging to artifacts in virtue of the material they are made of. Indeed, artifacts *may* be thought of as possessing an internal principle, because their matter possesses such a principle (or is itself composed out of matter that possesses one). For instance, a coat will fall to the ground if not hung from a nail; a balloon will fly upwards. However, such motions are not due to formal properties of the artifacts, but rather to the material constituting them. Thus, the corresponding principle belongs primarily or immediately to the matter. Since the matter constitutes the composite, its principle belongs to the composite as well, but only in virtue of belonging to the matter, thus non-immediately; therefore it does not belong to the artifact properly speaking. We may take then the two examples as illustrating two independent conditions for possessing a nature: essentiality and immediacy.

[7] The use of *sumbebēken* for characterizing the way artifacts relate to their matter may suggest that what is at issue here is, again, the distinction between *per accidens* and *per se* belonging. However, this distinction does not block the claim that the matter of artifacts is their nature, since the latter is conditionally necessary and not accidental to them. Hence, it is arguable that they possess *per se* the principle of change in their matter. Further, *Phys.* I.3, 186b18–20 introduced two senses of *sumbebēkos* (attribute): (a) accidental attributes (the principle of medicine belongs in this way to the patient); and (b) attributes in whose account the subject to which they belong is included (the principle of the matter belongs in this way to the artifact as a compound of matter and form). Cf. *Phys.* VIII.4, 254b8–10 and Simplicius *In Ar. Phys.* 1207.46–56.

The need for two independent conditions becomes relevant if we turn to the second part of *Physics* II.1, where Aristotle rehearses the competing arguments for identifying nature with either matter or form (193a9–b21). The argument in favor of matter suggests that the matter of natural beings belongs both essentially and immediately to them, while the matter of artifacts satisfies only the former condition. If so, then both conditions are needed for understanding the special role of matter in animate beings and its contribution to their nature. The argument proceeds from the claim that the nature of any material being is its first constituent, which in itself does not possess the arrangement of what it constitutes, as wood does not possess the arrangement of bed. Two characteristics of this constituent matter support the claim: First, matter, by being able to survive the loss of what it constitutes, has some ontological priority over the compound; second, if matter had a power to generate anything, it would generate something similar to itself rather than something similar to the compound, suggesting again that the principle of change belongs to it rather than to the arrangement of the compound (193a9–11, cf. 193a29). The first aspect turns on a criterion of survivability, while the second attributes a generative power to what survives. Satisfaction of both criteria grounds immediacy, but it is the second one that reflects a radical difference between artifacts and natural beings.[8]

The first criterion, survivability, does not seem to be a necessary condition concerning the contribution of matter that Aristotle endorses. For there are cases where Aristotle claims that the matter that survives the perishing of a compound is the same only in name or "homonymously."[9] This principle of homonymy seems to go against survivability, if the latter means that the matter persists as the same type of thing throughout in all cases. And even if Aristotle only endorses a weaker version of survivability, it seems that both artifacts and natural beings can satisfy it. Aristotle therefore needs a second criterion to further distinguish natural beings from artifacts.

[8] Even if we grant that wood, perhaps *qua* plant, generates wood, it seems weird to make the same claim for earth. Still, Aristotle argues in *On Generation and Corruption* that generation by reproduction of a specifically similar individual, as when human generates human, is on a par with the production of something generically the same, as when fire generates fire (I.5, 320b19–21). The argument here, however, does not require that natural entities literally reproduce themselves: It aims to show that if the matter of artifacts could generate something, what will be common between "parent" and "offspring" will be the material and not the arrangement. The fact that artifacts made out of, say, stone do not generate stone does not contradict this point.

[9] This principle applies to the matter of animate beings (*Metaph.* Z.10, 1035b25, *PA* I.1, 641a1), but possibly also, in a weaker way, to artifacts (see *Pol.* I. 2,1253a22, *DA* II.1, 412b15, *Metaph.* Z.10, 1034a22–1034b1, *Meteor.* IV.12, 390a12).

Aristotle's example of the wooden bed shows that the matter of artifacts fails to satisfy this second criterion, while the fact that the matter of natural beings does satisfy it was already suggested by Aristotle's claim that the underlying thing out of which a compound is constituted is the first thing out of which the constitution proceeds, which in animate beings is identified with their seed (or embryo: *Phys.* I.7, 190b2–3; I.9, 191a25–31). Now, the seed's power is analogous to the imagined generative power in the wood: It generates another being similar to the one it itself comes from. Thus, the nature of what gets generated can be partly identified with this matter that is its first constituent. Given this identity, the principle that immediately belongs to the first constituent will immediately belong to what it is the first constituent of. And, if so, the matter of natural things satisfies the criterion of immediacy as well, although properly speaking the generative principle merely resides in the matter (as we learn from Aristotle's *Generation of Animals*, it is the *formal* principle in the matter that generates; this specification is not pursued here). However, all this shows is that the matter of artifacts fails to satisfy the immediacy criterion, and Aristotle also wants to claim that artifacts also do not possess a nature because their efficient cause fails to satisfy the essentiality criterion (that is, their active and passive causes of changes are *always* separate).

II Aristotle's account of why artifacts cannot produce themselves

Artifacts and natural beings are contrasted in terms of the efficient cause involved in their production: "None of them [that is, artifacts] has in itself the principle of its own production" (*Phys.* II.1, 192a27–28). Since the production referred to here is a substantial change, the intended contrast seems to be that artifacts do not generate themselves, while natural beings do. However, even if this is in some sense true of natural beings, it is true only in a qualified sense. Strictly speaking, animate beings do not generate themselves. Thus, Aristotle's reason for denying that artifacts have a nature cannot merely be that they do not generate themselves. So what can it be? And whatever this reason is, will it also be able to exclude artifacts such as miraculous *automata* that can reproduce themselves as natural beings do?

The reason lies in the way Aristotle understands art as an exclusively human activity. This view surfaces in *Physics* II, but is formulated explicitly in other texts. For instance, in the *Nicomachean Ethics* Aristotle distinguishes art (*poiesis*) from action (*praxis*) by defining the former as "a productive disposition accompanied by true rational prescription (*logoi*)" (VI.4, 1140a10–11). Artifacts are produced by changes that stem from a disposition, i.e. some sort of psychological state in an agent that involves a capacity to attend to true *logoi*. And one can attend to such *logoi* in virtue of having a soul that is receptive to them in some way or another (*Int.* 1, 16a9–11). This in its turn suggests, first, that any craft contains an epistemic component, and, second, that to the extent that only human souls possess such an intellectual capacity, it is only humans that are able to invent, or internalize through learning, the principles that define a certain craft. Consequently, only humans can act as principles and efficient causes of artificial production.[10] This feature then differentiates artificial production from natural generation, in a way that puts the source of production in crafts *necessarily* in something separate from what is produced.

Aristotle comes to this view because of a conviction he shares with the materialist, namely that matter cannot organize itself so as to take the form of an artifact. The principles that reside in matter are not capable of effecting changes that will result in something that qualifies as the product of craft. At least not intrinsically and regularly so, for in some cases, such as the restoration of health, this may happen by accident. Aristotle makes this point in *Phys.* II.8, 198b15–17 when he claims that art generally either (a) completes what nature cannot bring to a finish, or (b) imitates nature. In cases under (a), what seems to be missing is an appropriate natural principle that can effect the necessary changes resulting in the corresponding end. Craft is described in *Metaphysics* A.1 as the result of the discovery of such principles that will realize specific ends, promoting utility or recreation.[11] What is essential for such a discovery is a piece of knowledge that is universal and explanatory, explaining what such and such material changes are good for. Cases under (b), on the other hand, are cases where the corresponding principle can be found in the domain of nature. A typical example is the restoration of health. Health can be the result of a spontaneous, accidental change, but it is more regularly and intrinsically produced by medicine. What differentiates these two cases is not the material changes in the body, but rather that in the latter case these changes are conceived, planned, and effected as something good for restoring the healthy state. Thus, medicine is described as comprising an intellectual, deliberative component, as well as a productive one (*Metaph.* Z.7, 1032b6–10, b15–23). The first component works towards discovering the principles

[10] See *Phys.* II.8, 199a20–23 and *Metaph.* A.1, 980b25–28. For the dependence of craft on practical wisdom and the tension this generates for Aristotle's analogy of craft with nature, see Broadie 1990.

[11] See *Metaph.* A.1, 981b13–17. The first inventors were admired because they were thought to be wiser, since they discovered something new, beyond the common perceptions. The object of their discovery must have been some universal and explanatory principle. See Cambiano 2012: 34.

that govern the production of health, while the second follows the opposite course by producing the healthy bodily state. What differentiates the two components is the intellectual grasp of the principles that govern the artistic process (absent in the spontaneous restoration of health) and that produce standardly beneficial results.[12] If this is so, then the dependence of crafts on knowledge of principles in the craftsman divorces the efficient cause of artificial production from what it produces, and entails that artifacts do not possess their principle *per se* or essentially (*NE* VI.5, 1140b6). Further, any other changes they can undergo *qua* products of such a principle will not belong to them essentially, for these changes will ultimately be explained by reference to the principle of their production.

To conclude, craft is primarily an intellectual capacity. The manual work necessary for artificial production is not a sufficient condition for the product to qualify as craft.[13] Master craft, i.e. art proper, must include, and is primarily identified with, universal and explanatory knowledge. In that sense it is necessarily separate (a) from the material changes directed towards the production of an artifact, and (b) from that particular artifact. For the case of miraculous *automata* this means that even if some craft could indeed produce artifacts that are able to effect their own changes, including reproducing themselves, these *automata* would not possess the principle of their changes in themselves *per se*, for their source is a principle that ultimately lies in human intelligence.

III Differentiating nature as a principle from the efficient cause

Artifacts, it was proposed, do not possess a nature because their principle of change is necessarily external. The principle of their matter does not belong immediately to them. And the principle producing their form and functions does not belong essentially, and need not belong immediately, to them (except accidentally), since it is something external. This may suggest

[12] *Physics* II.2 distinguishes two kinds of craft that involve knowledge: (1) one that uses the product; and (2) one that directs its production. (1) is directive in that it involves knowledge of the use and the form, while (2) is directive in that it involves knowledge of how the matter can acquire the qualities necessary for producing something appropriate for this use (194a36–b7). (2) also directs a further component, (3), the executing side of the craft, the actual manual labor needed. On this, see Pellegrin 2000: 1–26; cf. also *Pol.* III.11, 1282a2–5. *Metaphysics* A.1 distinguishes (1) from (3), which is relegated to the level of mere experience (*empeiria*) (980b5–12; b15–16), as lower-level manual workers act mechanically, like inanimate tools (981b2–5).

[13] This is confirmed by the fact that productive activity in other animals does not qualify as craft because it is not the result of inquiry or deliberation: see *Phys.* II.8, 199a20–21; cf. *HA* VII (VIII).1, 588a29.

again that in natural bodies the internal principle must be an internal efficient cause of their changes. However, several changes of natural bodies are caused by external efficient causes. For instance, the locomotion of simple bodies is initiated by external causes, and simple bodies and their mixtures suffer a number of natural qualitative or quantitative changes that are triggered by external efficient causes. Furthermore, animate beings seem to rely on external bodies for exercising some of their own natural capacities, e.g. nutrition, perception, and locomotion, and are thus at least in some sense moved by external efficient causes.[14] Of course, one could deny either (a) that the movements in question have *per se* and external efficient causes, or (b) that they are natural, i.e. depend on the nature of the being in question in an absolute sense.[15] Neither of these options, though, seems to fit with Aristotle's text.[16] A third option (c), which has recently been defended by Sean Kelsey, would be to deny the claim that possessing an internal efficient cause is necessary for possessing a nature that qualifies as a principle of change. I believe that option (c) is the most promising one, so let us explore Kelsey's interpretation further.

Kelsey's suggestion is to understand *principle* of change in terms of the following definition offered in *Metaphysics* Δ.1 (1013a10–13): "that in accordance with whose will what is moved is moved and what is changed changes." This definition of principle points to some kind of authority the principle has over what it is a principle of. In the *Politics* we find two types of authority relation between ruler and ruled, and Kelsey exploits them for distinguishing nature from craft, in a way that can accommodate the claim that some natural bodies are not efficient causes of their motion. The first authority relation, which Kelsey labels "*despotic*," is exemplified in the relation between master and slave, while the second, the "*non-despotic*" one, is exemplified in the relation between teacher and student (*Pol.* III.6, 1278b30–1279a8). The relevant difference between them concerns the location of the good that the authority relation aims at, what Kelsey calls the "*proper subject of the good produced*": Despotic authority relations aim at the good of the ruler,

[14] *Physics* VIII.4 argues that elements are not self-movers. For animate beings, see VIII.6, 259b6–20. See Johansen 2012b: 128–145 for a detailed defense of the claim that soul is part of the efficient causal story in nutrition, perception and locomotion in a way that does not threaten their possessing a nature.

[15] For the former view, see Waterlow 1982 and Graham 1999; for the latter, see Matthen 2001.

[16] Aristotle's claim that the movement of simple bodies is by nature and according to nature seems to rule out (b) (*Phys.* II.1, 193a1); the conclusion of *Phys.* VIII.4, 255b33–34, that they require external efficient causes of their movement, rules out (a). Alternative (c) is also followed by Scharle 2008: 171–172, who argues that elemental natures are principles of being changed and thus not principles as an efficient cause.

while non-despotic ones aim at the good of the ruled (Kelsey 2003: 76–77). Thus, even though changes involving non-despotic authority relations have an efficient cause, and this efficient cause is located in the ruler (which is the active component of the change and which is external to the proper subject of the change, namely the ruled or the passive component of the change), the ruled is still the subject and source of the changes undertaken, in the sense that the good of the relation is located in the ruled. Hence, a body can suffer a natural change, e.g. an elemental body can move naturally, just because the movement it suffers and the good produced is located in it, without possessing the efficient cause of that change internally.[17] A sufficient condition, then, for possessing a nature is to be the passive subject of a change undertaken in the context of a non-despotic authority relation where the good is located on the passive side of the change. In this way, Kelsey's interpretation can accommodate *natural* motions due to external efficient causes.

This proposal remedies the problem of natural changes that are not self-motions. Further, by interpreting nature in terms of a relation (and thereby classifying it in the category of relatives), it is easier to accommodate changes that are due to nature and at the same time necessarily involve external objects. For each of the capacities constituting the nature of a body, whether active or passive ones, will be related to a counterpart principle either in the material of the compound or to an external object.[18] However, I would like to amend the proposal in two respects.

First, the proposal seems rather restrictive by tying natural change to the subject of a good produced. There are natural changes that have external efficient causes, especially in inanimate bodies, where the result is not some good for the passive subject of the change. Rather, the result produced is a quality in the compound, the passive capacity for which belongs to the matter, but is ultimately determined by the form of the compound (*Meteor.* IV.2, 379b25–32; Aristotle mentions tears becoming rheum as an example). Aristotle explains that changes in inanimate bodies (such as solidification or liquefaction) involve three causes: the matter, the efficient cause, and the quality in the sense of form (IV.5, 382a28–30). The efficient

[17] Kelsey (2003: 79) writes:

> The suggestion then is that when Aristotle says that natural things have an internal principle of movement, he means that they are the proper subjects of their movements; by contrast, to say that artifacts have an external principle of movement would be to say that the proper subject of their movements is precisely not they themselves but something else distinct from them.

[18] On the classification of the natures of living beings under the category of relatives, see Johansen 2012b: 81–83.

cause is associated with the powers of heat and cold, while the dry and the moist are the passive powers that constitute composite bodies as matter, either directly or through the primary bodies, earth and water (IV.4, 382a1–7; IV.5, 382a31–32). Indeed, the passive and active factors of a change must belong to one genus, as the pairs wet–dry and heat–cold do (*GC* I.7, 323b31–32). True, we can also refer to what possesses these material properties, the compound constituted by the wet and dry, as the passive side of the change (*GC* I.7, 324a15–21). Indeed, an affection of its material properties brings about a change in the compound: For instance, fiery heat affects solid bodies, such as horn or iron, consisting of mostly earth and some moisture, and makes them soft (*Meteor.* IV.9, 385b6–11). Softness (or hardness) itself is a quality of the elemental material (*GC* II.2, 329b8–12), as well as of what this material constitutes. But the result of the change in the compound is the realization of a dispositional property, which is itself a characteristic differentia of bodies of its kind (*Meteor.* IV.4, 381b24–26; 8, 385a1–8), and the matter does not determine the limits within which the dispositional property is expressed. Otherwise, there would be no limit to the kind of change each body could suffer under the influence of heat or cold. Furthermore, it is questionable whether the expression of such dispositional properties is in every case something good for the subject. For although Aristotle seems to hold that even inanimate compounds have an end and function (*Meteor.* IV.12, 390a4–15), and this must be something good, not every change will result in something that accords with such a function. And yet such changes are due to the nature of the corresponding body. Thus, inanimate bodies seem to possess a number of dispositional properties as part of their formal natures, and such formal natures have priority as principles of the corresponding changes. If so, the inanimate body is the passive source of the change due to its material constitution, but the result of the change *ultimately* depends on its form, and this form can thus be identified as the internal principle.

Second, the proposal is also too generous. For it seems that being the proper subject of change, in the intended sense, applies not only to subjects of natural changes, but also to subjects of spontaneous changes. Typically, changes that result spontaneously resemble natural ones. They bring about an end-state, standardly achieved by an efficient cause aiming at that end-state, but in their case the efficient cause was not aiming at this end. So, for instance, the healing of a patient, standardly the result of the exercise of medical craft, may also be the product of a spontaneous process caused by environmental conditions. The difference between natural and spontaneous cases lies in the relation of the efficient cause producing that change and the good that results for the entity suffering the change. In spontaneous cases, the efficient cause and the end produced are related *per accidens*, while in standard, natural cases, the relation is *per se*. Now, if an entity suffers a natural change only in virtue of being the proper subject of the good the change produces, then it is not clear how to differentiate between causes that relate *per se* and causes that relate *per accidens* to the good produced, and hence between standard and spontaneous cases, for in both cases, the proper subject of the good produced is the entity that constitutes the passive side of the change, e.g. a generated organism. Both types of process, it seems, will count as natural.[19]

For an account of nature that can accommodate the above difference we need to turn to *Physics* II.1, 193a30–b18, where Aristotle gives three arguments for the claim that nature must be identified with form rather than matter. Aristotle there specifies three different aspects of forms, being an actuality, an efficient, and a final cause, but we only need to focus on the latter two, since they apply exclusively to the forms of living beings.

The second argument revisits the claim that nature must be identified with what persists throughout generation and corruption. But the thought implicit earlier, and equally present here, is that the nature of an entity is the element common between it and what generated it, since its persistence might be taken as evidence for locating the efficient cause of the change in it. What is common between generating and generated animal is that they belong to the same kind. And the nature of each kind is identified, sc. 193b3–4, with the corresponding form. So the nature of an animate being includes its form in its role as an efficient cause in reproduction. However, this nature acts as an efficient cause for generating a different specimen, and thus it is an *external* principle of what is generated and no part of its own nature. It seems, though, that the third argument qualifies this claim. In this last argument Aristotle contrasts changes produced by nature with those produced by craft. The latter do not lead to the craft itself but to a different form in the subject suffering that change, e.g. medicine starts a change that leads not to medicine but to the recovery of health. Natural generations, by contrast, proceed to nature, and thus what grows should be identified with what it will become, its form. The exact reference of Aristotle's mention of the thing that grows (*phuomenon*) is not certain,[20]

[19] The proposal creates problems for other changes as well: For instance, any change of place that an animal undergoes, where it is the proper subject of the good produced by that change, e.g. Daedalus flying from Crete to Icaria, would qualify as natural.

[20] The word is used primarily for plants (e.g. *DA* II.1, 413a25, *GA* I.23, 731a8, and *HA* VIII.19, 601b12), but could also refer to animal fetuses (e.g. *GA* II.3, 736a34–35).

but *Metaph.* Δ.4, 1014b20–22 informs us of the following: Those things are said to grow that derive augmentation from something else by contact and organic unity, or by organic adhesion, as in the case of embryos. It is thus plausible that Aristotle here has in mind animal embryos formed out of semen and catamenia (as well as the seeds of plants). What defines organic unities is that there is something common to both parts of the unity that makes them grow together and makes them become one with respect to continuity (*Metaph.* Δ.4, 1014b23–25, cf. *Phys.* v.3, 227a10–16).

If this is the stage of generation Aristotle has in mind here, it is one he touches upon in his detailed account in *Generation of Animals*, where we are told that seeds and fetations not yet separated from the parent possess nutritive soul only potentially. They do so actually when they start drawing nourishment to themselves (II.3, 736b8–10), and this happens once their first principle has become distinct, in the form of the heart in blooded animals or in the form of shoot and root in plants (II.4, 739b34–740a4). Previously, the growing thing was only potentially separate, while actually it was a unity with the parent, as the comment on growth from *Metaphysics* Δ suggests. But now that comment is connected with the following sense of nature: "the source from which the primary movement in each natural object is present in it in virtue of itself" (1014b18–20). If the embryo that is not yet actually separated is a unity with what generates it, and, additionally, if the principle of its primary movement belongs to what generates it in virtue of itself, it follows that the principle must, somehow, belong to the embryo. Further, the principle of this movement is, by the lights of the second argument, the form in the generating substance. So the embryo has as its principle a form that is also the form of the unity in which it participates. This should not lead us to the thought that the growing thing possesses its own internal efficient cause. It indicates, though, that by forming a unity with what generates it, the embryo shares in the form of that entity. And in *that* sense it, arguably, possesses an internal principle of change. To conclude, although the third argument aims to establish that the growing thing's nature is its final cause, the association of the growing thing in conception with its efficient cause as the form in the parent also suggests, together with the second argument, why the growing thing's nature is its end, and also why it is always an *internal* principle that determines the growing thing.[21]

[21] See *PA* I.1, 641b32–35 (and Lennox 2001b comments ad loc.) on a similar account of why the end can be called a nature, building on the dual nature of the seed involved in natural generation: "for the seed is a seed in two ways, as that *from* which and that *of* which; that is, it is a seed both of what it came from . . . and it is a seed of what will be from it."

Therefore, in order to differentiate natural from spontaneous generation, we must require that the parent's form effecting the change is shared by the growing thing in the relevant way. They are not mere passive subjects of the change, but they share the same form with that which acts as their efficient cause. To be sure, even if we amend the proposal in this way, the proposal results in a merely sufficient condition for possessing a nature, not in a necessary one. For, as argued earlier, inanimate beings are not, at least not universally, efficient causes of the changes they undergo. From this we should conclude, not that for all natural beings efficient causation must be excluded from our understanding of their principle of change, but rather that nature must be understood non-uniformly. The remaining question, of course, is whether, despite their differences, the principles in question share sufficient similarity, such that they can legitimately be grouped together under a general conception of what it means to possess a nature. The most difficult case our interpretation needs to accommodate as a case of *natural* motion is the motion of the elements.

IV The natural motion of the four elements and their mixtures

Aristotle denies that elements are self-movers, and holds that their local motion has an external efficient cause (i.e. the active power that moves them is always external), while they themselves possess merely a passive principle of change (*Phys.* VII.4, 255a6–18; 256a1–2).[22] Despite this, he holds that their motion is natural, and this leads to the following *aporia* (VIII.4, 254b33): The elements move naturally in one direction on the axis of up and down, and yet do not qualify as self-movers.

The resolution of this *aporia* comes in two stages. The first introduces a distinction between causes, the second a distinction between senses of potentiality. At the first stage, Aristotle suggests that, as movers may *move* naturally or move unnaturally, so with things *moved*, in some cases they are moved naturally, while in others they do so unnaturally. We should ask, though: What causes does Aristotle have in mind here? The point about natural versus unnatural movers concerns efficient causation; but the point about naturally versus unnaturally movable things concerns, it seems, formal causation: Movables move naturally when they are "potentially in their own actuality" (VIII.5, 255a24–30).

[22] The interpretation offered here is indebted to Bodnár 1997 and Gill 2009. For other interpretations, see Matthen 2001, Scharle 2008, Katayama 2011. I offer a more elaborate defense of my reading in Stavrianeas, "The nature of the elements" (in press).

This potentiality cannot be their potentiality to move (intransitive), for they are actually moving. Rather, it is a potentiality that belongs to them while they are moving to their proper places. And to be in their proper place is – at least part of – their form. However, even if we grant that the actuality they are potentially in is their principle in the sense of form, each moving portion is not in its proper place yet. So in what sense does it possess this principle already? The second stage of the resolution may be read as addressing this point, as Aristotle suggests that in order to grasp what moves the elements, we need to specify the particular sense in which the potentiality/actuality distinction applies to their movement:

> (1) One who is learning a science knows potentially in a different way from (2) one who while already possessing the knowledge is not actually exercising it. Wherever (3) something capable of acting and something capable of being acted on are together, what is potential becomes actual, e.g. the learner becomes from one potential something another potential something (for one who possesses knowledge of a science but is not actually exercising it knows the science potentially in a sense, though not in the same sense as before he learnt it). And (4) when he is in this condition, if something does not prevent him, he actively exercises his knowledge: otherwise he would be in the contradictory state of not knowing. (*Phys.* VIII.4, 255a33–b5)

Sections (1) and (2) divide two senses of potentiality, while (3) and (4) two corresponding senses of actuality. For instance, the first-level potentiality to become hot (1) belongs to a portion of an element that is not hot, but cold, such as earth. Once a cold portion is transformed into something hot, such as fire, the first-level potentiality becomes actual (3), and from then on the generated portion of fire possesses the actuality of being hot. By the same token it also possesses a second-level potentiality (2) to heat, if not impeded. The exercise of this second-level potentiality constitutes a second-level actuality, on a par with the exercise of the power to theorize (4), i.e. with the exercise of a capacity (255b6–7). It is in this latter sense that the movable is potentially in its actuality, so let us see how this schema applies to elemental motion and resolves the *aporia*.

The power of fire to be light, i.e. to be up, belongs as a first-level potentiality (1) to something that is actually heavy, e.g. earth. Once the latter is transformed into something light, i.e. earth into fire, this potentiality is fulfilled and the element must, according to this schema, possess a capacity (2) that is similar to its capacity to heat. However, the two capacities behave differently with respect to their transition to their respective actualities (4).[23] The capacity of fire to rise higher whenever exercised includes a local change and resembles the actuality of a first-level potential, as potential, i.e. as something incomplete, while the exercise of its capacity to heat does not involve such a change. In the light of this difference, it remains puzzling why Aristotle classifies the exercise of the capacity for locomotion as a transition from a second-level potentiality to the corresponding actuality (a transition that standardly does not involve change), rather than to the transition from a first-level potentiality.

Part of the answer is found in what differentiates the two levels of potentiality–actuality pairs: The learner becomes from one potential something another potential something, while the knower will exercise his capacity, if nothing impedes, i.e. whenever he wills to exercise it, for otherwise he would be in the contradictory state of not knowing. Transitions at the first level consist in moving from one state to another state (possibly its contradictory state).[24] Now, in one sense, this is true of elemental locomotion, which is a change to the natural place from its contrary one. However, unlike other local changes – say, a stone thrown upwards – movement to the natural place is not accidental to the elements. Rather, it is a constant and regular feature, and it expresses what each element is.[25] If this change, then, consisted in the destruction of a contrary, its result would be the annihilation of what is essential to the moving entity. Far from this being the case, the element in moving to its proper place remains unchanged and gets united with its like. So its potentiality to be in its proper place by becoming active preserves its nature by bringing it closer to its form. In this respect, the locomotion of the elements resembles cases at the second level of the actuality–potentiality distinction.[26] Therefore, an activity of this kind may be attributed to a principle that something possesses in itself *per se*, even if it has an external efficient cause.[27]

[23] Indeed, the distinction concerning causes in the first stage of the resolution of the *aporia* prepares us to expect just that: The capacity of fire to heat is a capacity of fire as a natural mover, while its capacity to move up is a capacity of fire to be moved naturally.

[24] Cf. *DA* II.6, 417a20ff.

[25] See Gill 2009: 151 on this.

[26] On the other hand, the fact that this motion is an activity of some kind justifies, to some extent, the thought that it is not a mere capacity for a corresponding actuality or activity, but that it is rather itself an actuality.

[27] This is confirmed by Aristotle's illustration in *On the Heavens*: "to ask why fire moves upward and earth downward is the same as to ask why the healable, when moved and changed *qua* healable, attains health and not whiteness" (*Cael.* IV.3, 310b15–17; cf. 310b27–30). The healable that is already moved *qua* healable, i.e. as something that has the potentiality to be healthy, is already on its way to being healthy, in the sense that it already possesses the form of health (*Metaph.* Z.7, 1032b25–27; see also *DA* II.2, 414a10). Similarly, the fire moving up is partly realizing its form in moving towards it, as it occupies successively different locations that are gradually nearer to its proper place, even though none of them *is* its proper place.

And there is a further reason for holding that elemental motion is similar to second-level activities and to the exercise of natural capacities. This becomes clear if we consider the qualification Aristotle adds concerning such capacities, namely that they are active *if not impeded*, and its application to the respective capacities of fire to heat and to be up. All that is needed for impeding the capacity of fire to heat is that no suitable subject is found in its surroundings; then its capacity is inactive. In any other situation, fire is active and heats the proximate bodies. No change in the fire is involved in its transition from inactivity to its exercise. The capacity of fire to be up, on the other hand, is impeded, we are told, whenever fire is not in its proper place. This may be so either because it is held there by an external force, or simply because it was generated there. So as soon as the impediment is removed or the portion generated, the corresponding capacity is active. True, one may be tempted to think that the portion is in some sense impeded while still moving, but this is not quite the sense in which it is impeded when it is held immobile by force. Further, it does not seem right to say that fire is inactive when moving to its proper place in the way the capacity of fire to heat may be inactive. In the former case it rises higher, while in the latter there is no heating going on.[28] What is common in the two cases, and similarly to the exercise of other natural capacities, is that there is no internal obstruction that can form an obstacle to their exercise; no process within the element can stop its capacity from being constantly active. In other words, Aristotle's qualification "if nothing impedes" should be understood as "if nothing *external* impedes." And this seems to be a standard feature of capacities.[29] To be sure, natural capacities in animate beings form part of a complex network of potentialities whose exercises may interfere with each other. Moreover, they belong to ensouled bodies, where a formal and a material side may be antagonizing each other. Regardless, Aristotle's point is that, considered on its own, each capacity is *internally indefeasible*. No impediment can come from a normally functioning capacity such as to block its being exercised. This

[28] Indeed, Aristotle concludes the *aporia* by saying that the portion of fire is already light and will realize its proper activity as soon as it is generated (VIII.5, 255b10–11). And he states as much a few lines below: "if what hinders it [that is, the portion of air] is removed, it realizes its activity and continues to rise higher" (255b20–21).

[29] The exercise of a capacity seems to be compromised only by something external – see *Metaph.* Θ.7, 1049a7; 1049a13–14; and *MA* 8, 702a17. This means that when the external conditions are right, the capacity itself is sufficient for explaining the changes its exercise will be involved in. Thus, it makes sense to say that what possesses such a capacity possesses in itself a principle for the corresponding changes, which can be understood as its internal principle of change, its nature.

feature, then, offers one more reason to view elemental motion as governed by an internal principle.

At this point, however, the similarity between capacities in animate and inanimate beings breaks down. The former are capacities for complex activities, while the latter are for simple and uniform changes. The transition to activity does not depend on a complex process involving changes in other material beings (such as in the activity of nourishing oneself) or in parts of the entity itself (such as in the activity of moving one's own body locally). We could qualify such activities as *homogeneous* or *non-plastic*. For this reason, too, it may be that a passive principle is sufficient for governing elemental motion, as well as changes of inanimate beings in other categories.[30]

In sum, elemental motion is crucially similar to transitions from capacities to their corresponding activities. It is not a standard local change, but part of the realization of elemental formal nature. Since, as we just saw, this capacity differs from capacities in animate beings by being active from the start, its *per se* efficient cause must be something external.[31] However, this does not make it a capacity to be changed by something else, for it requires no external stimulus for becoming active, nor is it a capacity to be changed by itself *qua* other, due to the simplicity of elemental body. What it is is a capacity to be changed in itself *qua* itself, and hence it satisfies in one way *Physics* II.1's definition of nature.

V Conclusion

We may conclude, then, that we should not insist on an exclusive understanding of what a principle of change is, or on constraining our interpretation of nature as a principle to the role of efficient causation in all natural phenomena. The absence of an internal efficient cause does not disqualify a change as natural, as in the case of elemental locomotion, nor does its presence alone qualify a change as natural, as in the case of miraculous *automata*. Rather, what matters are the specific capacities that characterize natural kinds and whose exercise regularly produces changes that preserve the nature of the entities of those kinds. In these changes, entities can play active or passive roles in conformity to their form.

[30] Note that the model applied to the motion of the elements holds similarly of their qualitative and quantitative alterations: *Phys.* VIII.5, 255b12–13.

[31] Simplicius *In Ar. Phys.* 1220.20.

With respect to *Physics* II.1, then, it seems neither necessary nor desirable to read it as proposing a complete and finished definition of what nature is. Rather, Aristotle offers up an initial account of what differentiates natural beings as a first conception or general description, while providing further developments and qualifications in the remaining, more specific, natural treatises concerning what is distinctive of each natural kind. The main aim of *Physics* II.1 is to delineate the domain that natural bodies populate and that natural science investigates.

CHAPTER 4

Aristotle on chance as an accidental cause

James Allen

I

"Chance" stands in for the genus designated αὐτόματον by Aristotle, which he divides into two species, τύχη, luck or fortune, which is operative in the realm of practice, and its counterpart in the natural world, the automatic or spontaneous, called, like the genus, αὐτόματον. The terminological distinction is stipulative and corresponds neither to Aristotle's own usage nor that of other authors. He defines luck and the automatic as causes by accident, and of the efficient type (*Phys.* II.6 198a2–7; cf. II.5, 196b23, 30, 197a5–6, 12–14; II.8, 199b23–25). The definition and the distinction both come from his fullest discussion of chance, which occupies chapters 4–6 of *Physics* II.

He has good reasons for tackling the subject here, in the treatise he dedicates to fundamental questions of natural philosophy. Chance is accounted a cause, in the sense that people appeal to it to explain what happens, not only in the practical sphere, but in the natural world as well. His point of departure in *Physics* II.4 is the question whether and how chance can be accommodated in the framework of the four causes (II.4, 195b31–36; cf. *EE* VIII.2, 1247b1–9). Chance is also regarded – rightly, in Aristotle's view – as indefinite (ἀόριστος), unpredictable (παράλογος), obscure (ἄδηλος), and uncertain (ἀβέβαιος),[1] and he aims to supply an account that will show how chance occurrences can have causes while retaining the distinctive character that sets them apart from their non-chance counterparts. His other main object is to combat on the basis of this account the errors of the natural philosophers who would exclude teleology from physics, replacing explanation in terms of final causes with appeals to chance. And, although he touches on it only briefly in the *Physics*, Aristotle

[1] παράλογος (197a18; *EE* VII.14, 1247a33); ἀόριστος (199a20; *EE* VII.14, 1247b12; *Metaph.* D.30, 1025a25; E.2, 1027a6); ἄδηλος ἀνθρώπῳ (197a10); ἄδηλος (*Metaph.* K.8, 1065a32); ἀβέβαιος (197a30–31).

expects that his account will also undermine another widely held view, namely that lucky occurrences are the effects of unseen divine agency, so that they are mysterious only from a limited human perspective (II.4, 196b5).[2]

Of course, if chance did not exist, there would be no problem to solve, but Aristotle has little patience with the well-known argument that there is no such thing as chance because it is always possible to specify a definite cause for everything that has happened (196a5–6). He observes that, though familiar to *everyone*, this argument has not prevented *anyone* from regarding some things as due to chance, others as not (196a11–17). His brisk way with this argument is a clue that an issue that one might have expected to loom large in an examination of chance will receive little attention. On a certain conception of chance, i.e. roughly what we mean when we speak of "objective chance," events are by chance because they are not determined. No specification of the conditions in which they occur, however complete, is incompatible with their failure to occur, so that, one might suppose, there is no cause because of which they occur when they do (as opposed perhaps to an explanation of the conditions that make it possible or likely for them to occur). The argument that "to be caused" is thereby "not to be by chance," with the converse implication that "to be by chance" is "to lack a cause," suggests that this idea was not inconceivable, but Aristotle nowhere confronts it directly.

One reason is because Aristotle is taking sides in a controversy that was already under way, and – for all the differences between them – on the side of Plato.[3] In this controversy the opposed alternatives are not *by chance* or *by necessity*, but *by chance* or *for the sake of something*; chance and necessity can even be combined.[4] Their adversaries are philosophers who would explain the beneficial arrangements in which the natural world abounds as the result of chance, which they present as an alternative to the form of explanation favored by Aristotle's party, according to whom it is for the sake of the good they do or make possible that matters are so arranged.[5] One opponent identified by Aristotle is Empedocles, who, he says, held

[2] It receives more attention in *EE* VIII.2, where, in the course of discussing the popular view that happiness is good fortune (εὐτυχία), Aristotle rejects the idea that lucky people are the beneficiaries of divine favor (1247a23–29). Epicurus too treats this view as widespread, maintaining that the person who has taken his philosophy to heart will not believe, as the many do, that luck is a goddess (*Men.* 134). It is credited to the Pythagoreans by Aristoxenus (DK I 478, 23–33).

[3] And Socrates, on whom see Sedley 2007.

[4] At *Int.* 9, 18b5–6, Aristotle says, without mentioning causation, that if everything is by necessity, nothing is by chance.

[5] On beneficial arrangements in nature, see Cooper 1982.

that animal organs are mostly products of chance (*Phys.* II.4, 196a23–24; cf. II.8, 198b23–32). Others, whom he does not name, have, he says, made chance the cause of both this and other worlds (II.4, 196a24ff.). They resemble the "impious" people, likewise unnamed, against whom Plato has the Athenian argue in a passage in *Laws* X that prepares the way for a proof of the gods' existence (888eff.). They maintain that the greatest and most beautiful things in the world, including the whole heaven, everything in it, animals, plants, and seasons, are products of chance and nature (889a), and that they arise in accordance with chance from necessity (κατὰ τύχην ἐξ ἀνάγκης). By appealing to chance and nature, they deny that these things are products of mind (νοῦς), god (θέος), or art (τέχνη) (889c). According to Plato's Athenian, they have inverted prior and posterior; it is the soul (ψύχη), its faculties, and motions that are prior, while nature – if by this are meant bodies, their powers, and motions – is posterior. But the way in which those who couple nature with chance conceive of nature is misguided. The Athenian's opponents want to describe as nature the genesis of the primary beings, but this honor, he maintains, belongs rather to soul, which can best and most correctly be said to be by nature (892c).

As it figures in this controversy, then, chance is the random factor which, by bringing together bodies in the right conditions and proportions, supplies the initial conditions that allow them, acting in accordance with the necessities imposed by their material natures, to give rise (among other things) to the greatest and most beautiful of existing things – a kind of cosmic luck. The question whether the random element introduced by chance is due to indeterminism or simply has (determining) causes that are beyond the scope and ambition of natural philosophy appears not to arise in the controversy, in which Plato and Aristotle's object is simply to prove that chance, conceived as the not-further-to-be-explained source of the material combinations that give rise to beneficial arrangements in nature, cannot replace teleology.

To be sure, Aristotle and Plato are on the same side only with substantial qualifications. Plato's preferred account of the natural world assigns priority to soul or mind, ultimately divine mind. Aristotle's god neither creates nor administers the world. The things that are and come to be for the sake of something in the realm of nature do so without the benefit of deliberation or design. What is more, the intended scope of final causal explanation in Aristotelian natural philosophy remains a matter of controversy. Nonetheless, on the fundamental point that at least some natural beings and their doings are for the sake of something, he agrees with Plato.

Although Aristotle parts ways with his ally when it comes to the priority of art and mind, his account of chance pays at least as much attention to luck and the practical sphere, where mind is operative and which he tackles first in *Physics* II.5, as it does to the automatic, which he takes up in II.6. This may be an instance of proceeding from what is prior and more familiar to us to what is less so, but there is a risk in adapting an account of luck tailored to the realm of agency, where it is easy to contrast beneficial occurrences from which the for the sake of something is absent with the those in which it is present. On my interpretation, Aristotle regards chance occurrences as something like byproducts of final causation, which are only intelligible as exceptions to the teleological rule. As a result, his account of chance threatens to beg the question in dispute between him and his predecessors, according to whom *nothing* in nature ever occurs for the sake of *anything*. This is especially plain in Aristotle's own argument from priority, with which his account ends and in which it seems it was meant to culminate.

> But since the automatic and luck are causes of things of which mind or nature might be causes, when something comes to be as a cause of these same things *per accidens*, and since nothing that is *per accidens* is prior to that which is *per se*, it is clear that no cause *per accidens* is prior to that which is a cause *per se*. Hence the automatic and luck are posterior to both *mind* and *nature*; so however much the automatic may be the cause of the heavens, mind and nature are necessarily prior causes both of many other things and of this universe. (198a5–13; trans. Charlton modified)

It is clear from the broader context that, unlike those who couple nature with chance, Aristotle takes nature, which he puts on a level with mind here, to be, like it, operating for the sake of an end. The burden of his argument is that automatic outcomes are possible in nature only if it contains occurrences that are for the sake of their results as well.[6]

In what follows I shall be chiefly concerned to examine Aristotle's account of chance, on the basis of which he takes himself to be entitled to the argument for the priority of *per se* causation for the sake of something and the consequent impossibility of chance causation *per accidens* apart from it. But I shall conclude by returning briefly to the question whether Aristotle begs the question against his opponents. I conclude that he does, but that he implicitly acknowledges this when he returns to

[6] This is how the passage is understood when it is summarized in *Metaphysics* K.8 (1065b26ff.) and by a number of commentators. Aristotle has already made this view explicit in chapter 2 and will do so again in chapter 7 (194a25ff., 198b4). Cf. M. R. Johnson 2005: 102, 104; Sedley 2007: 192–194.

confront their appeal to chance as an alternative to final causal explanation in *Physics* II.8, whose argument does not beg the same question.

II

Aristotle begins his treatment of luck in II.5 by drawing attention to two features, both of which belong likewise to the automatic (196b10–22).

1) Luck is not a cause of things that come to be always or for the most part, but is, in a sense still to be specified, infrequent.
2) It is actions done for the sake of something that have lucky outcomes.[7]

These features are presumably intended to illuminate the notion of accident as it applies to luck. Though not the only available interpretative strategies, two opposed approaches can be distinguished by which of the two features they emphasize. One takes being an accident, or an accident of the relevant kind, to be at bottom a matter of the (low) relative frequency with which the initiating event, under some specification, is followed by its lucky or automatic outcome.[8] In my view it is the second feature that will help us get a better fix on how the notion of accident applies to chance, while the infrequency condition is best viewed as an additional constraint on chance accidents, whose more fundamental nature is specified in other terms.

The second feature has proved to be a puzzle, however. Both in *Physics* II and elsewhere, *not* coming about for the sake of what results is said to be a mark of luck or chance.[9] For example, in a passage from *Posterior Analytics* II.11 to which we shall return, Aristotle roundly asserts that "nothing that is by chance comes to be for the sake of something":[10]

> Among things that come to be from thought some never arise from chance or from necessity, e.g. a house or a statue, but for the sake of something, while others such as health or safety also arise from luck . . . but nothing by chance comes to be for the sake of something. (95a3–6)

[7] Note that both outcomes and the actions that bring them about, i.e. what it was lucky that one did, are said to be "by luck" (cf. *Phys.* II.8, 199b18–22).

[8] Judson 1991b: 81, 90 takes Aristotle to hold that the conditions that chance events come to be accidentally and that they do rarely are the "same" or "equivalent." My disagreement has not prevented me from profiting from his article, especially his account of how the notion of the "for the most part" is brought to bear on chance (92–93).

[9] *Phys.* II.5, 196b34, 197a16; II.6, 197b19, 23; II.8, 199b21–22; *APo.* II.11, 95a5–9; *Rh.* I.10, 1369a32–34. *Protrepticus* fr.11 Ross. Cf. Lennox 1984: 250–251 and Judson 1991b: 77 with n.8.

[10] ἐν δὲ τοῖς ἀπὸ διανοίας τὰ μὲν οὐδέποτε ἀπὸ τοῦ αὐτομάτου ὑπάρχει, οἷον οἰκία ἢ ἀνδριάς, οὐδ' ἐξ ἀνάγκης, ἀλλ' ἕνεκά του, τὰ δὲ καὶ ἀπὸ τύχης, οἷον ὑγίεια καὶ σωτηρία . . . ἀπὸ τύχης δ' οὐδὲν ἕνεκά του γίνεται.

Yet Aristotle places chance occurrences squarely among those that are for the sake of something both at the start of and throughout the account in *Physics* II.

> Anything which might be done as the outcome of thought or nature is for the sake of something. Whenever something like this comes to be by accident, we say that is by luck. (196b17–24)

A little further on, in connection with his favorite example of a man who went to the market with another end in view but succeeded in collecting a debt by luck, Aristotle says:

> Plainly then luck is a cause by accident among those things that are for the sake of something that are according to choice. (197a5–6; cf. 5, 196b29–31)

And other passages say much the same thing (cf. 196b23–24, 32–33, 197a6, 197b18). Since the philosophers against whom Aristotle is chiefly concerned to argue, at least as he understands them, maintain that *nothing* in nature comes to be for the sake of *anything* and propose coming to be by chance as an alternative to coming to be for the sake of something, we should expect the fact, which Aristotle takes it to be, that everything that happens by chance is – somehow – for the sake of something to be more than a merely verbal coup: It should deliver a real sting.One solution that is favored by a number of ancient commentators and modern scholars holds that a lucky occurrence must be for the sake of something, to be sure, but something other than the outcome by bringing about which it qualifies as lucky.[11] That is, for the action A whose lucky outcome is X, there is a Y not identical with X for whose sake A was done. Although this may be true according to Aristotle, it is not his point, which is rather that the action is *somehow* for the sake of the effect by bringing about which it qualifies as lucky. Consider a passage in *Physics* II.8 that looks back to the doctrine of chapters 4–6.

> That for whose sake and what is for the sake of this may come about by luck; for example, we say that a stranger came by luck, paid the ransom and went away, when he did so as if he had come for the sake of this though he did not come for the sake of this – and this by accident. (199b18–22)[12]

[11] Porphyry reported by Simplicius *In Ar. Phys.* 336, 28–29; Philoponus *In Ar. Phys.* 268, 30ff., 274, 10ff.; Themistius *In Ar. Phys.* 51, 16–20, 52, 10ff.; Bonitz 1969: 58; Torstrik 1875: 429, 445–448, 466, 468; Zeller 1879: II.2 335–336, 947–948. But cf. Irwin 1990: 520–521, who thinks the solution works for luck but not the automatic.

[12] τὸ δὲ οὗ ἕνεκα, καὶ ὃ τούτου ἕνεκα, γένοιτο ἂν καὶ ἀπὸ τύχης, οἶον λέγομεν ὅτι ἀπὸ τύχης ἦλθεν ὁ ξένος καὶ λυσάμενος ἀπῆλθεν, ὅταν ὥσπερ ἕνεκα τούτου ἐλθὼν πράξῃ, μὴ ἕνεκα δὲ τούτου ἔλθῃ. καὶ τοῦτο κατὰ συμβεβηκός . . .

Here paying the ransom is "that for whose sake" and coming to the marketplace is "what is for the sake of this," even though the stranger did not come to the marketplace for the sake of paying the ransom. When Aristotle takes lucky actions to be actions for the sake of something, then, he means that every lucky action A whose lucky outcome is X belongs among actions that are, *somehow*, done for the sake of X.

The more recent tendency has been to distinguish senses of "for the sake of X," so that a chance event may be for the sake of X in one sense, but fail to be for its sake in another. This is on the right lines so far as it goes, but I shall argue that one of the ways of distinguishing senses goes astray from the outset, while the other can be better grounded and pursued further. The proposal with which I disagree takes the sense of "for the sake of X" in which an event comes to be for the sake of its automatic or lucky outcome to be that it actually has that outcome.[13] This seems to me less like another sense of the expression than its reference or extension in that sense, however. According to the second proposal, the other sense of "for the sake of X" is something like "such as might have come to be for the sake of X."[14] This is better, but it is still misleading, I maintain, to speak of "senses" of the expression "for the sake of something." Rather, I suggest, the difference in meaning that Aristotle exploits here is, at bottom, a difference in verbal aspect.

As we know, in Greek and English, among other languages, the simple present is used for purposes other than saying that something is going on at the time of the speaker's utterance. In English, where one typically uses the present progressive to do this, it would often simply be incorrect. "Paul drives to work" means that Paul drives to work usually or habitually. But "Paul drives to work now" is incorrect, unless "now" means something like *these days*. One says "Paul is driving to work now." But the simple present is used not only to represent an action as habitual in this way, but also to express another kind of general proposition as well. If, on a hike in the woods, one of us idly observes to the other that "trees burn," the effect should be different not only from the one elicited by "trees are burning," but also from the response to an assertion that trees habitually burn. Statements like these are sometimes

[13] Judson 1991: 77 with n.11, who adopts this interpretation, calls this the non-explanatory sense. He takes himself to disagree with Ross 1936, but it is not clear that Ross, who believes that, in the relevant sense, *heneka tou* means actually having a result that is end-like (39, 517, 518, 520, 523), is not of the same opinion. Cf., however, Ross 1924: II 187 (ad *Metaph.* Z.7, 1032a29).

[14] Cf. Simplicius *In Ar. Phys.* 335, 20–26 and 336, 27–30. Cf. Lennox 1984: 251–252.

called generic propositions.[15] This way of speaking is not confined to assertions about independently specified subjects. It can also be used to specify a subject with the aid of a relative phrase: e.g. "things which burn" or "things that fly" do not mean things that are burning or are flying (or even do so usually).

This use of the present tense, which is familiar to us in propositions about substances and natural kinds, is not confined to them. Teachers of philosophy should have no difficulty imagining a situation in which it would be appropriate to say "propositions are true and arguments valid" or "the things that are valid are arguments, not propositions," although there is, of course, no dearth of invalid arguments or false propositions. It is also available for use in assertions about actions and occurrences. Consider statements like the following, which would be out of place in an ordinary conversation. "Paul did something that is done in order to hurt others (or alternatively 'that people do,' 'that one does'), but he did not do it in order to hurt anyone." Plainly we are talking about an agent, a particular action of his, and a kind of action to which the action belongs. What is asserted is, roughly, that Paul performed an action (or, if you prefer, an action of a kind) that is such as to be done in order to hurt someone without doing it in order to hurt anyone on this occasion. If different senses are in play here, they are not different senses of "in order to hurt someone" but of "is done" or the whole verbal phrase "be done to hurt someone." The same action can be (something that is) done to hurt someone, viewed as an instance of a type, and not done in order to hurt someone, regarded in all its concrete particularity, as done by this agent, on this occasion, out of these motives.[16] It is a point of this kind, I suggest, that Aristotle is making about

[15] Cf. Moravcsik 1994: 229–244. The terminology is somewhat fluid: Sometimes "generic" designates the genus of which the habitual is a species.

[16] "That for the sake of which" (τὸ οὗ ἕνεκα) is "one of the items in which or about which an action is" (*EN* III.1, 1110b33–a1), and ignorance regarding one or more of "the particular circumstances of the action and the objects with which it is concerned" can render an action involuntary (1110b30–a2). All interpreters agree that Aristotle is not describing a self-deceived agent ignorant of his own intentions. To avoid this, some interpreters have taken τὸ οὗ ἕνεκα to mean the result or issue of the action in three relevant passages from Aristotle's ethical works (*EN* III.1, 1111a2–19; V.8, 1135b12–16; *EE* II.9, 1223b1–5). Gauthier and Jolif 1968: II.1, 185 ad *EN* 1111a5; Judson 1991b: 77–78; and Stewart 1892: I 240. This sense is otherwise unattested (except perhaps in the *Physics*) and goes unrecognized by lexicographical authority. To judge by their translations and commentaries, others are unpersuaded: see Crisp 2000: 39–40; Dirlmeier 1956; Rowe and Broadie 2002: 125; explicitly Grant 1866: II 13n.18 and Taylor 2006: 148. I suspect that the kind of ignorance Aristotle has in mind is about the actions by means of which to bring about the agent's end. His meaning would have been clearer had he written: "the agent was in error regarding *that for the sake of which* in supposing that so acting was a way of bringing it about." The culprit would then be the extreme grammatical compression that results from Aristotle's determination to force all the possible kinds of error into a single telegraphic construction.

chance events, most explicitly when he summarizes his results in chapter 6.

> It is clear, then, that among the things that come to be for the sake of something without qualification, when they come to be not for the sake of what results, the cause being external, we then use the expression "by chance" of them.[17] (197b18–20)

The meaning of "without qualification" (ἁπλῶς) here has long been a puzzle.[18] I suggest that Aristotle is using the term in the well-attested sense of "in abstraction from the particular qualifications or circumstances, or with those circumstances or qualifications mentally removed" (cf. *Top.* II.II, 115b29–35). He means occurrences or actions that one would say are for the sake of X when they are viewed in abstraction from the particular circumstances in which they are done or take place. A good parallel is furnished by a passage from the discussion of the voluntary in *Nicomachean Ethics* III.I, where Aristotle describes actions like throwing goods overboard as involuntary without qualification, but voluntary in certain circumstances (1110a18, b5; cf. 1110a10). Translators have turned to phrases like "in the abstract" to bring out the fact that it is actions that are not worthy of choice when viewed in isolation from the circumstances in which they are performed that are involuntary without qualification, even though there are circumstances in which they are performed voluntarily.[19] So, too, actions and processes that come to be or are done for the sake of X, when viewed in abstraction from the circumstances in which they actually occur, can take place without doing so for the sake of X. When they do, Aristotle is saying, and all goes well, they and their results come about by chance.

Aristotle sometimes conveys the idea of things that come to be or are done for the sake of something in the abstract by means of another construction, in which the adverbial phrase "for the sake of something" is joined to the definite article, which has the effect of turning it into an adjective and allows the whole phrase to serve as a substantive (196b35–36,

[17] ὥστε φανερὸν ὅτι ἐν τοῖς ἁπλῶς ἕνεκά του γιγνομένοις, ὅταν μὴ τοῦ συμβάντος ἕνεκα γένηται ὧν ἔξω τὸ αἴτιον, τότε ἀπὸ τοῦ αὐτομάτου λέγομεν. The point of calling the cause "external" is a puzzle. See Judson 1991b: 94 with n.52.

[18] The commentators tend to treat ἁπλῶς as equivalent to καθόλου (universally) and take Aristotle to be talking about things that come about for the sake of something both in the sphere of luck and in that of the automatic (Philoponus *In Ar. Phys.* 288, 17–20; probably Themistius *In Ar. Phys.* 55, 8–9; Ross 1936: 522). Judson (1991b: 93) takes ἁπλῶς to mean "in the most general sense" and this to be the non-explanatory sense, which means actually having the result in question.

[19] "Voluntary in the abstract" (Ross); "an sich betrachtet" (Dirlmeier 1956); "abstraction faite de telles circonstances" (Gauthier and Jolif 1968). Cf. Irwin and Fine 1995: 111n.91, who connect the use of ἁπλῶς here with that in *Nichomachean Ethics* III.I.

197a6; cf. 199b19). In one, admittedly difficult and much disputed, passage in chapter 6, we find "when what is for the sake of another thing does not come to be for the sake of this" (ὅταν μὴ γένηται τὸ ἕνεκα ἄλλου ἐκείνου ἕνεκα), which, I suggest, is once again meant to express the idea that chance occurs when an occurrence that takes place for the sake of something in the abstract takes place without on this occasion doing so for its sake (197b23).[20] Although there is no verb, it is natural to understand one, and when translating to supply it, as I have done in the less controversial passage that I already cited from *Posterior Analytics*:

> Among things that are from thought some never arise from chance or from necessity, e.g., a house or a statue, but for the sake of something, while others such as health or safety also arise from luck . . . but nothing by chance comes to be for the sake of something. (*APo.* II.11, 995a3–6)

Here, being for the sake of something is a characteristic said to be absent from every episode of chance. Yet some things that come about from thought – the Greek simply has the definite article joined to "from thought" – also come about by chance.[21] If we assume, as we are surely entitled to do, that on every occasion when something comes about from thought in the sense intended, it comes about as the result of an action undertaken for its sake, then it might appear that we are faced with a contradiction. Yet I doubt that readers have ever been troubled by this passage (cf. *Metaph.* Z.9, 1034a9ff.). It should now be plain that when Aristotle classes chance occurrences among "things that come to be for the sake of something" (196b17) he does not mean that individual episodes or particular occurrences each take place for the sake of something.

III

How does the idea that lucky agents do something that is done for the sake of an end (in the abstract) without doing it for the sake of that end

[20] The context is an obscure etymological argument that turns on the phonetic similarity between μάτην, "in vain," and αὐτόματον. Charlton 1970: 48, followed by Judson 1991b: 77n.9, emends 196b34–36 to yield another example: ἦλθε δ' οὐ τούτου ἕνεκα, ἀλλὰ συνέβη αὐτῷ ἐλθεῖν, καὶ ποιῆσαι τοῦτο τὸ τοῦ κομίσασθαι ἕνεκα· Bonitz 1961: 58, excises the phrase τοῦ κομίσασθαι ἕνεκα, with which there is already a problem: see Ross 1936: 520 and Irwin 1990: 520n.12.

[21] The widespread view about affinity between art and chance, *technê* and *tuchê*, that Aristotle shares is an expression of the same outlook. These are in a way concerned with the same things (*EN* VI.4, 1140a17; *Phys.* II.5, 197a7). Aristotle can, without contradiction, treat art, deliberation, and the like, on the one hand, and chance, on the other, as excluding each other because an instance or episode of one cannot at the same time be an instance or episode of the other, and also as productive of the same things by the same means (cf. *EN* III.3, 1112a22; VI.4, 1140a17).

contribute to the understanding of luck and chance? The point, as I have already suggested, is to help us get a fix on how the notion of accident applies to chance, and the sting will come in the argument from priority that is based on the opposition between the *per accidens*, so conceived, and the *per se*, conceived in the corresponding way. Very roughly, Aristotle's formulation serves to situate luck in a world in which there are things – call them "ends" – for the sake of which beings – call them "agents" or "people" – do things, and things that they do for the sake of these ends. There is, if you will, a repertoire of actions that are done for the sake of ends and that are available to agents acting for the sake of those ends. Again, speaking very roughly, when someone performs an action drawn from this repertoire, but does not, on this occasion, perform it for the sake of the end for the sake of which it is done (in the abstract), he performs the action and achieves this end by accident and by luck. Were he to perform the action for the sake of the end, however, and succeed, it would not be by accident or by luck.[22]

This help is welcome because of the way in which Aristotle introduces the notion of accident and the corresponding idea of the *per se* (II.5, 196b25–27, 197a14–15). He observes that, just as there is both being *per se* and being *per accidens*, something may be a cause of, or be responsible for, something *per se* or *per accidens* (196b24ff.). E.g. of a house the cause *per se* is a house builder, but *per accidens* the pale or the musical. This does, to be sure, illustrate what it means to be a cause by accident. It is because the builder, who is the cause *per se* of the house, happens to be or is *per accidens*,

[22] Aristotle's examples are all of inadvertently lucky agents, e.g. someone who goes to the market with another end in view and by chance meets and is paid by a debtor. Yet other cases of luck will be like gambling and winning, where the agent wants the outcome and is trying to bring it about. Aristotle considers cases like these in the discussion of good fortune in the *Eudemian Ethics*, where he mentions throwing dice (VIII.2, 1247a27, b17) and people who want something good act with a view to succeed, but do so despite reasoning poorly (1247b28–32). A full discussion of this kind of luck would require an essay of its own, but two observations are apposite. First, to avoid being by accident, the non-chance achievement of an end may require more than the presence of a wish to bring it about, resulting in an action that does in fact bring it about. Paradigm examples of *per se* activity of the relevant kind involve the exercise of a faculty for the sake of an end by the selection of an action suited to bring it about, which is at the opposite extreme from doing something – anything – at random in the hope of hitting upon the means to secure one's end. (There will, of course, be many – harder – cases between the two extremes.) It may be possible to understand the relevant notion of accident so that, in addition to cases of inadvertent luck, it also covers cases of this latter kind. Roughly speaking, to avoid doing something that is done for the sake of an end (in the abstract) by accident it is not enough to do it out of a wish to attain that end: one must also do it *because* it is suited to the end, and this will require that one somehow recognizes what makes it so. Second, perhaps Aristotle is less concerned with this kind of luck in *Physics* II because he does not think there is a simple analogue in the natural world to the untethered intentions, wishes, or hopes found in the practical realm.

say, a flute player that the flute player can be said to be the cause of the house *per accidens*, but in this case, one assumes, the house was not built by accident, and its building owes nothing to chance. It seems that it will always be possible to find a description under which an event that is plainly not due to chance has an accidental cause.[23] To see how chance occurrences are by accident, we need to direct our attention away from substances or objects and towards their faculties or powers and the actions or processes in which these are realized or displayed.

In *Nichomachean Ethics* v.8 Aristotle says that when people perform actions that are just or unjust involuntarily, they do not act justly or unjustly, except *per accidens*, because they perform actions that happen to be, or are by accident, just or unjust (1135a15–19; cf. *Metaph.* E.2, 1027a3–5). The construction rendered "happen to be" contains the verb *sumbainein* from which the expression that we translate as "accident" is formed. In a more literal translation, which also preserves more of the original grammar, Aristotle says: "they perform actions to which being just or unjust happens" or "with which being just or unjust coincides." We should draw two lessons. First, it is possible to perform an action in two ways, accidentally and non-accidentally – sometimes Aristotle says "simply" or "without qualification" (ἁπλῶς). In this passage, the person who does what is unjust voluntarily does it simply; the person who does it involuntarily does it by accident. And, second, the person who performs an action by accident does so by performing another action simply or without qualification, the one to which the first action – the one he or she does by accident – happens to be identical or with which it is *per accidens* the same. Here, for example, the involuntarily and accidentally unjust actor does something simply, not knowing that in so doing he is inflicting undeserved harm by accident if he is acting out of ignorance; or non-accidentally does something that prevents a greater evil by doing which he unavoidably but accidentally inflicts undeserved harm if he is acting under compulsion.

It is also important to note that what counts as accidental for one purpose need not do so for others. So, for example, elsewhere Aristotle is willing to say that someone who thinks that it is necessary to have the virtues for the sake of external goods does noble deeds *per accidens* (*EE* viii.3, 1249a14–16). Since he does them because advancing his own selfish interests requires doing noble deeds, what he is doing simply, properly speaking or non-accidentally, is advancing his own selfish

[23] Cf. Judson 1991b: 90.

interests. However, there is no doubt that he does the noble deeds intentionally (if not with the intention of acting nobly) and voluntarily.

What of the *per se* with which the *per accidens* character of luck and the automatic is contrasted? Here are the relevant passages from *Physics* II.5 in which the expression "cause *per se*" is introduced. Note the shift from agents to the same agents in activity and the actions they perform, and that all the examples are of agents acting with a view to an end (cf. II.3, 195b5). This is not because *per se* causation is confined to cases that evidently involve final causation. A hot or a cold body could *per se* be the cause of the heating or cooling of another body. But the cases in which we are interested, where luck is involved, have outcomes that are ends, the *per se* causes of which, if any, would act for their sake.

> E.g., of a house the cause *per se* is a house builder, but *per accidens* the pale or the musical. The *per se* is determinate, the *per accidens* indeterminate. (196b26)

And, twenty-two lines later:

> E.g., of a house the cause [*per se*, without qualification] is a house builder, but *per accidens* a flute player, and of someone having come collecting the money, not having come for the sake of this, [the causes by accident are] infinite in number, e.g., wishing to see someone, or prosecuting or defending or in order to watch. (197a14ff.)

Collecting the money plays the part of "that for whose sake," and the action of having come (to where the debtor is) that of "what is for its sake." The "*se*" in "*per se*," the *itself* in virtue of which, refers to the agent, but not simply the agent, or even the agent in so far as he or she possesses a certain power or faculty for bringing about the result at issue, but to the agent as actually using or exercising it to that end.

Normally when we discover that, e.g., a flute-player who is the accidental cause of, say, a cure, is also a doctor, we are entitled to rule out luck or chance as the cause, but not always. Suppose a doctor is throwing darts to relax after hours and, quite unintentionally, hits someone suffering from fever, as it happens drawing precisely the amount of blood that he would have prescribed be drawn had he been practicing the medical art, and, as a result, cures the person struck by the dart. A doctor, exercising the medical art, would be the *per se* cause of the corresponding action and outcome, those of which and for which the medical faculty or power is a faculty or power. When, as in the present case, the doctor does this action and causes this outcome without so acting – by performing *simply* another action, throwing darts, with which being something that is done for the sake of a cure in the abstract, drawing such and such an amount of blood, happens to coincide or is by accident the same – he is not their *per se* cause, not their cause as a doctor, and so the person with the fever is cured by accident and as a matter of luck.

To judge by these examples, then, the agent is the cause *per se* or in virtue of itself of an outcome when it exercises one of its faculties – the faculty of bringing about results like this – with a view to that end, by selecting from the repertoire of actions open to it one that is for its sake in the abstract.[24] It is this action that the agent does simply or without qualification. The result, when all goes well, is not by accident, and hence not by luck. By contrast, the agent brings about the same result by luck and is the cause of it by accident when, as we should say, he does the same thing with the same result, not, however, simply, but by accident. Although not the *per se* cause of the lucky outcome this time, the agent is the *per se* cause of something else, namely of the action that he performs simply or without qualification for the sake of another end, possibly exercising another faculty, by performing which he performs by accident the action that is for the sake of the lucky outcome in the abstract, which happens to be or is by accident the same as the first action.

Cases of luck and chance more generally are, Aristotle insists, also characterized by low relative frequency (II.5, 196b10–13, 36; II.8, 199b24–25). In Aristotle's example, by going to the market, say, to buy fish, I place myself in proximity to my debtor only rarely. But are the two characterizations of luck, being accidental, as I have interpreted it, and being rare necessarily coextensive? Aristotle's language is compatible with more than one interpretation. It seems conceivable that there could be cases in which by doing one thing *per se* one could hardly fail to do something else *per accidens*, as a concomitant. Elsewhere, of course, Aristotle is happy to allow that a substance is always and necessarily attended by some of its accidents (*Metaph.* Δ.30, 1024a30–34; *Apo* I.7, 75b1); and some cases where the items accidentally unified are actions are also like this – doing noble deeds by accident in the case of the agent acting for the sake of external goods cited above, for instance. Perhaps, then, the rarity requirement functions as an additional constraint, so that for an occurrence to be by luck, it must be both by accident in the specified way and rare. Alternatively, rarity might be a further requirement on being an accident of the type relevant to chance (cf. Phys. II.8, 199b24–25). And the gap between these two alternatives could be reduced by further reflection on the place of concomitant effects. To a

[24] Cf. Judson 1991b: 92–93, whose conclusion, reached by a different route, mine resembles.

well-informed agent they come as no surprise, and they are part of what he does intentionally even if not with the intention of bringing them about.

If we restrict our attention to the practical realm, Aristotle's account promises to do justice to the desiderata that he set up at the outset. Luck is neither a break in the causal order nor a case of divine meddling in mortal affairs by the gods on behalf of their favorites. Every lucky outcome has a cause, namely, it seems, an agent acting, and acting for the sake of an end, albeit one different from the one by bringing about which he is lucky. It is merely that in the Aristotelian universe everything, including every action, is what it is and, unavoidably, an indefinitely large number of other things by accident as well. As a result, in doing what we do simply or that of which we are the cause *per se* we cannot fail to do by accident and be the cause of an indefinitely large number of other things as well (cf. II.5, 196b28–29). Sometimes, one of these will be one that (in the abstract) is done for the sake of an end that we value and brings it about. This, roughly speaking, is luck. As an inescapable byproduct of agents acting with a view to ends, then, there is nothing at all surprising about luck in general; it is, rather, to be expected. At the same time, individual episodes of luck will be unexpected, obscure to human understanding, and of an indefinite character that sets them apart. Whereas the explanation for a non-lucky outcome, say, a cure, will be the exercise by an agent of a faculty for curing directed to this end, that of a lucky cure will now be the exercise of this unrelated faculty to this unrelated end, now to that end (cf. *EE* VII.14, 1247b12ff.).

The priority argument at the end of *Physics* II.6, which concludes that mind and nature are necessarily prior causes both of many other things and of this universe, plainly applies to the natural world, but we will be in a better position to understand it if we continue, for a little while longer, to restrict our attention to the realm of practice by considering a version applying to luck. To do something by luck and to be lucky as a result is to be the cause *per accidens* of an action that is done or that people do for the sake of that result. Speaking in this way is only intelligible if the practical realm contains agents who are *per se* causes of many of their actions and their results by performing the former for the sake of the latter. Without this, luck so conceived would be unintelligible, and, whatever happened and however it came about, it would not be lucky or by luck.[25]

[25] I think this is the kind of explanatory priority that Aristotle has in mind, although every action for the sake of an end in the abstract of which an agent is the cause *per accidens* is itself *per accidens* the same as an action of which the agent is the *per se* cause and which could, perhaps, be seen as prior to the other.

What, then, of automatic occurrences in the realm of nature? Much that is at home in the practical realm will have no place there – intention, deliberation, choice, art – but if Aristotle's account applies *mutatis mutandis* to nature as well, for there to be *per accidens* automatic outcomes there will have to be in the natural world a repertoire of processes that occur for the sake of ends (in the abstract), but that can on occasion take place without doing so for the sake of those ends. And this can be the case, it would seem, only if there are episodes of *per se* causation understood along the same lines as in the practical sphere, i.e. episodes in which beings function as *per se* causes for the sake of ends. Plainly, on this account, luck and *mutatis mutandis* the automatic have a place only in a domain where explanation in terms of final causes is at home. But any such assertion about the domain of nature would be vehemently rejected by Aristotle's opponents, who are nevertheless happy to appeal to luck or chance. I conclude that the account of chance on which the priority argument is based does beg the question in dispute.

IV

Were the argument of *Physics* II.8 based on the same account without modification, it would suffer from the same defect and be open to the same objection. But on closer inspection, it becomes clear that it is not. Aristotle begins with the example of rainfall, which, by falling, makes the crops grow (198b16–23). Although it remains a matter of controversy whether he is expressing his own opinion or making a concession to his opponents for the sake of argument, Aristotle presents this process as one that does not take place for the sake of its beneficial consequences, but is rather explicable entirely in terms of material necessity (198b18). It just so happens, or is an accident (συμβαίνειν, 198b21, 23), that they coincide with it. The point is to make available the idea of an accidental unity whose two terms are, on the one hand, processes stemming from material necessity – heating, cooling, condensing, and the like – and, on the other, those bringing about beneficial outcomes like plant growth. When they say that things like bodily organs, e.g. the jaws, are products of chance, Aristotle now takes opponents to be committed only to the view that these processes come to be or take place *as if*, or just as they would if, for the sake of the good they do (198b29ff.). To give the notion of chance a possible purchase on the occurrences whose explanation is disputed, an *as if* teleology is sufficient; the alternative to accident now is merely the notional possibility of *per se* causation for the sake of an end.[26] In effect, Aristotle has conceded, if only for the sake of argument, that chance is conceivable even in a domain from which final causation is completely absent. He does not argue, as he did before, that if some occurrences in the realm of nature are by chance, and therefore caused *per accidens* as lucky events in the sphere of agency are, there must be others that are caused *per se* by end-directed powers acting for the sake of their results.

Instead, he relies for his main argument on the infrequency requirement, lent support by the sheer implausibility of supposing that complex beneficial arrangements like the organs of an animal body could be the product of accident. The argument is framed by a stark alternative. Either the processes that give rise to, e.g., the jaws, do so by coincidence, i.e. occur only as if for the sake of the outcome and, it is implied, by chance, or they come to be for its sake. The infrequency requirement is used to eliminate the first alternative. Animal parts like the jaws arise as they do always or for the most part, therefore not by chance, and therefore, since this is the one remaining alternative, for the sake of something. The argument, which has been the object of much study, is hardly free from problems. The rainfall example with which the chapter begins seems to have left room for at least the notional possibility of regular coincidences in nature – whether this is Aristotle's settled opinion or not – where the beneficial result, e.g. plant growth, and the processes that give rise to it, e.g. heating, cooling, condensation, are not related as that for the sake of which and what is for its sake respectively (except with the qualification *as if*), and all the explanatory work is done by material necessity. An argument that, whatever might be said of rainfall and crop growth, bodily parts like the jaws cannot be the products of such coincidences is conceivable, but none is supplied. And Aristotle does not seem to pay adequate attention to the fact that those thinking along the same lines as Empedocles, as he is viewed here, seem to have envisaged original, one-in-a-million accidents (which would satisfy the requirement for infrequency) that give rise to arrangements that are not only beneficial in the short term, but are also self-sustaining or self-replicating, and that, once in place, can be relied upon to work always or for the most part. Nonetheless, the argument does not beg the question begged by the priority argument of *Physics* II.6.

There follows a series of supporting arguments based on the resemblance between natural and artistic processes. In both, a set of events that must occur when and in the precise sequence that they do if the benefit that

[26] I owe the idea of an *as if* teleology to Wieland 1970: 259 and passim.

results is to be produced can be observed occurring regularly and predictably (199a8ff.). Aristotle invites us to appreciate, e.g., that if a house were to come to be by nature, it would come about just as it does now by art (199a12; cf. 199b28).

It is only now, after he has established to his satisfaction that nature is a cause for the sake of something, that Aristotle alludes, in a passage that I have already cited, to the idea that played such an important part in chapters 4–6:

> That for whose sake and what is for the sake of this may come about by luck; for example, we say that a stranger came by luck, paid the ransom and went away, when he did so as if he had come for the sake of this though he did not come for the sake of this – and this by accident. (199b18)

Although it is possible that *Physics* II.8 contains Aristotle's second thoughts, I suspect that the two arguments are complementary. The priority argument belongs to an exposition of Aristotelian doctrine. The main argument of *Physics* II.8 meets Aristotle's opponents half way, and, by showing, if it succeeds, that they cannot be right even so, provides crucial support for one of the assumptions on which the account of chance already expounded rests, namely that there are final causes in nature. With this principle secured, Aristotle now re-states and re-affirms his account. An outcome comes about by chance when an action or process that is for its sake, i.e. belongs to a repertoire of actions that are done, or processes that are set in train, for its sake (in the abstract), is done or set in train by accident, meaning without being done or set in train for the sake of the lucky or automatic outcome that is its issue.

Appendix: The etymological argument of *Physics* II.6, 197b22–31

The point that Aristotle is trying to extract by etymological means, I suggest, resembles that about the relation between thought or art and chance. Just as the same things come about by thought or art as come about by chance and *vice versa*, so occurrences (of the kind) that take place in vain when they are done for the sake of an end but without success come about by chance when they are done by accident but with success, and of course neither by chance nor in vain when they are done non-accidentally for the sake of the outcome at issue and with success. It is meant to lend support to the idea expressed in the immediately preceding passage, that when something that is for the sake of something in the abstract comes about without coming about for its sake in actual fact, we speak of chance

(197b18–20). The next remark (197b20–22), that those chance or automatic events that are objects of choice are by luck, is a parenthesis (cf. II.5, 196b18–19). The main topic remains the conditions under which it is proper to speak of the automatic.

Aristotle now turns to the expression μάτην, "in vain." The ὅτι in the next sentence is usually taken to mean *because*:[27]

> The expression "in vain" is a sign *because* it is said when what is for the sake of another thing does not come to be for the sake of that thing. (197b22–23)[28]

But these are not the conditions under which the expression "in vain" is correctly applied. They restate the conditions under which the expression "by chance" is used. Attempts have been made to emend the passage so that it means: "the expression 'in vain' is used when something is for the sake of something else and what it is for the sake of does not come about."[29] I suggest that instead we take ὅτι to mean "that" and to introduce a clause specifying that of which the expression "in vain" is a sign:[30]

> And the expression "in vain" (μάτην) is a sign *that* it [namely, ἀπὸ τοῦ αὐτομάτου] is said when what is for the sake of another thing does not come to be for the sake of that thing.

Aristotle explains the meaning of μάτην with two examples (197b23–29).[31] If one walks for the sake of relaxation and relaxation does not ensue, one has walked in vain. It is correct to say that something occurs in vain, he continues, when it is by nature for the sake of something else, and that for whose sake it was and which it is by nature such as to bring about does not take place (197b25–27). To occur in vain, something must:

[27] Trans. Lacey 1997.

[28] σημεῖον δὲ τὸ μάτην, ὅτι λέγεται ὅταν μὴ γένηται τὸ ἕνεκα ἄλλου ἐκείνου ἕνεκα.

[29] Simplicius knew of an alternative reading that yields this sense (*In Ar. Phys.* 349.4–6), which is adopted by Pacius 1596: 58; Torstrik 1875: 464; Carteron 1926: 1.31; and Wicksteed and Cornford 1929: 34. Prantl 1854: 82 proposed an emendation to the same effect, adopted by Bonitz 1969: 39; Hamelin 1931: 134; and Charlton 1970: 49. Ross 1936: 523 believes the received text can yield the required sense: "the MS reading offers no great difficulty if we remember that ἕνεκα του may mean 'producing an end-like result.' The phrase means 'when that which was intended to produce an end-like result does not produce it.'" He is followed by Wagner 1989: 48; cf. 472.

[30] Although committed to a different account of why "in vain" is a sign, Philoponus construes this sentence as I do (289: 28–29 Vitelli). Ross 1936 also treats it as a possibility (523 ad ll. 22–23).

[31] οἷον εἰ τὸ βαδίσαι λαπάξεως ἕνεκά ἐστιν, εἰ δὲ μὴ ἐγένετο βαδίσαντι, μάτην φαμὲν βαδίσαι καὶ ἡ βάδισις ματαία, ὡς τοῦτο ὂν τὸ μάτην, τὸ πεφυκὸς ἄλλου ἕνεκα, ὅταν μὴ περαίνῃ ἐκεῖνο οὗ ἕνεκα ἦν καὶ ἐπεφύκει, ἐπεὶ εἴ τις λούσασθαι φαίη μάτην ὅτι οὐκ ἐξέλιπεν ὁ ἥλιος, γελοῖος ἂν εἴη· οὐ γὰρ ἦν τοῦτο ἐκείνου ἕνεκα.

(i) be for the sake of X in the abstract
(ii) actually be done for the sake of X
(iii) fail to bring X about.

Aristotle emphasizes condition (i), as the next example, of the man who bathes with a view to bringing about an eclipse, makes plain. It would be absurd, he says, for this man to say that he had bathed in vain, absurd because bathing is not for the sake of eclipses, meaning not that the agent did not want an eclipse, but that (in my terminology) bathing is not something done for the sake of eclipses in the abstract.

Now comes the phrase that is supposed to bring what we have learned about μάτην (in vain) into relation with the automatic (τὸ αὐτόματον) (197b29–30).

> So then the αὐτόματον even according to its name when it (αὐτό) should come to be in vain (μάτην).[32]

That the expressions αὐτόματον and αὐτό μάτην resemble each other is plain; but the remark is otherwise obscure in the extreme. The absence of a main verb in the translation is not an oversight; it is also missing from the text, and Aristotle's next remark does not help (197b30–32):

> For (γὰρ) a stone fell not for the sake of hitting someone, but it fell automatically because it might have been made to fall by someone and for the sake of hitting.[33]

This is not an example of something happening in vain, but seems to illustrate the contention for which the etymology is supposed to provide evidence, that is, once again, I suggest, that the expression αὐτόματον is used when an occurrence that is for the sake of X in the abstract does not occur for its sake (197b22–23).[34]

Two explanations have been proposed, each with serious defects. According to the traditional favorite, an automatic occurrence brings about an automatic outcome *instead* of the outcome for whose sake it was, and therefore occurs in vain in relation to the latter but not in relation to the former, its chance result.[35] The problem is that, though true of some

[32] οὕτω δὴ τὸ αὐτόματον καὶ κατὰ τὸ ὄνομα ὅταν αὐτὸ μάτην γένηται·

[33] κατέπεσεν γὰρ οὐ τοῦ πατάξαι ἕνεκεν ὁ λίθος· ἀπὸ τοῦ αὐτομάτου ἄρα κατέπεσεν ὁ λίθος, ὅτι πέσοι ἂν ὑπὸ τινὸς καὶ τοῦ πατάξαι ἕνεκα.

[34] Noticed by Bonitz 1969: 39, who, however, did not think it was a sufficient ground for retaining the ms. reading at 197b22–23.

[35] Alexander, *On the Soul* II, 178.24–29 (in Bruns 1887); Alexander apud Simplicium *In Ar. Phys.* 349.13–35; Themistius *In Ar. Phys.* 55.19–56.7; Pacius, 468; Ross 1936: 524; Guthrie 1946: 70–76; Waterfield 1996: 242.

episodes of chance, it is not true of others, in which the chance outcome occurs *in addition* to the outcome for whose sake the event takes place.[36] According to the alternative, μάτην in the crucial phrase αὐτό μάτην (at 197b30) means not "in vain" but, as it also can, "without a goal or purpose".[37] Automatic outcomes would then be the result of events that took place without a goal, or at least without the goal of producing this result. The problem is that, in the context of the bathing and walking examples, μάτην *does* mean "in vain."[38] What is more, Aristotle means his etymological digression to furnish support for a contention about the automatic. Which contention is especially clear if one retains the manuscript reading at 197b22–23 and understands it, as I do, to mean the expression "in vain" is a sign *that*, but, even if one does not, the most likely contention remains the same, namely that X is an automatic outcome when an event that is for its sake in the abstract brings it about without, on this occasion, occurring for its sake. On neither proposal does the etymological excursus obviously support this or any other relevant contention.

I suggest that just as the same occurrences come about by art as come about by chance and *vice versa*, so occurrences that take place in vain come about by chance and *vice versa*. Take an event that is for the sake of X in the abstract: Have someone perform it for the sake of X, but without bringing X about, and it was done in vain. Change the story by making it bring X about without being done for the sake of X, and it brings X about by chance.[39] If this interpretation is on the right lines, Aristotle's point is sound, even if his etymology is not. It is not clear that the phrase as it appears in the ms. at 197b29–30 can bear this weight as it stands, but there are plausible emendations that would allow it to do so. Although it would hardly make matters crystal clear, an original present subjunctive (ὅταν

[36] Cf. Guthrie 1946, who, although he favors this explanation as the best available solution, sets out this problem very clearly. It had already been noticed by Philoponus *In Ar. Phys.* 290.17–20 and 290.25, and Simplicius *In Ar. Phys.* 348.27–31.

[37] Philoponus *In Ar. Phys.* 290.9–25; Prantl 1854: 82, Wicksteed and Cornford 1929: 34, Wagner 1989: 472.

[38] Torstrik 1875: 463 and esp. Guthrie are very clear on this point. Even Wagner (1989: 472), who favors this interpretation, acknowledges that it requires μάτην ("zwecklos") first to mean "vergeblich" and then to mean "ohne Zweckbestimmung."

[39] Torstrik 1875: 463–464, who thinks much of the passage about the relation between the expressions μάτην and αὐτόματον is nonsense and very likely not due to Aristotle himself, nevertheless comes closest to this interpretation. He notes that one of the points that the passage could have made but does not is that both vain and chance events are by nature for the sake of something else τὸ πεφυκὸς ἄλλου ἕνεκα. In my view, the passage does make this point by taking it to underlie the similarity between αὐτό μάτην and αὐτόματον.

αὐτό μάτην γίνηται), altered under the influence of the nearby aorist subjunctives at 197b19, 197b23, and 197b34, might make it easier to view the temporal clause a filter for event types at 197b29, so that Aristotle's meaning would be, roughly, "So then the αὐτόματον even according to its name [is found] when it (αὐτό) comes to be in vain (μάτην) [i.e. is the sort of thing that comes to be in vain or is such as to come to be in vain]." Alternatives worth considering include: an original ὅτι ἂν or ὅσ' ἂν with γένοιτο (in the latter case with the singular αὐτό μάτην understood as though in inverted commas and explained as the result of Aristotle's desire to exhibit the phonetic similarity between αὐτόματον and αὐτό μάτην).[40]

In its favor, the point this interpretation would have Aristotle make here is one that he has already been at pains to emphasize, namely that X is an automatic outcome when an event that is for its sake in the abstract brings it about without, on this occasion, occurring for its sake.

[40] Cf. *Phys.* II.5, 196b22; II.6, 198a6; and esp. II.5, 197a35: ὅσ' ἂν γένοιτο ἕνεκά του.

CHAPTER 5

Man from man but not bed from bed: Nature, art and chance in Physics II

Margaret Scharle

The first argument in *Physics* II.8 serves as the foundational text for understanding the domain, extent, and character of Aristotle's natural teleology. On the basis of this text, most scholars think that Aristotle's natural teleology applies exclusively to biological things (plants and animals) and that the elements (earth, air, fire, and water) either are not teleological or are teleological only in so far as they play a role in biological processes. In addition, some scholars think this text shows natural teleology to operate not only within an individual living thing, but also to extend throughout the *scala naturae*, with lower things (like elements) existing for the sake of higher things (like animals and plants, and ultimately humans). With what they take to be the domain and extent of natural teleology confirmed by this text, scholars look outside the *Physics* to deepen their understanding of the character of natural teleology (as well as related concepts such as cause, end, nature, chance, and necessity) through careful consideration of its application in particular explanatory contexts. Those convinced of the restriction of natural teleology to individual biological things seek clarification predominantly in the biological works, such as *Generation of Animals* and *Parts of Animals*, while those seeking, in addition, better understanding of a supposed commitment to an overarching teleology across the *scala naturae* turn also to such works as *Metaphysics* XII and even the *Politics*.[1]

In a previous paper I argued that the role of nature in *Physics* II.8's first argument for natural teleology has been widely misunderstood, and as a

I thank Sean Kelsey and Devin Henry for written comments on an early draft of the chapter, and Mariska Leunissen for her comments on the penultimate draft. I also thank Kellyn Bardeen for her excellent editorial assistance and philosophical insight.

[1] Gotthelf and Lennox 1987 and Lennox 2001a are paradigmatic examples of the former approach, while Sedley 1991 is such of the latter. Notable exceptions include M. R. Johnson 2005, Leunissen 2010, and Quarantotto 2005.

result Aristotle has been interpreted with an overly biological focus.[2] I suggested a new reading of the winter-rain example that appears in the argument and argued that water is teleological *on its own*, independent of biological processes. If I am correct in my interpretation of the text, we should be looking not only at the biological works, but also at the elemental works such as the *Meteorology, On Generation and Corruption,* and *On the Heavens* to understand the character of natural teleology and related concepts.

In general, there are two desiderata for a proper interpretation of the first argument in *Physics* II.8: First, the interpretation must show that the premises and conclusion are ones Aristotle himself would accept; and second, since the argument is meant to engage an opponent, its interpretation must offer a satisfying account of the dialectic in the passage.[3] In my previous work, I showed my new interpretation to best satisfy the first desideratum, but did not speak to the second. In this chapter, I argue that my interpretation best satisfies the second desideratum as well and, more importantly, suggests a unified interpretation of the dialectic across the whole of *Physics* II.

I The argument of *Physics* II.8

In *Physics* II.8, Aristotle considers a puzzle from an opponent to his natural teleology. The challenge makes use of some of Empedocles' views, yet, as Alan Code points out, the problem is not posed by Empedocles and may not be one that he would have endorsed.[4] I follow Code in referring to the opponent of *Physics* II.8 as the "Empedoclean" opponent, who presents what I call the "statement of the problem":

> There is the difficulty: what prevents nature from acting neither for something nor because it is better, but as Zeus rains – not in order that the corn may grow, but of necessity. (For what was taken up must become cold, and what has become cold, having become water, must come down. When this has happened, it turns out that the corn grows.) Similarly also, if someone's corn on the threshing floor is ruined it does not rain for the sake of this, so that the corn may be ruined, but this simply results. Why then should it not

[2] Scharle 2008.

[3] In arguing for the importance of the dialectic in *Physics* II.8, I do not mean to suggest that Aristotle's method in the *Physics* is dialectical as opposed to scientific. I see the *Physics*' use of dialectical and scientific methods as complementary and integrated, but I will not pursue the complex issue here. For the purposes of my interpretation, "dialectic" need not be understood in any more technical a sense than engaging an interlocutor, as opposed to talking past him.

[4] Code 1997: 127.

> be the same with the parts in nature, e.g. that our teeth should come up of necessity – the front teeth sharp, fitted for tearing, the molars broad and useful for grinding down the food – since they did not arise for this end, but it was merely a coincidental result; and so with all other parts in which we suppose there is purpose? Wherever then all the parts came about just what they would have been if they had come to be for an end, such things survived, being organized by chance in a fitting way; whereas those which grow otherwise perished and continue to perish, as Empedocles says his "man-faced ox-progeny" did. (*Phys.* II.8, 198b17–34)[5]

In the course of the chapter I will discuss what this challenge amounts to. But first I will focus on Aristotle's response to this challenge with the passage I call "the winter-rain argument":

> Such are the arguments (and others of the kind) which may cause difficulty on this point. Yet it is impossible that things are this way. For these things [e.g. animals] and all things that are by nature, come to be in this way either always or for the most part, and nothing from luck or chance does. For it does not seem to be from luck or from coincidence that it rains often in winter, but if in the dog-days; nor that there are heatwaves in the dog-days, but in winter. If, then, things seem to be either from coincidence or for the sake of something, and if these things are not able to be from coincidence or from chance, they would be for the sake of something. But clearly all such things are by nature, as these speakers themselves would say. The "for the sake of something," then, is in things which are and come to be by nature. (*Phys.* II.8, 198b35–199a8)

Many commentators suggest that Aristotle presents nothing in the winter-rain argument that does anything more than baldly beg the question – asserting, more than arguing, the position of natural teleology against his Empedoclean opponent. I believe a more compelling argument can be constructed from this passage by bringing in texts from outside the *Physics* to sharpen our understanding of the shared ground between Aristotle and his Empedoclean opponent. The structure of the argument, I suggest, is as follows:

1. (Accepted Premise): "Things [e.g. animals] seem to [come to] be either from coincidence or for the sake of something."
2. (Disputed Premise): "These things [e.g. animals] are not able to [come to] be from coincidence or from chance."
3. (Sub-Conclusion): "These things [e.g. animals] would [come to] be for the sake of something." (1, 2)

[5] Trans. Barnes 1984; modified. All translations are from Barnes 1984, unless otherwise noted.

4. (Accepted Premise): "All such things are by nature, as these speakers themselves would say."
5. (Conclusion): "The 'for the sake of something,' then, is in things which are and come to be by nature." (3, 4)

As presented, the argument is valid. In order to establish his conclusion, however, Aristotle must be able to convince the Empedoclean opponent to accept each of the premises, so that the conclusion will follow. After considering Premises (1) and (4), I will focus on what is, perhaps, the most problematic premise, Premise (2).

As Leunissen notes, Premise (1)'s use of δοκεῖ suggests mutual agreement.[6] There is a trivial reading of the agreement, argued by Code and Charles, in which the disjunction is trivially true because the opponent agrees to one of the two disjuncts.[7] On my view, Aristotle positively moves his opponent to accept Sub-Conclusion (3), that "these things [e.g. animals] come to be for the sake of something," by asserting the disjunction in the strong sense required for the entailment of (3) (that the rejection of one disjunct entails the acceptance of the one remaining). If this were the case, he would not be simply arguing against the disjunct that the opponent accepts and asserting his own: He would be using the rejection of the opponent's position to drive the acceptance of his own. By moving the Empedoclean opponent to accept Sub-Conclusion (3), Aristotle can make use of Accepted Premise (4) to reach the final Conclusion (5). If Charles and Code are correct that all Aristotle can do is show one of the disjuncts to be false, then the fanfare that Aristotle makes over Premise (4) as shared ground – "as these speakers themselves would say" – would be for naught; without the opponent's acceptance of Sub-Conclusion (3), the mutual agreement on Premise (4) would serve no purpose in the dialectic.

On my view, the Empedoclean opponent maintains that tooth growth and the generation of whole animals *appear* to be teleological: We suppose that there is purpose in these cases (ἐν ὅσοις δοκεῖ ὑπάρχειν τὸ ἕνεκά του; *Phys.* II.8, 198b28–29).[8] The opponent suggests that things might not be as they appear and offers his own alternative: Maybe they just came to be that way by chance, in a similar way to things that are not, in fact, teleological (like rain resulting in corn growth or rain resulting in corn rot). The opponent thereby concedes that phenomena like tooth development in animals and the generation of whole animals *might* be teleological, and the

[6] Leunissen 2010: 29. [7] Charles 1991: 113; Code 1997: 129.

[8] I acknowledge Sean Kelsey for suggesting this view to me.

question on the table is whether they, in fact, are – either they are teleological (as they appear to be) or they are not (because the opponent has presented a successful alternative). This is what Premise (1) formalizes. In admitting that the phenomena appear teleological, the opponent accepts that the onus is on him to unseat the presumption in favor of teleology.

So interpreted, Premise (1) does not claim that *for any occurrence whatsoever*, it happens either by chance or for the sake of something. Not even Aristotle thinks that.[9] Rather, Premise (1) states that *the phenomena in question* happen either for the sake of something (as they appear to) or by chance (because the opponent has offered a successful counter).[10]

Let us now consider Premise (4). Aristotle directly states that the Empedoclean opponent would accept the truth of this premise, that all such things that are at issue in the dialectic are "by nature" (φύσει). This is important because Aristotle and his Empedoclean opponent disagree as to which types of natures exist, so the meaning of the term "by nature" must be interpreted in a way that is neutral to this philosophical disagreement. The Empedoclean opponent denies the existence of natures above the level of the four elements: earth, air, fire, and water. Homogenizing the views of Aristotle's materialist predecessors – including Empedocles – *Physics* II.1 attributes to them the view that the four elements are "the whole of substance, all else being its affections, states, or dispositions" (193a25).[11] Aristotle's own view, by contrast, is that each natural thing – each animal, in this case – has a nature of its own, where nature is an inner source of movement and rest that belongs to the thing "primarily in virtue of itself" (πρώτως καθ' αὑτὸ) (*Phys.* II.1, 192b22), and that animal generation (propagation of the species) comes about caused by the source of movement and rest that is the animal nature.

9 The production of bile is a case in point: Bile is produced neither for the sake of something nor coincidentally, but simply as a necessary byproduct of the teleological operations of the liver (*PA* IV.2, 677a12–18). The nature of the animal does not aim to produce bile: In fact, the healthier the liver and the more pure the blood, the less bile is produced.

10 This premise is suggested by the Empedoclean opponent himself in the statement of the problem: When it comes to the phenomena under consideration, the two options are "either from coincidence or for the sake of something." Sauvé Meyer 1992: 796–797 points out that although the initial disjunction is misleadingly stated as "not in order that . . . but of necessity" (198b18–19), the gloss of that disjunction (198b19–23) does not mention necessity. The reason the opponent gives for denying that the phenomena are teleological in the restatement of the position at 198b27 is that they are coincidental (198b27), even though necessity is mentioned in this passage. Ultimately, the final statement of the opponent's position (198b27–32) does not mention necessity at all, thereby suggesting that the opponent agrees to the disjunction for the phenomena in question.

11 See also *Metaph.* V.4, 1014b35–1015a3.

Without begging the question, Aristotle cannot build into (4) the assumption that animal generation is *by the nature of the animal*, because in the Empedoclean opponent's view animal generation is *by the nature of the elements*. However, in *Physics* II.1, Aristotle carefully marks off the distinctions between something that "has a nature" and something that is "*by* nature" (φύσει). The locution "*by* nature" is introduced as a description of that which is *by* an inner source of movement and rest, which has a much wider scope than "*has* a nature," for it includes not only things that *have* a nature, but, more generally, that which is the result of nature (e.g. the natural activities of things that have nature), whether the nature of an animal or the nature of the elements.

Premise (4) therefore remains neutral between Aristotle's own view and that of the Empedoclean opponent, who denies that animals "have" a nature (for they do not have a nature that belongs to it primarily in virtue of itself), yet admits that animals are "by" nature in the sense that they are by the nature of the elements. In order to reach his conclusion, Aristotle needs nothing stronger than the neutrally stated Premise (4).

The greatest difficulty in interpreting Aristotle's argument lies in showing why he thinks the Empedoclean opponent could be moved to accept the truth of (2) – so I will spend the bulk of the chapter explaining and defending this premise. The text seems to offer the following claims in support of (2):

A. (Accepted Premise): "It does not seem to be from luck or from coincidence that it rains often in winter, but if in the dog-days; nor that there are heatwaves in the dog-days, but in winter." *(There is a regularity in winter rain, and in summer heatwaves, which cannot be due to coincidence.)*

B. (Disputed Premise): "These things [e.g. animals] and all things that are by nature come to be in this way either always or for the most part." *(Animals and things that are "by nature" come to be with the same type of regularity as winter rain and summer heatwaves.)*

C. (Disputed Premise): "Nothing from luck or chance does [i.e. nothing from luck or chance comes to be in this way either always or for the most part]." *(Nothing that comes by chance comes to be with the same type of regularity as winter rain and summer heatwaves.)*

Premise (2) (Conclusion): "These things [e.g. animals] are not able to [come to] be from coincidence or from chance." *(Therefore, animals and things that happen "by nature" cannot come to be from chance.)* (A, B, C)

The question of how A, B, and C are meant to support Premise (2) is not obvious, although the gloss after each quotation offers a preview of the interpretation I will argue. To begin, I want to consider whether the Empedoclean opponent would accept the truth of these claims. Let us consider Premise (A). Most scholars now think Aristotle means that while summer rain produces corn growth coincidentally, winter rain produces corn growth teleologically.[12] In my previous paper, I offered a detailed argument against this interpretation by showing that it would lead to a conclusion Aristotle himself would not accept.[13] Importantly, this interpretation also fails to be dialectically satisfying. Again, Aristotle's use of δοκεῖ here, as in his statement of Premise (1), suggests mutual agreement.[14] But the case of rain's production of corn growth was originally intended by the Empedoclean opponent as an obvious example of chance, and it is not at all clear why pointing out this *seasonal* connection would move the opponent to change his mind. After all, the opponent has *already* said that "cold" is responsible for the rain (*Phys.* II.8, 198b14), which is close to acknowledging its seasonality.

On my reading, Premise (A) presents winter rain, taken on its own, as an example of a non-coincidental phenomenon, and this is simply a restatement of the opponent's assertion: "For what was taken up must become cold, and what has become cold, having become water, must come down" (*Phys.* II.8, 198b19–20). Premise (A) simply makes this mutual agreement precise by clarifying that the process takes place when it is cold, and that it is typically cold in the winter, and not in the summer.[15] Premise (A) additionally notes that we would not say that winter rain happens by coincidence, but only summer rain.

Let us now consider Premises (B) and (C). The statement of the problem shows the Empedoclean opponent to reject either or both: Although nowadays species reproduce true to type, and thus nowadays come to be, for instance, with a set of teeth with molars in back and sharp teeth in front (either always or for the most part), nonetheless it was not always so, the Empedoclean opponent would say. The species we see now were once less

[12] I argue against the full range of alternative interpretations in Scharle 2008: 148–167. Most recently, Leunissen 2010: 10–48 has suggested that Premise (A) refers to the fact that farmers use winter rain to grow their crops. Although I do not have the space to pursue the point, I think her otherwise persuasive interpretation strains the text in requiring that winter rain not be among the things whose nature is for the sake of something.

[13] Scharle 2008: 151–167. [14] Leunissen 2010: 29.

[15] *Pace* Leunissen, who argues, "*If the argument is to be rhetorically effective*, it seems that there must be some non-accidental way in which Aristotle thinks winter rain (even if not itself caused teleologically) serves the growth of crops" (2010: 30; my emphasis).

common and only became common because the arrangement of their parts was conducive to survival. So, if (B) is interpreted as a claim about what has always or for the most part happened *throughout all time*, the Empedoclean opponent will simply deny its truth: One would only think (B) is true if one were focused myopically on the current era and ignored the fact that things did not always or for the most part come to be as they do nowadays. However, if (B) is interpreted as a claim about what nowadays comes to be always or for the most part, (B) will be true, but (C) will be false. For the Empedoclean opponent claims that what nowadays happens always or for the most part nonetheless comes to be by coincidence. In putting forward B and C in support of (2), it is hard to see how Aristotle does not just baldly beg the very question at issue.

Some commentators simply concede that he does so. For example, Cooper holds that Aristotle's argument relies on his view that the species of organisms are eternal and thus did not come to be as Empedocles hypothesizes.[16] Other commentators push Aristotle's question-begging back a step. For example, Judson holds that Aristotle's argument relies on the claim that the proper account of the generation of animals must make reference to the fact that the arrangement of teeth, for example, "serves the life of the organism."[17] Thus, Aristotle shows that the Empedoclean opponent's account of the arrangement of teeth is impoverished because it maintains that the arrangement of teeth is merely coincidentally beneficial. But, again, this issue lies squarely within the disputed ground. Similarly pushing the question-begging back one step, Code maintains that "all the opponent must admit is that it is always or for the most part the case that if in *human* development a front tooth is formed, then it is suitable for biting."[18] But, as we have seen when examining Premise (4), the Empedoclean opponent denies the existence of natures above the level of the four elements, and thus thinks that an animal is simply a coincidental arrangement of earth, air, fire, and water, and there is no robust, non-arbitrary kind "Human" by which to distinguish certain occurrences of tooth formation from others. Further, the Empedoclean opponent might press that even if the designation of a kind is not wholly arbitrary, the designation is completely *ad hoc*: If you can designate the kind "Human" in part *by reference to the fact* that things in this category have teeth suitable for biting and chewing, then it will be true, in a trivial sense, that tooth formation in a human regularly leads to teeth suitable for

[16] Cooper 1987: 246–253. See also Charlton 1970: 123, who claims the argument is "inconclusive."
[17] Judson 2005: 352. [18] Code 1997: 131.

biting and chewing. But the kind of regularity within the "kind" designated in this *ad hoc* fashion is not the kind of regularity that (C) claims is not the result of coincidence.

In what follows, I will suggest that Aristotle uses the uncontroversial shared ground of (A) as the fulcrum of his argument against the Empedoclean opponent: Because the Empedoclean opponent will uncontroversially agree to (A), he will have to concede (C) and (B), and thus (2). In order to appreciate the impact of this strategy, we first have to understand the origin of the challenge posed in the statement of the problem.

II The origin of the problem

In *Physics* II.4–6, Aristotle argues that both *thought* and *nature* enjoy a priority over chance: "spontaneity and chance . . . are posterior (ὕστερον) to intelligence and nature" (*Phys.* II.6, 198a9–10), in the sense that movements brought about by either thought or by nature are teleological, while those that might have been brought about by either, but in fact were not, are coincidental (*Phys.* II.5, 196b22–23). For example, "the stone that struck the man did not fall for the sake of striking him; therefore it fell spontaneously, because it might have fallen by the action of an agent and for the sake of striking" (*Phys.* II.6, 197b29–32). Here the fall is coincidental, because it might have fallen by thought – for the sake of striking the man – but in fact did not.

It is just this priority of nature and thought over chance that misleads the Empedoclean opponent in the statement-of-the-problem passage. The opponent's case of rain accidentally resulting in corn growth – "Zeus rains, not in order to make the corn grow" (*Phys.* II.8, 198b18–19) – is perfectly analogous to the II.6 stone case: replacing "striking the man" with "corn growth," Aristotle can give the same description: "it did not fall for the sake of [growing corn]; therefore it fell spontaneously, because it might have fallen by the action of an agent [i.e. Zeus] and for the sake of [growing corn]." Coincidence for Aristotle is the lack of either *thought* or *nature*, and in the case of both the stone and the rain, the coincidence is because of the lack of *thought*, emphasized by the opponent's appeal to "Ζεὺς" in stating the problem. The opponent seizes on what we might call the "lack-of-agency" model of coincidence (that is, coincidence by lack of thought) to formulate his understanding of the "lack-of-nature" coincidence supposedly exhibited in the case of animals. The opponent asks: "Why then should it not be the same with the parts in nature?" (*Phys.* II.8, 199b24), and ultimately with the generation of whole animals (*Phys.* II.8,

198b29–32)? On my reading, the winter-rain argument responds to this question by claiming that the "lack-of-agency" model is the wrong model of coincidence.[19] In the winter-rain argument, Aristotle re-orients the Empedoclean opponent by offering a "lack-of-nature" case of coincidence (summer rain) and a *natural* case of teleology (winter rain) to serve as the exemplars of the model on which to reconsider the opponent's case of animal generation. Given that the case of animal generation fits the model exemplified by winter rain, not summer rain, the opponent should conclude that animal generation (and thus the development of functional teeth as part of that generative process) is similarly teleological.

This reading demands a distinction between what I am calling the "natural" model of teleology and what I am calling the "agency" model of teleology. *Physics* II.4–6 claims that regularity is a hallmark of both models: Both things that come to be by nature and things that come to be by thought come to be regularly, while things that come to be by coincidence do not (*Phys.* II.5, 196b10–11, 20–21; 197a3–4, 20, 31–32; 197a33–35). Agents, by the teleological direction of their thought, and natural things, by the teleological direction of their natures, both reliably produce their respective ends, unless something impedes: The builder regularly builds houses, and the plant regularly grows roots. But there is something distinctive about the type of regularity exhibited by natural things at the level of generated wholes, a distinction Aristotle makes in his argument in *Physics* II.1 against his materialist predecessors (including Empedocles): "man is born from man but not bed from bed" (193b9–10). The type of regularity exhibited by natural things is that of continuous generation, for nature is an internal principle of "production" (ποιήσεως) (*Phys.* II.1, 192b30). And this is the type of regularity with which winter rain comes to be.

III Regularity on the natural model of teleology

Aristotle's refined view, as my previous paper argues in depth, is that water moves into its natural place *naturally and teleologically only upon being generated by the sun.*[20] Although it is true that water moves into its natural

[19] *Pace* Cooper 1987: 245n.5, who maintains that the art–nature analogy plays no role in the winter-rain argument, and therefore claims, given the fundamental importance of the winter-rain argument, that one must "reject the suggestion" that the analogy is "central and fundamental to Aristotelian natural teleology."

[20] For the full argument for this view, see Scharle 2008: 150–181. As I mention there (n.70), the sun is the efficient cause of three teleological cycles of evaporation and condensation, and water's coming to be and movement as part of these cycles is imitative of the divine and teleological.

place whenever it falls unobstructed (whether as rain or otherwise), winter rain is the *only* rainfall that forms a cycle that imitates the divine. *Metaphysics* XII.10 (1075a19–22) suggests that elemental movements are teleologically directed towards the prime mover via their imitation of the heavenly bodies: Water's rectilinear motions can imitate the circular motion of the heavenly bodies, which in turn imitates the activity of the prime mover. The only way for a rectilinear motion to imitate circular motion is for it to be part of a cycle that "reverts again to the beginning . . . Hence it is by imitating circular motion that rectilinear motion too is continuous" (*Cael.* II.10, 337a7). The rectilinear movement of water in winter rain, then, is imitative *because* (expressed by ὥστε) it occurs on the heels of another rectilinear movement (of air) with which it composes a cycle. *Meteorology* I.9 confirms that winter rain composes part of an imitative cycle of generation: "[W]e get a circular process that follows the course of the sun . . . When the sun is near [i.e. in the summer] the stream of vapor flows upwards; when it recedes [i.e. in the winter], the stream of water flows down [as winter rain]" (346b35–347a1; see also *Meteor.* II.4, 359b34–360a3).[21] The fact that water's natural movement is efficiently caused by the sun ensures that the natural movement of water will occur on the heels of air's upward rectilinear movement. Only in this way does water's movement imitate circular movement, which imitates the prime mover. In contrast, whereas winter rain is properly caused by the recession of the sun, which ensures its coordination with air to form an imitative cycle, summer rain is caused by the "recoil" (ἀντιπερίστασις) of hot and cold (*Meteor.* I.12, 348b8–10, 349a5–9) and is thus "violent" (ὕδατα λαβρότερα, *Meteor.* I.12, 348b11, 348b23, or ῥαγδαῖα, 349a7). Summer rain is therefore neither imitative nor teleological – a rectilinear dead-end.[22]

In general, *Metaphysics* XII.10 shows that things are in "joint-arrangement" (συντέτακται) (*Metaph.* XII.10, 1075a15) with one another to the extent to which their activities approximate that of the prime mover by their imitating the circular motion of the heavenly bodies: Winter rain

[21] See also *Metaph.* IX.8, 1050b28ff.

[22] Most recently, Leunissen 2010: 30n.57 has resisted my interpretation of winter rain in *Physics* II.8 because she "take[s] it that the crux for Aristotle in *Ph* II.8 is to show that regular natural phenomena have regular beneficial outcomes due to the fact that *nature is an efficient cause that acts for the sake of something.* Under Scharle's interpretation, however, it is the retraction of the sun in the winter that is the efficient cause that makes the water return to its natural place, but this efficient cause itself never – neither in the winter, nor in the summer – acts for the sake of this outcome." My response is to deny that natures are always efficient causes (see Scharle 2008: 171–173) and to maintain that the stated conclusion of *Physics* II.8 is that nature is for the sake of an end (198b17–18, 198b10–11, 199a7, 199b32–33), not that nature is an *efficient* cause for the sake of an end.

imitates the circular motion of the heavenly bodies by moving rectilinearly on the heels of air's upward movement in the summer (thereby forming the generative cycle of water and air), while animals and plants do so by generating another of their kind (*GC* II.10, 336b27–337a8).[23] Aristotle argues, "coming-to-be and passing-away will, as we have said, always be continuous (συνεχῆς)" (*GC* II.10, 336b25), for God "fulfilled the perfection of the universe by making coming-to-be uninterrupted; for the greatest possible coherence would thus be secured to existence, because that coming-to-be should itself come-to-be perpetually is the closest approximation to eternal being" (*GC* II.10, 336b32–337a1). We have now arrived at a precise articulation of the type of regularity exhibited on the natural model of teleology: that coming-to-be itself comes-to-be regularly. This regularity is initiated by the sun – the sun's circular motion ensures not only that sublunary elemental transformation and locomotion will come to be in an imitative pattern, but also that animals generate another individual of their own kind in an imitative pattern: "since the upper movement is cyclical, the sun moves in this determinate manner; and since the sun moves thus, the seasons (ὧραι) in consequence come-to-be in a cycle, i.e. return upon themselves; and since they come-to-be cyclically, so in their turn do the things whose coming-to-be the seasons initiate [e.g. plants and animals]" (*GC* II.11, 338b3–5; see also *Cael.* II.3, 286a13–286b2).[24] And it is important for my interpretation that Aristotle has this relationship between the sun and generable things clearly in mind in *Physics* II itself when he says, "man is begotten by man and by the sun as well" (*Phys.* II.2, 194b13).

So far I have argued that the same type of regularity is exhibited in both elemental and animal generation. I now want to show that this regularity comes in degrees according to an entity's rank on the *scala naturae*. *Generation of Animals* II.1 lays out the ranking as follows: Beginning at the top of the *scala naturae*, Aristotle places the heavenly bodies (both living and eternal), then living things (living but not eternal), and finally the sublunary elements (neither living nor eternal) (731b24–732a1). This passage understands rank according to goodness, while the texts I consider next suggest that ranking is alternatively calibrated according to the *degree of regularity* exhibited in the entity's imitative activity. These two descriptions of ranking ultimately come to the same thing: Since the prime mover

[23] In fact, *On the Soul* II.4 maintains that, for living things, generating another of their own kind is "the most natural" (φυσικώτατον; 415a26) act to which all other natural activities are subordinate (415b1–2).

[24] On my view, the seasons initiate coming-to-be in the sense that animals make use of the seasons for the sake of growth. See Scharle 2008: 161–165.

is the best thing in the cosmos, the more closely something approximates its activity, the better it is; and the more closely it approximates the prime mover, the more regular and uninterrupted its activity.

The introduction to the *Meteorology* confirms that the sublunary elements exhibit a "regularity less (ἀτακτοτέραν) than" the heavenly bodies moving in a circle (*Meteor.* I.I, 338a20–b4). This contrast in degree of regularity is also at work in *Metaphysics* XII.10's household analogy: Taken together with the introduction to the *Meteorology*, the analogy suggests that the heavenly bodies are to the sublunary elements as the freemen are to the slaves and beasts. Thus, we should expect that the heavenly bodies "have least license to act as they chance to, but all or most of what they do is arranged (τέτακται)," while the sublunary elements "can do a little towards what is communal, but act mostly as they chance to" (1075a19–22).[25] Although the circular motion of the heavenly bodies always imitates the activity of the prime mover, not all sublunary elemental movements imitate the activity of the prime mover, but only those movements, such as winter rain, that take the sun as their efficient cause.

On the Soul confirms that this regularity varies in degree even within the stratum of living things: "since then no living thing is able to partake in what is eternal and divine by uninterrupted continuance (for nothing perishable can forever remain one and the same), it tries to achieve that end in the only way possible to it, and success is possible by the more and the less (τὸ μὲν μᾶλλον τὸ δ᾽ ἧττον)" (II.4, 415b4–6). Thus, Aristotle seems to think that the heavenly bodies exhibit the highest degree of the type of regularity at issue, followed by animals and plants, which exhibit a lower degree of regularity (and greater and lesser degrees within this stratum), and, at the lowest level, sublunary elements exhibit the lowest degree of regularity.

Physics II offers two anti-materialist arguments that clearly capitalize on the degrees of regularity Aristotle articulates. First, *Physics* II.4 (196a25–b4) argues against Democritus' view that animals and plants come to be by mind or nature and not by chance, but that the heavenly spheres did come to be by chance. Aristotle argues that the movements of the heavenly spheres are much more regular than the generation of plants and animals, so if Democritus agrees that the regularity with which plants and animals come to be cannot be due to mere chance, *a fortiori* it is so in the case of the heavenly spheres.[26] Democritus cannot consistently maintain that the

[25] Trans. Sedley 2000: 328. [26] See also *PA* I.I, 641b10–23.

generation of animals is not due to chance while arguing that the movements of the heavenly spheres are.

Physics II.8 itself offers an analogous argument, which I will call the "olive-headed-vines argument," found just thirty lines down from the winter-rain argument and levied against the same Empedoclean opponent:

> Again, in plants too we find that for the sake of which, though the degree of organization (διήρθρωται) is less. Were there then in plants also olive-headed vine-progeny, like the "man-headed ox-progeny," or not? An absurd suggestion; yet there must have been, if there were such things among animals. (199b10–13)

Once again, Aristotle takes as a premise a claim about the degrees of regularity exhibited in the generation of things at different levels of the *scala naturae*: The generation of animals is more regular than the generation of plants, so that if the generation of animals is due to chance, as Empedocles suggests, *a fortiori* the generation of plants is, in which case plant generation should exhibit the botanical analogue of man-faced ox-progeny (e.g. olive-headed vines).[27] But Empedocles never mentions anything like olive-headed vines, which shows his inconsistent application of the notion of chance.

IV Reconsidering the winter-rain argument

I now submit that *Physics* II.8's argument for Premise (2) is yet another instance of the *a fortiori* argument pattern exhibited in both *Physics* II.4's argument against Democritus and the II.8 olive-headed-vines argument against the Empedoclean opponent. In fact, Aristotle suggests this in the introductory passage directly preceding the statement of the problem: Here Aristotle complains that his predecessors posit the existence of causes like love or *nous* that could operate teleologically, but fail to use them in the proper explanatory contexts. As we have seen in the *a fortiori* arguments, they "touch on it" to explain some phenomena, but fail to use it to explain the phenomena that require it even more (καὶ γὰρ ἐὰν ἄλλην αἴτιαν εἴπωσιν, ὅσον ἁψάμενοι χαίρειν ἐῶσιν; *Phys.* II.8, 198b13–16). The origins of this complaint can be found in Plato's *Phaedo* (97b–99d), where Socrates expresses disappointment in Anaxagoras' positing *nous*, but failing to make use of it as a teleological cause of phenomena that require one.

Recall the sub-argument at issue in the first argument of II.8:

[27] Simplicius' commentary on *Physics* II agrees with Alexander that this is another *a fortiori* argument.

A. (Accepted Premise): "It does not seem to be from luck or from coincidence that it rains often in winter, but if in the dog-days; nor that there are heatwaves in the dog-days, but in winter." *(There is a regularity in winter rain, and in summer heat waves, which cannot be due to coincidence.)*

B. (Disputed Premise): "These things [e.g. animals] and all things that are by nature, come to be in this way either always or for the most part." *(Animals and things that are "by nature" come to be with the same type of regularity as winter rain and summer heatwaves.)*

C. (Disputed Premise): "Nothing from luck or chance does [i.e. nothing from luck or chance comes to be in this way either always or for the most part]." *(Nothing that comes to be by chance comes to be with the same type of regularity as winter rain and summer heatwaves.)*

2. (Conclusion): "These things [e.g. animals] are not able to [come to] be from coincidence or from chance." *(Therefore, animals and things that happen "by nature" cannot be from chance.)* (A, B, C)

While the *Physics* II.4 argument against Democritus relied on the claim that the heavens display a greater degree of regularity than animal- and plant-generation, and the II.8 olive-headed-vines argument relied on the claim that there is a greater degree of regularity exhibited in animal generation than in plant generation, here in the II.8 winter-rain argument Aristotle suggests that animal generations exhibit the same type of regularity as elemental generations in the form of winter rain. And just as the *Physics* II.4 argument showed Democritus mistakenly to attribute to chance the heavenly motions, even though they are more regular than the generations he does not attribute to chance, and just as the olive-headed-vines argument shows the Empedoclean opponent mistakenly to attribute to chance animal generations, even though they are more regular than the plant generations, he does not attribute to chance (at least in so far as he does not posit the existence of anything like olive-headed vines), so too this winter-rain argument shows the Empedoclean opponent mistakenly attributes animal generations to chance even though they exhibit the same type of regularity whose degree is the same as (if not more than) the elemental generations exhibited in winter rain, which the opponent does not attribute to chance. Not only does my interpretation make better sense of the dialectic between Aristotle and his Empedoclean opponent, it reveals that Aristotle thought two of his formidable materialist opponents – Democritus and Empedocles – fell prey to the same kind of error, simply from different ends of the *scala naturae*.

So understood, Premise (A) supports (C) directly by showing that the type of regularity exhibited in winter rain – a kind of continuity of generation – is the kind of regularity that cannot be chalked up to chance. And if the regularity in (B) is this same type of regularity that (C) claims cannot be due to chance, then neither can the regularity by which animals are generated be chalked up to chance. By accepting (A), then (C), then (B), the Empedoclean opponent reaches conclusion (2).

But why would the Empedoclean opponent admit that there is any continuity at all in animal generation such that it displays the same type of regularity as winter rain? Even though Aristotle thinks the species are eternal, and thereby disagrees with the Empedoclean opponent's version of "natural selection," they both share the view that animal generation is itself continuous: Even the Empedoclean opponent thinks that *the animals that generate* do so continuously, for he claims that the ones whose parts were unsuitably arranged die and continue to die (*Phys.* II.8, 198b29–33). As Aristotle highlights in the next set of arguments, Empedocles thought that animals are generated from seeds: Aristotle even quotes Empedocles' poem that "what was 'undifferentiated first' was seed (σπέρμα)" (*Phys.* II.8, 199b8–9). As they do for Aristotle, Empedocles' seeds serve as the link between one generation and the next, and can thereby suggest that he is committed to a kind of continuity of generation, even if it is not eternal. Aristotle wants to point out that in so far as the Empedoclean opponent thinks this, there is, after all, a regularity found in animal generation that cannot be simply coincidental, for this regularity is *the same type of regularity* – a kind of continuity in generation – that the Empedoclean opponent agreed could not be due to mere coincidence in the case of winter rain. Water's falling in winter (when the sun recedes) ensures (for the most part) its subsequent evaporation (when the sun returns); in so doing, winter rain metaphorically "sows the seeds" of the next winter's rainfall. For although it may rain in the summer, this kind of rain does not form a generative cycle with air that will lead to another iteration of rain the following winter. Likewise, although men may give birth to ox-faced progeny, those are not the ones that will, in turn, generate. And even if there are plenty of episodes of coincidental rain that randomly come to be here and there out of season, and even if there are plenty of random ox-faced monstrosities that come to be, nonetheless *the rain that forms part of the generative cycle with air* will (for the most part) come to be again, and *the animals that give birth to animals that survive and generate* will (for the most part) generate again. This is the distinctive regularity found on the natural model.

At this point one might worry that my interpretation pushes Aristotle's question-begging back just one step, in a way that is similar to Code's interpretation. It is worth returning to Code's interpretation in order to distinguish mine. As I note above, Code's interpretation requires that "the opponent must admit that it is always or for the most part the case that if in *human* development a front tooth is formed, then it is suitable for biting," and more generally, "for any given natural kind K, tooth formation of a K" regularly results in functional teeth.[28] I argued that this assumption would violate Premise (4), understood as neutral between Aristotle's own position in which there are robust kinds above the level of the elements, and the Empedoclean opponent's view in which there are not. By contrast, my interpretation of the argument does not violate Premise (4), interpreted neutrally. I argue that Aristotle first sets out the example of winter rain as the natural model of teleology to which the Empedoclean opponent's own version of animal generation conforms. The opponent admits that the *animals that generate* do so continuously and thus must be explained teleologically, just as winter rain must be. Once Aristotle has identified this set of animals – the ones that generate according to the natural model exemplified in winter rain – he can then say that *in those things* tooth formation regularly leads to functional teeth, while remaining neutral as to whether it is the elemental natures or the natures of the animals that are responsible for the continuity. That is to say, Aristotle thinks he must first establish the need for teleological explanation at the level of the generation of whole animals – for this is the level that exhibits the distinctively natural pattern of teleology – and then consider the teleology of the formation of their parts as part of the generative pattern.

In my reading, all the Empedoclean opponent needs to accept is that winter rain exhibits a regularity that cannot be due to coincidence, and that animal generation exhibits the same type of regularity. Thus, animal generation, like winter rain, requires a teleological explanation, and given that these phenomena are by nature (as opposed to by thought), nature will be the teleological cause. To accept this, the Empedoclean opponent need not accept (i) Aristotle's particular teleological explanation for the regularity, although *Metaphysics* XII.10 later makes an argument to this end, nor must the opponent accept (ii) Aristotle's appeal to natures above the level of the elements. If the Empedoclean opponent thinks he can show that elemental natures are robust enough to explain the continuity of animal generation, so be it (as far as this argument is concerned).[29]

[28] Code 1997: 131. [29] *Pace* Judson 2005: 349.

That said, Aristotle may think that he also has given the Empedoclean opponent reason to doubt his ability to explain the continuity of animal generation by appeal solely to the four elemental natures. Although Aristotle does not spell this out in the winter-rain argument, if he can convince the Empedoclean opponent that winter rain displays not only the same type of regularity, but that winter rain is *less regular* than animal generation (as Aristotle himself thinks), then he would have given the Empedoclean opponent reason to doubt that elemental natures could explain the continuity of animal generation. In other words, if elemental processes (exhibited in winter rain) are less regular than biological generative processes, the former could not explain the latter. But, again, Aristotle need not argue this in order to reach his conclusion.

V Reconsidering *Physics* II.8's place in *Physics* II

With my interpretation that winter rain is teleological in its own right, I am in the unusual position of claiming that Aristotle extends natural teleology even to the level of the elements. This novel approach allows me to construct an argument I believe to be more interpretatively satisfying and more dialectically satisfying than other existing readings. In addition to this, my interpretation suggests a unity for the set of arguments in *Physics* II.8 and for the whole of *Physics* II. Let me conclude with a brief sketch of this unified picture.

Aristotle's conception of nature in *Physics* II is presented as opposed to both a Platonic tradition, in which nature *is* art, and a pre-Socratic one, in which nature is not properly distinguished from chance. Rejecting the pre-Socratic tradition, both Plato and Aristotle insist that the world's order and structure are explained teleologically and are not due to chance; however, Aristotle disagrees with Plato, who holds that nature would have to be art in order to explain such order. Overall, *Physics* II insists that art and nature are analogous in being the source (ἀρχή) of order, yet distinct in that nature is an internal source of this order, while art is an external one (*Phys.* II.1, 192b9–34). In arguing against his pre-Socratic predecessors' conception of the natural world as operating by chance, however, Aristotle's nuanced conception of the relationship between art and nature is at once a help and a hindrance.

Consider Aristotle's response to the pre-Socratic predecessors' argument in *Physics* II.1 for the claim that nature is matter (193a13–25). Aristotle depicts the pre-Socratics as taking lessons about nature from artificial examples: Just as matter is the nature of the bed, so too, they think, is matter the nature in the natural cases of bronze, bone, and wood. In modeling nature after art, the predecessors relegate much of nature to chance, for if *solely matter* (earth, air, fire, and water) is nature, then everything else is "its affections, states, and dispositions" (*Phys.* II.1, 193a24–25). But in response to this materialist challenge, Aristotle argues that form "is more (μᾶλλον) nature" than matter (*Phys.* II.1, 193b6) by sharply distinguishing art from nature: "man is born from man but not bed from bed" (*Phys.* II.1, 193b9–10). This statement reiterates Aristotle's introductory claim that nature is an *internal principle* of production (ποιήσεως; *Phys.* II.1, 192b29).

This sharp distinction between nature and art gets blunted by *Physics* II.4–6's insistence, against his pre-Socratic predecessors, on nature and thought's allied priority over chance (*Phys.* II.6, 198a9–10). I have argued that it is this shared priority of nature and thought that led the Empedoclean opponent of *Physics* II.8, once again, to mistakenly model nature after the agency we find in art. Here we find the opponent erroneously attempts to model a "coincidental" development of animals (which Aristotle goes on to argue against) on the lack-of-agency coincidence between rain and corn growth. Through the winter-rain argument, Aristotle restores nature's priority over chance by returning to *Physics* II.1's sharp distinction between art and nature: "man is born from man, but not bed from bed." The distinctive regularity by which natural things generate could not be found in the case of Zeus's raining for the sake of growing crops, and the example of winter rain is meant to highlight the distinctive feature of natural generation in terms that are mutually acceptable to both Aristotle and his opponent. Aristotle is then in a position to move the opponent, through an *a fortiori* argument, to see that animals generate in this distinctively natural way as well.

This unified reading of *Physics* II makes sense of *Physics* II.8's second argument to the conclusion that nature is prior to art (199a9–33) as an attempt to block, once and for all, the opponent's tendency to take lessons about nature from artificial examples, and to be misled, thereby, into thinking that nature operates by chance. In so doing, Aristotle brings together the arguments of *Physics* II.1 (that form is more nature than matter) with the first argument of *Physics* II.8 (that nature is for the sake of something) to conclude that "form must be the cause in the sense of that for the sake of which (ἡ οὗ ἕνεκα)" (*Phys.* II.8, 199a33); in other words, that form is for the sake of form, or "man is born from man."

CHAPTER 6

In defense of the craft analogy: Artifacts and natural teleology

Charlotte Witt

In the *Physics* Aristotle distinguishes artifacts from natural beings, noting that the latter have an internal origin of motion and rest, which artifacts lack. Still, it is striking that artifacts appear frequently in the passages of the *Physics* that introduce ideas like matter and form, which suggests that artifacts are ontologically similar to natural substances in some important respects.[1] Further, artifacts figure centrally in Aristotle's arguments for natural teleology, which also suggests that he finds significant similarities between artifacts and natural beings. Nonetheless, many commentators argue that artifacts do not even count as substances for Aristotle, which must mean that artifacts and natural beings are of a different ontological type and radically dissimilar. Some think that artifacts lack the unity of true substances, being only accidentally one (Kosman 2013); others think that it is an artifact's inability to reproduce that is key (Katayama 1999); and still others think it is the fact that artifacts are not self-regulating (Shields 2008) and lack intrinsic ends (Shields 2008 and Katayama 2009). The idea that artifacts and natural beings are radically dissimilar casts doubt on the role that artifacts play in Aristotle's argument for natural teleology, perhaps leading some scholars to downplay the significance of these arguments or to omit them altogether.

In this chapter I focus on the idea that artifacts lack intrinsic ends, since this claim is directly relevant to the legitimacy of Aristotle's use of artifacts in his arguments for natural teleology. For, if artifacts lack intrinsic ends it is difficult to see what role they could play in an argument meant to establish that natural beings have intrinsic ends. But it turns out that

[1] Of course, Aristotle's use of artifacts in relation to natural beings (and substances) is found in many texts. For a discussion of texts other than the *Physics*, see Katayama 1999 and Shields 2008. In this chapter I am interested in the relatively narrow question of what Aristotle's argument for natural teleology suggests about the ontological status of artifacts.

Aristotle's use of artifacts in his argument for natural teleology in *Physics* II.8 casts doubt on the claim that artifacts lack intrinsic ends in the relevant sense. Indeed, I think that one of Aristotle's arguments, the argument from mistakes, presupposes that artifacts have intrinsic ends or proper functions, a view we find him explicitly endorsing in another text. But if Aristotle does make a distinction between the intrinsic ends or proper functions of an artifact and its possible uses, then it is perfectly appropriate to use an analogy between artifacts and natural beings in arguing for intrinsic ends in nature, or natural teleology. Moreover, Aristotle's legitimate use of artifacts in his argument for natural teleology, in turn, casts doubt on the claim that artifacts are not substances, and are either of a completely different ontological type or not beings at all.

I Introduction: The craft analogy

Aristotle's discussion of nature and natural teleology in the *Physics* opens with a distinction between artifacts and natural beings. Things that exist by nature have an inner origin of motion and rest, which includes not simply locomotion, but also growth and diminution and alterations. "On the other hand, a bed and a coat and anything else of that sort, *qua* receiving these designations – i.e. in so far as they are products of art – have no inner origin of motion and rest" (192b16–18). It might be reasonable to expect that, if this were a basic ontological distinction, artifacts and art would both recede from the subsequent discussion of nature and natural teleology. Or, at the very least, we might think that artifacts would not play a central role in Aristotle's explanation of, and argument for, natural teleology. Alternatively, we might expect Aristotle to warn us of the ontologically inadequate and foreign character of his favorite examples. But it turns out that Aristotle uses craft examples frequently, and draws freely on the structural likenesses between nature and craft in his discussion of nature and in his argument for natural teleology. Borrowing the term from David Sedley (2010), I will call both Aristotle's use of craft examples and his references to specific similarities between craft and nature Aristotle's "craft analogy."

To modern readers the craft analogy might seem mistaken, even misleading. In particular, Aristotle's use of the craft analogy to argue for natural teleology might seem confused. After all, absent a Platonic cosmic craftsman, it just seems wrong to argue from the existence of function and purpose in artifacts to the existence of function and goal-directedness in natural beings, or to think that processes of craft creation will tell us anything useful about processes of natural generation.[2] Some scholars think that Aristotle used the craft analogy simply for pedagogical reasons because crafts and artifacts are more familiar to us (or to his audience) than nature and natural beings, but he didn't really think there were important similarities between the two that could be legitimately used in an argument by analogy for natural teleology. I discuss the pedagogical interpretation of Aristotle's craft analogy and its implications further in section II.

Alternatively, one might take the craft analogy at face value, which is that Aristotle thought there were important similarities between craft and nature that he could legitimately exploit in his argument for natural teleology. If that is right, then Aristotle could not have thought that artifacts were ontologically in a different category from natural beings, or not beings at all. In section III I focus on Aristotle's arguments for natural teleology in *Physics* II.8, and I develop the idea that Aristotle's use of artifacts to explain and to justify natural teleology is cogent and not misguided or mistaken. The basic assumption underlying Aristotle's use of artifacts in the argument for natural teleology is the mimetic relationship between art and nature: the idea that art "imitates" nature. Two features are particularly important. First, Aristotle points out that both artistic creation and natural generation consist of a sequence of stages, and that their ordering and the relationship of the stages to one another are themselves ordered by their being for the sake of the end. A second important feature shared by artistic creation and natural generation is the possibility of mistakes or errors. Aristotle argues that, like art and artistic creation, nature and natural generation are subject to defect and error.

> If then in art there are cases in which what is rightly produced serves a purpose, and if where mistakes occur there was a purpose in what was attempted, only it was not attained, so must it also be in natural products, and monstrosities will be failures in the purposive effort. (199a31–b4)

But this argument and the notion of a mistake or a monstrosity presuppose that *both* artistic and natural processes have intrinsic ends. To see this, it is helpful to distinguish two points: first, whether or not a being has an intrinsic end, and, second, where the intrinsic end originates or comes from. Artifacts have intrinsic ends, with external origins in the artist's intentions and motions; natural beings have intrinsic ends, with internal origins in that the end develops and unfolds through internal

[2] See Sedley 2010 for a useful discussion of Platonic and Aristotelian teleology.

processes.[3] But if artifacts have intrinsic ends, then one reason for holding that artifacts and natural beings are ontologically different, or that artifacts are non-beings, is mistaken.

In section IV I address the apparent tension between the idea that artifacts have intrinsic ends, and Aristotle's definition of them as having an external origin of motion and rest. If an artisan creates an artifact (and is its external origin in that sense), then doesn't the form or end have an external origin? In what sense, then, does an artifact have an intrinsic end?

II The pedagogical interpretation of the craft analogy

A preliminary issue is whether the craft analogy *is* central to Aristotle's explanation and justification of teleology, and, if so, in what way?[4] It is useful to distinguish two questions. First, is the craft analogy central to Aristotle's explanation and justification of natural teleology? Since examples drawn from crafts, like building, and arts, like medicine, appear throughout *Physics* II.8 it is clear that Aristotle uses them in his exposition of natural teleology. So much should be uncontroversial. But, we might wonder, in what way are crafts important? How does Aristotle use them? According to the pedagogical interpretation of the craft analogy, Aristotle uses examples drawn from the crafts to illustrate points about causation and teleology in nature and natural beings because they are more familiar (or accessible) to us than natural beings and processes, and not because they share any important ontological features with natural beings.

> Aristotle's extensive and careful use of the model of arts in his discussion of the teleological aspects of natural things is due to the fact that technology is more familiar to humans than other natures, not because he thinks that technology is fundamental to nature. On the contrary he holds that art imitates nature. (Johnson 2005: 9)[5]

In support of the pedagogical interpretation there are a number of texts in which Aristotle articulates a distinction between principles that are familiar to us or more knowable to us and principles that are more knowable by

[3] It is worth noting that Aristotle often identifies the male parent (or the motions in the semen) as the moving origin of animal reproduction. So, even in the case of natural beings, the original source of the generative motion is external to the developing offspring.

[4] For differing views on this issue, see Cooper 1982 and Sedley 2010.

[5] "Aristotle himself uses the art model. It will become clear that Aristotle does not intend it to explain anything, but merely to give us a notion of how causality itself works on the basis of facts most readily available to ourselves (i.e. how intentional agency and the crafts can produce definite results)" (M. R. Johnson 2005: 126).

nature or without qualification.[6] Indeed, the *Physics* opens with this observation on the method of the inquiry:

> The natural way of doing this is to start from the things which are more knowable and clear to us and to proceed to those which are clearer and more knowable by nature; for the same things are not knowable to us and knowable without qualification.

Aristotle illustrates this point using the example of universals (more knowable to us) and individuals (more knowable by nature). It is worth pointing out that the examples Aristotle uses to illustrate his distinction, namely universals and individuals, fits very well with the subsequent discussion of the principles invoked by the natural philosophers (and Plato) to explain change. But it is not equally obvious how to apply Aristotle's epistemic distinction to the case of artifacts and natural beings, to art and nature, since the paired items seem to be of the same ontological type. Artifacts and natural beings are individuals; art and nature are universal principles. Moreover, it is not intuitively obvious that artifacts are more familiar to us than natural beings. Why and in what way is a bed, for example, more familiar than a goat? And in what way is the production of a house or a temple more understandable (to someone who is not a builder) than the generation of an animal? Surely, animal reproduction would be better understood by a farmer than how to design and build a temple. Finally, the pedagogical interpretation holds that Aristotle's argument for natural teleology moves from the allegedly clear and easy example (artifacts) to the confusing and difficult case (natural beings). But we don't actually find this direction of argument in the text. Instead, Aristotle's exposition and argument weave back and forth between artifacts and natural beings in a free-flowing manner.

So it just isn't clear how to apply Aristotle's methodological principle – of moving from what is more knowable to us to what is more knowable by nature – to the case of artifacts and natural beings; nor is it clear that Aristotle intends us to do so. This gives us reason to doubt the relevance of Aristotle's methodological principle to his use of the craft analogy in his argument for teleology. Moreover, even if we are meant to read this text in the light of Aristotle's distinction between what is more knowable to us and what is more knowable by nature, the pedagogical interpretation leaves open the question of whether Aristotle thinks – in addition – that there are important ontological similarities between artifacts and natural beings and

[6] See e.g. *Phys.* I.1, 184a19–20; *APr.* II.23, 68b35–37; *Metaph.* VII.3, 1029b3–12; and *EN* I.3, 1095b1.

significant structural similarities between goal-directed processes in craft and in nature to which he can appeal in his argument for natural teleology. In other words, the pedagogical interpretation is compatible with the claim that the craft analogy plays a central and legitimate role in Aristotle's explanation and justification of natural teleology. The pedagogical interpretation, in itself, does not determine whether the craft analogy is used simply for purposes of illustration and teaching or whether it is meant as a piece of argument that points to, and presupposes, genuine and important similarities between craft and nature. A close look at Aristotle's argument for natural teleology reveals that there are good reasons to endorse the second alternative.

III Art imitates nature

In *Physics* II.8 Aristotle makes several arguments for natural teleology. The first (and most extensive) argument does not make use of the craft analogy (198b1–199a8).[7] But the two following arguments, plus an important disclaimer, do employ the craft analogy. Looking just at the number of arguments in this chapter, it is fair to say that the craft analogy is central to Aristotle's case for natural teleology. First, Aristotle uses the craft analogy to make what I will call *the structural argument*: the idea that in a natural process of development, each stage in the sequence of stages is for the sake of the end or goal of the process (199a9–19). Next, Aristotle uses the craft analogy to make what I call *the argument from mistakes* to show that there are intrinsic ends in natural beings and natural processes (199a34–b8). In order for the argument from mistakes to work, however, Aristotle must think that there are intrinsic ends in art and artistic processes as well, since he employs the craft analogy to make his point. Finally, Aristotle clarifies the connection between deliberation and goal-directedness, again employing the craft analogy. While the clarification is not part of a direct argument for natural teleology, it does support the legitimacy of the craft analogy in the justification of natural teleology by defusing an obvious criticism, namely that goal-directedness requires conscious thought or deliberation (199b26–33).

These uses of the craft analogy in Aristotle's argument for natural teleology establish that he thought there were important similarities

[7] For an interpretation of this argument, see Cooper 1982 and Leunissen 2010: 16–18. Cooper downplays the role of the craft analogy in Aristotle's argument for natural teleology, and Leunissen is skeptical about its significance in Aristotle's argument.

between craft and nature that could be legitimately exploited to establish the existence of ends in nature. Before entering into the details of the arguments, it is useful to explore the conceptual basis for these similarities. In the *Physics*, Aristotle says that art "imitates" (and sometimes "completes") nature (*Phys.* II.2 194a21–22 and II.8, 199a15–17). Initially it might seem that the term "imitation" has connotations of Platonic metaphysics, and these connotations might seem to reinforce the idea that art and nature are radically, ontologically dissimilar, like Platonic forms and particulars. It would then seem misguided for Aristotle to use the craft analogy in his argument for natural teleology. However, it is unlikely that Aristotle is referring to Platonic imitation, since Aristotle claims not to understand Plato's relationship between forms and particulars, and, in particular, Aristotle seems puzzled by the metaphoric language Plato uses in describing the relationship (*Metaph.* I.6, 987b13–14).

There are two contexts outside of the relationship between art and nature in which Aristotle makes fairly extensive use of the notion of imitation: in his theory of tragedy and the "mimetic arts," and in his account of elemental transformation. Aristotle says that elemental transformation imitates the eternal movement of the celestial circles. Johnson (2005: 147) is helpful in pointing out that we need not interpret Aristotle as personifying the elements; rather, Aristotle holds that elemental transformation resembles eternal celestial motion in that it is continuous. In the *Poetics* Aristotle uses the term "mimesis" to express what tragedy is: "the imitation of an action that is serious," and he refers to painting, tragedy, epic, and the like as "the imitative arts" (8, 1451a31). "Tragedy is essentially an imitation not of persons but of action and life" (8, 1450a16). However, it is likely that the term has an importantly different connotation in the *Poetics* from that in the *Physics.* Halliwell (2002: 26) distinguishes:

> Aristotle's specific conception of the mimetic arts (mimetic *qua* forms of intentional representation) from his quite separate and general principle of "the imitation of nature" by human productive craft (*technê*), the latter counting as "mimetic" not in terms of representation but by virtue of *analogy* or analogousness to natural processes of production.

Technê does not imitate nature in the sense of represent nature; rather, craft imitates nature because it is like nature or resembles nature in certain respects. These similarities or resemblances are what the craft analogy depends upon, and in order to see what similarities are important for Aristotle's defense of natural teleology, we need to return to the *Physics.*

Let's begin with an argument from early in the *Physics* that uses the craft analogy and the notion of mimesis in order to make an epistemological point. This will allow us to appreciate that Aristotle uses the craft analogy in contexts other than his argument for natural teleology, which underlines the breadth of the analogy. In a discussion of whether the natural scientist should study both matter and form, Aristotle draws a comparison with what a doctor or a house-builder should know. This comparison begins with the claim that "art imitates nature":

> But if, on the other hand, art imitates nature, and it is the part of the same discipline to know the form and the matter up to a point (e.g. the doctor has a knowledge of health and also of bile and phlegm, in which health is realized, and the builder both of the form of the house and the matter, namely that it is bricks and beams, and so forth); if this is so, then it would be part of natural science also to know nature in both its senses [as matter and as form]. (*Phys.* II.2, 194a22–27)

Notice first that in this text the notion that art imitates nature has a metaphysical meaning; both natural beings and artifacts are composites of matter and form. Then notice that the shared hylomorphic structure of artifacts and natural beings is taken to have an epistemological significance. Aristotle uses the craft analogy and in particular the relationship of imitation to further an argument and to establish a conclusion. After all, if the imitation relationship did not hold between artifacts and natural beings, then there would be no way to infer from some property that is true of artifacts to the corresponding property in natural beings. We will see that the craft analogy in Aristotle's argument for natural teleology follows the same pattern. First, there is a claim about a feature of crafts or craft production, then the imitation relationship between craft and nature is invoked (or clearly presupposed), and finally a conclusion is drawn about natural beings or their development.

The structural argument (*Phys.* II.8, 199a8–32) opens with the claim that where an end exists, all the preceding steps are for the sake of that end. Aristotle concludes the structural argument: "if, therefore, artificial products are for the sake of an end, so clearly also are natural products" (199a18–19). Notice the direction of argument – it runs from a claim about artifacts to a claim about natural beings. Aristotle assumes that we will agree about the teleological structure of artistic processes, and, since art imitates nature (explicitly stated at 199a15–17), the same structure will also be found in natural processes: "The relation between the later to the earlier items is the same in both artificial products and natural products" (199a19).

Aristotle devises a thought experiment about a house to express the structural similarity between craft and nature: If a house were a natural product, then it would be produced in the very same way by nature (meaning via the same teleological sequence) as it actually is by art. He adds that things made by nature, if they were made by art as well, would come to be in the same way. Notice how strong Aristotle's point is here; the teleological structure of artistic productions and natural generations are the same, not just similar in some respects.

Aristotle does not mention the obvious difference between the two teleological sequences, namely that in an artistic creation the artisan is the external origin of the motions that constitute the teleological sequence, whereas in a natural generation after the initial motion (from the male parent) the origin of the development is internal. This omission is all the more noteworthy since Aristotle differentiates artifacts and natural beings precisely on this point. It seems that, for Aristotle, the similarity in the teleological structure of artistic creation and natural development is independent of the causal origin of the respective processes. It seems therefore that the teleological character of the artistic process with which Aristotle compares the natural process is not simply determined by the intentions of the artist alone. This is a very important point, to which I return in section IV.

The argument from mistakes (*Phys.* II.8, 199a32–b7) turns on another similarity between artifacts and natural beings, namely that, like artistic creations, natural generations sometimes go awry and the result is a mistake (*hamartia*) or a monstrosity (*terrata*). The terminology here is interesting. Like imitation (*mimêsis*), the notion of a mistake (*hamartia*) appears in Aristotle's theory of tragedy in the *Poetics* to refer to the kind of error made by a tragic character like Oedipus, who unknowingly killed his father and married his mother. The vocabulary of monstrosity (*terrata*), in contrast, finds its home in Aristotle's biological writings and refers primarily to the deformed offspring that result from botched processes of reproduction. Aristotle reasons from the presence of mistakes in artistic processes and products to botched natural developmental processes and offspring. It sometimes happens that "the literate man makes a mistake in writing and the doctor pours out the wrong dose" (199a32–35). So, too, in nature we find defective seeds or processes of development that result in monstrosities like the "ox-progeny." Hence in nature (as in art) there are goal-directed processes that do not reach their goals or ends. We tacitly recognize the existence of goals or ends in nature when we recognize deformed or monstrous outcomes. The notion of a mistake or monstrosity presupposes that another outcome ought to have happened; it presupposes that there is an end or form the product ought to have realized. But notice that these presuppositions also hold in the case of artifacts. The notion of a mistake also presupposes that there is a form or function the artifact (or outcome of the artistic process) ought to have or ought to serve, even if in fact the artifact is defective and cannot serve that function. The function of the artifact has a normative component; it is not simply what it is doing or what it can do. Consider the defective house that has lost its roof in a hurricane; it no longer actually provides shelter for humans and animals, and it no longer can serve that function. Nonetheless it is a house – albeit a defective one. It – so to speak – ought to be able to provide shelter for humans and animals. The argument from mistakes presupposes that artifacts have proper functions, which are sometimes not realized owing to a defective process of creation. Similarly, Aristotle reasons, natural processes can also produce deformed animals and plants.[8]

In order for the argument from mistakes to be a cogent argument for natural teleology, Aristotle must think that mistakes in art and in nature presuppose that there are proper ends to be realized. And these proper ends are not simply whatever the product can be used for or is being used for.[9] That is, in order that the notion of a mistake do the work that Aristotle assigns it in this argument, he ought to distinguish the proper end of an artifact, which it realizes through the appropriate process of creation and is internal to it, from the many uses an artifact might be put to. But if this is correct, then Aristotle ought to recognize that artifacts, like natural beings, have intrinsic ends, or, in other terminology, proper functions.

And, indeed, we find that Aristotle does recognize that artifacts have proper functions in a text from the *Politics*. Aristotle contrasts kinds that are by convention alone, like money, that have no intrinsic value or use-value, with artifacts like a shoe that have a proper function. Money is by convention alone; it has no intrinsic value or use-value.[10]

> Others maintain that coined money is a mere sham, a thing not natural but conventional only, because, if the users substitute another commodity for it, it is worthless, and because it is not useful for any of the necessities of life, and, indeed, he who is rich in coin may often be in want of necessary food. (*Pol.* 1.9, 1257b10–14)

[8] For a discussion of deformed animals in Aristotle, see Witt 2012.

[9] For a further discussion of the proper ends of artifacts, see Witt 2015.

[10] Aristotle understands money to have exchange-value rather than use-value, although there are tensions in his theory; see Meikle 1994.

The context for this comment is Aristotle's discussion of wealth acquisition, which has a proper use in the household, just as a shoe has a proper function or use in relation to protecting the foot. Although shoes can be used to acquire wealth or to barter, this is not their proper use (*Pol.* 1.9, 1257a5–10). Money, in contrast, has only an exchange value, and – unlike the shoe – money has no proper function (*oikeian kresin*) at all. Aside from the conventions governing money and its use, it has no intrinsic value or proper use. So, not only is the distinction between what an artifact can be used for and its proper function presupposed by the argument from mistakes, but we also find Aristotle explicitly drawing that distinction in another text.

The idea that artifacts have intrinsic ends or proper functions appears to be inconsistent both with the idea that artisans create them (and so their ends have an external origin in the artist's intentions) and with the idea that artifacts serve the ends of their creators (and these ends are extrinsic to the artifact). I address the first inconsistency in the next section. The second inconsistency might mean one of two things. The idea might be that in any given case an artifact serves the purpose of its creator (or user), or the idea might be that all artifacts, taken collectively, serve the purposes of human beings (or other tool-using animals), and in that sense they do not have intrinsic ends. The first claim blurs the distinction drawn above between the proper function of an artifact (in relation to which the language of mistakes is appropriate) and what it is used to do or can be used to do. We can use an ashtray as a hammer, but it is still an ashtray after all. Similarly, we can use a horse to pull a plow, but that is not its natural function. This is the sense in which artifacts, like natural beings, have intrinsic ends or proper functions. The second claim, that artifacts taken together serve the collective purposes of human beings (or other tool-using animals) is correct, but it is also compatible with the idea that artifacts have intrinsic ends in the sense specified above, i.e. proper functions. And this is the sense required for the argument from mistakes for natural teleology to be cogent.

IV Art does not deliberate

Recall that Aristotle differentiates between artifacts and natural beings because artifacts have an external origin of motion and rest in contrast to the internal principle that natural beings have. Animal reproduction and artifact production seem to differ on this point; the developing animal embryo has an internal source of change whereas each stage in the production of the artifact is shaped by the activities of the artisan under the guidance of her intentions. This seems to be a crucial point of difference between artifacts and natural beings, and one that is directly relevant to the issue of intrinsic ends. The artisan's individual psychological state (her intention) is the external origin of the movements that produce the artifact with its proper function, and this seems to be a significant and relevant difference between artistic production and natural generation, namely that the intrinsic end of an artifact has an external origin in the artisan's intentions. The craft analogy falters due to this difference, which seems to create a fatal ambiguity in the notion of an intrinsic end. That an artifact has a particular proper function is causally dependent upon the individual psychology of the creator. In contrast, that the natural being has a particular proper function or intrinsic end is not causally dependent upon an external source in the same way.[11]

This is a reasonable criticism of Aristotle's use of the craft analogy, since it points to another apparent ambiguity in the notion of an intrinsic end. And it is a criticism that resonates with those contemporary readers of Aristotle who think that Aristotle's craft analogy is confused, since the idea of creating an object with purpose or function makes no sense independent of the individual psychology and intentional motions of an artisan. However, in a few difficult comments, Aristotle eliminates the apparent ambiguity by explicitly bracketing the process of artistic deliberation and individual psychological intention from the craft analogy.

> It is absurd to suppose that purpose is not present because we do not observe the agent deliberating. Art does not deliberate. If the ship-building art were in the wood, it would produce the same results by nature. If, therefore, purpose is present in art, it is present also in nature. (*Phys.* II.8, 199b27–30)

One way to interpret this comment is that Aristotle means that a true artisan does not need to deliberate about how to create an artifact; she is so adept and practiced that she simply acts without deliberation or reflection. Indeed, Aristotle remarks earlier in the chapter that non-human animals make things without art, inquiry, or deliberation (199a20–21).

However, it is not likely that this is what Aristotle means, because in other contexts he emphasizes the process of means–ends reasoning used by human artisans. It is more likely that Aristotle is directing our attention to the central role of craft knowledge as what is responsible for the existence of the artifact, an entity with a proper function or intrinsic end. That "art

[11] It is worth remarking again that in one sense a natural being does have an external origin of motion, namely the male parent. But the developing natural being is not causally dependent on the male parent throughout the process of development as the artifact is dependent upon the artificer during the entire process of creation.

does not deliberate" means that craft is the knowledge of what is to be made and how to make it, which is to be distinguished from an artisan's individual psychological process of reasoning out how to achieve a particular end-result and the subsequent series of motions of which she is the origin. So Aristotle distinguishes between the individual psychological process of the artisan, who deliberates, and is the external origin of the product and the craft itself, which is (as it were) a stationary body of knowledge. Hence, Aristotle's use of the craft analogy in his argument for natural teleology does not use an ambiguous notion of intrinsic end or proper function, because the terms of the comparison do not include the origin of the motion or the individual psychological process that an artisan might employ in determining what to do.

Where does this leave us? On the one hand, Aristotle distinguishes beings that are products of art from natural beings in terms of the locations of their moving causes or principles of origin. In this context, the individual psychological processes of the artisan and the motion of the male parent (or sometimes the motions in the semen) are the relevant items. However, when Aristotle deploys the craft analogy in his argument for natural teleology, he explicitly brackets the issue of the origin of artistic creations and natural generations, which makes sense, since this is a point of disanalogy between art and nature. With the issue of the moving cause off the table, the worry about a possible ambiguity in the notion of an intrinsic end or proper function is defused, since the origin of the intrinsic end is not relevant to Aristotle's craft analogy and, hence, not relevant to his argument for natural teleology.

V Conclusion

Readers of *Physics* II must be struck by Aristotle's copious use of the craft analogy in his explanation of what nature is, and in particular its central role in his arguments for natural teleology. Yet scholars argue that artifacts are not substances, and even not beings at all, which would undermine the value of the craft analogy especially in Aristotle's arguments for natural teleology. In this chapter, I have challenged one important thesis central to the ontological demotion of artifacts, namely the claim that they lack intrinsic ends and proper functions and in this way are importantly different from natural beings.[12] The idea that artifacts have intrinsic ends

[12] Of course, the question of whether or not artifacts have intrinsic ends is only one issue that is germane to the topic of their ontological status. For a discussion of other important topics, see Katayama 1999, Kosman 2013, and Shields 2008.

and proper functions is of central importance for the cogency of using the craft analogy to argue for natural teleology. Since my interpretation grants intrinsic ends and proper functions to artifacts it supports the cogency of Aristotle's arguments for natural teleology that make use of the craft analogy. It emerged that when Aristotle uses the craft analogy to argue for natural teleology, he does so by bracketing the causal question of the principle or origin of the creation or generation, and in this way eliminates a difference between craft and nature that might seem to undermine his use of the craft analogy. Hence, my interpretation provides a partial response to those who might think that the craft analogy is mistaken or misleading on the grounds that it occludes a crucial difference between art and nature, namely the role of the artisan or creator.

CHAPTER 7

The origins of Aristotle's natural teleology in Physics II

Robert Bolton

I The background

Aristotle discovered natural science. Or so he believed, at any rate, as we can see most clearly from the opening book of his *Metaphysics* (A.3–7). There, in presenting the historical background to what he offers in the *Physics*, he argues at some length that none of his predecessors, not even those who by tradition were known as students of nature (οἱ φυσιολόγοι) or naturalists (οἱ φυσικοί), had any adequate or workable conception of the subject. The earliest of these *physikoi*, says Aristotle, were interested chiefly in the nature of matter or in the ultimate material constituents of things, those constituents from which other things come to be and into which they perish while they themselves, in their substance, persist unchanged (A.3, 983b6–18). They offered no proper or adequate account of why (*dia ti*) the ultimate material constituents are modified in the ways that they are, or of why and how they manifest regularity and order in their patterns of alteration (984a19–25, b11–22). Some, such as Parmenides, offered no account of the patterns of change in nature in so far as they denied the reality of change altogether (984a29–b4). Others, such as Plato, believed that natural change was too irregular and fluctuating to be a suitable subject for genuine science or scientific explanation (A.6, 987a32–b1; cf. *Metaph.* M.4, 1078b12–17). Still others, such as Empedocles and the Atomists, offered, at best, only partial or inconsistent explanations for the regular patterns of natural change (A.4, 985a10–29, b19–20).

In addition, Aristotle claims, none of the earlier thinkers made any clear or proper use of the *essence* or *form* of natural objects, in definition or in explanation concerning them. Plato only made gestures in this direction, as

Earlier versions of this chapter were presented at Toronto and at USC. I am grateful to those present, and especially also to Mariska Leunissen, for very helpful comments and suggestions.

did the Atomists in their use of the shapes and positions of the atoms to explain their patterns of interaction (A.7, 988a34–b1; A.4, 985b10–19 with H.2, 1042b11–15, 1043a5–7, 14–21). Finally, and most importantly for our purposes here, Aristotle claims that no one at all before him had any real understanding of natural teleology, or of those goals for the sake of which things come about *in nature* (A.7, 988b6–8). The main lesson that Aristotle wants us to draw from his historical survey is, of course, that genuine natural science must bring into play each of his so-called "four causes" of what it studies – material, efficient, formal, and final. As he presents it in *Metaphysics* A, this is the heart of a proper conception of natural science, a conception that he supposes he was the first to properly grasp (cf. *Physics* II.1–3, *Metaphysics* H.4).

As one might expect, the accuracy of Aristotle's survey of the details and of the deficiencies of the approaches to causality of his predecessors has often been questioned, most especially perhaps his claim that none of his predecessors had identified the final cause. This claim has struck many recent commentators as surprising and even, as some have said, "startling," particularly in view of the extensive use that Plato makes of teleology in the *Timaeus*.[1] Plato's specific accounts, in the *Timaeus*, of the functions and goals of various natural entities are often alluded to and sometimes endorsed by Aristotle himself in his own biological works.[2] So why does Aristotle think that even Plato, like his other predecessors, had no grip at all on the final cause, that is, on the way in which goals figure as genuine causes (*aitiai*) in nature? This is a difficult problem which was recognized already in antiquity, for instance by Alexander, whose own brief attempts at a solution, however, take inadequate account of the special difficulty posed by the *Timaeus*.[3] The problem clearly requires resolution if we are to properly understand what is distinctive about Aristotle's own teleology in his *Physics* and how it differs from approaches offered by his predecessors. Let us begin, then, by considering how it is that Aristotle defends his judgment that his predecessors totally neglected the final cause in nature. This defense is laid out in *Metaphysics* A.7, in the following passage:

> That for the sake of which actions and transitions and changes take place they [the earlier thinkers] do speak of as a cause in a certain fashion but not in this way, that is, not in the very way in which this is in nature a cause. For

[1] See already Ross 1924: 179 and 1936: 38; Cherniss 1944: 454.

[2] Cf. e.g. *Tim.* 91e–92a with *PA* III.10, 686a25–31, and *Tim.* 76c–d with *PA* II.14, 658b2–10. See also *Philebus* 53–54, and Sedley 2010.

[3] Alexander, *In Metaph.* 59.28ff. See now also M. R. Johnson 2005.

> those who speak, for instance, of intelligence (νοῦς) or of friendship (φιλία) do posit these causes as something good. But they do not at all speak as though anything is or comes to be for the sake of these things. Rather, they speak as though changes [to good results] *originate from* these things. In the same way, those who speak of the one (τὸ ἕν) or of being (τὸ ὄν) as an entity of this kind [as a cause which is good] say that it is a cause of the substance (οὐσίας) [of certain things] but not that it is for the sake of this [good] that anything is or comes to be. (988b6–14)

In this passage, Aristotle first concedes that some of his predecessors did indeed identify certain things that functioned as causes for them as things that are good, and responsible for good in some way. The latter point is emphasized also in an earlier related passage in A.3, 984b11–22. In A.3, as in *Physics* II.3, Aristotle himself says that the goal or final cause of a thing is the *good* of that thing in some way (983a31–32, 195a23–26.) But this, Aristotle contends, is not sufficient for any of these presumed causes to count as final causes or as that for the sake of which something is or comes about in nature. A cause may be good and of something good, Aristotle indicates, but still be only a cause *from which* change originates – that is only an efficient cause, as the operations of intelligence and friendship were for those who offered them as causes. Aristotle has in mind here especially Anaxagoras and Empedocles as those who, respectively, made use of the operations of *nous* and *philia* as efficient causes. This is clear from many other passages, for instance in *Metaphysics* Λ.10, where he develops further this complaint against them (1075b1–16, cf. *PA* I.1, 640b5ff.). Earlier, in A.3, Aristotle claims that Anaxagoras in particular identified his cause of what is good, or well, in nature with an efficient cause (984b20–22). But, as he also indicates in A.3, and as we shall shortly see in some detail, he does not have only these two figures in mind in A.7 as those who erred in the way they did (984b18–22).

One main thing, then, that is very clear from the first part of our passage in A.7 is that, in Aristotle's view, a final cause in nature, properly understood, cannot be an efficient cause in the sense that it cannot be something that serves somehow to literally originate change, even regular change to some good result. Any conception of a goal according to which it does serve somehow to literally originate change, even to a good result, will fail to capture the way in which goals serve distinctively to explain things in nature. This is a point that Aristotle also emphasizes elsewhere, for instance in another important background passage for our purposes in *On Generation and Corruption* I.7, where he says:

> The productive agent [i.e. the efficient cause] is a cause in the sense of that from whence change originates. But that [goal] for the sake of which

> something obtains is not a productive agent, which is why health [the doctor's goal in healing] is not a productive agent [of the healing] except metaphorically (κατὰ μεταφοράν). For when the productive agent is present [e.g. when the doctor is healing] the patient is coming to be in respect of something [e.g. health]. But when the [resulting] stable states are there [e.g. health], the patient is no longer coming to be but is [e.g. healthy, so that the efficient cause is no longer present], and the forms [e.g. of health, produced by the doctor in the patient] and ends (τέλη) are such [resulting] stable states. (324b13–18)

Here Aristotle emphasizes again that final causes are not efficient or productive causes. He allows that a final cause, in some cases, may in a metaphorical or non-literal way be called a productive cause, as health or the form of health is when it figures in the desire and aim of a doctor to bring about health.[4] But the health that *figures in* the efficient cause in this case, as only an intended object, is not a true literal final cause or goal of the doctor's healing efforts since, Aristotle says, the actual goal is the stable resulting state of health which is not present until the efficient causal process is finished. For Aristotle, then, the doctor's desire in healing is not for the *actual* concrete resulting state, since no such particular state is yet actual, and, we may add, none may ever be actual. His desire is rather, we should say, that a certain *type* of state or form – say a certain balance of hot and cold – should be realized in a certain actual body, and there is no reference in this description of his desire for the *particular* final state which, Aristotle says, is the true end. If the doctor's desire is never satisfied, that does not alter the content of his desire, which is the same whether the desire is satisfied or not. So reference to that actual final state and true end which exists only when the desire is satisfied cannot be required for, or included in, the correct specification of the content of the desire itself, or thus of the efficient cause of healing.

In *Poetics* 21 Aristotle helps us to understand better what he means when he says in *On Generation and Corruption* 1.7 that health is an efficient cause of the doctor's healing only metaphorically. He says there: "Metaphor consists in giving a name to something that belongs [strictly] to something else, where the transference is from the more inclusive (*genos*) to the specific (*eidos*), or from the specific to the more inclusive, or from specific to specific, or by analogy" (1457b6–9).

It is useful here to see, first, that the terms *genos* and *eidos* in this passage do not mean *genus* and *species*. In one of Aristotle's examples in the text,

[4] See also *Metaph.* Z.7, 1032b6–10, Λ.10, 1075b8–10, and further below.

the entity *large number* stands as *genos* to the number 10,000 (τὸ μυρίον) as *eidos*, so that one can use the term "10,000" as the Greeks did, non-literally, to mean "a large number" (1457b11). We use the term "a million" in this same non-literal way today when we say, for instance: "There are a million things wrong here." We mean literally: "There is a large number of things wrong here." But the number 10,000, or a million, is not a *species* of the kind (*genos*) large number, in Aristotle's normal sense of his term for species. These are, rather, particular paradigm instances of large numbers. So the terms *genos* and *eidos* are used in *Poetics* 21 to refer in a very general way to the more and the less inclusive. In the case in question in *On Generation and Corruption* I.7, the desired goal, health, is a productive or efficient cause metaphorically, in that it is a less inclusive part or feature of the strict, more inclusive, efficient cause, namely the doctor's deliberated *desire for health*, and it stands for this whole efficient cause in any proper reference to health itself as an agent. The metaphor or figure of speech in question here is an example of what we now call synecdoche – the use of a part to stand for the whole. The actual final cause or goal itself, then, is not ever literally an efficient cause or productive agent, although the desire to achieve a goal of a certain type may well be. In *Physics* II.7 Aristotle does say that "often" efficient and final causes "go towards one," but only "in kind" (198a24–27). His example is that of a human being, since a human being is the goal of human generation, and a human being – another human being the same in kind – is the efficient cause of this. Here the efficient and final causes are not literally identical. Indeed, in *Metaphysics* A.3 Aristotle says plainly that the final cause is "opposed to" the efficient cause (983a31).

In the second part of our passage in A.7, Aristotle turns more explicitly to Plato, though without mentioning him by name (988b11–14). He introduces certain features of Plato's so-called late ontology or "unwritten doctrines," features that he has earlier described in A.6 and 7, in particular the doctrine that "the one" is the cause of the substance of certain other things. In Aristotle's earlier report, "the one" is in fact for Plato the cause of the substance of his forms (987b18–21, 988b4–6). However, although "the one" may be good, as it is in Plato's account, and may serve as a source of goodness for the things that share in it, "the one" is not the goal or that for the sake of which any of its participants are or come to be. So for this reason there is no final causality in this case at all but only at most, as Aristotle says earlier in A.7, a gesture at a type of formal causality (988a34–b6). For Plato, "the one" is doubtless something by reference to which someone can *understand what it is* for something to be or to come to be good, by virtue of its unity, but that is not sufficient to make it a final cause. This can be true of a purely formal cause.[5]

II Aristotle on Plato's teleology in the *Timaeus*

In this attempt – Aristotle's only attempt in *Metaphysics* A – to show that no earlier thinker, including Plato, had introduced genuine final causes in nature there is no direct reference to the *Timaeus*. But how surprising, or "startling," is this? Although there is no explicit objection to the adequacy of the teleology of the *Timaeus* as such or by name in A.7, there is, arguably, a clear indication there of what Aristotle's main objection to it would be. As we have seen, there is a direct objection in A.7 to the attempt at natural teleology of Anaxagoras, and, as commentators have frequently noted, the introduction and use of intelligence or *nous* by Anaxagoras, in order to account for the goodness and order of things in nature, are revived by Plato himself in the *Timaeus*. In a section of the *Phaedo* well known by Aristotle, where Plato first mentions Anaxagoras' approach, it is rather quickly put aside, purportedly because of Anaxagoras' failure to adequately exploit it, but with the wish expressed that someone might make proper use of it (97b–99d; cf. Aristotle, *Metaph.* A.4, 985a18–21, *Phys.* II.8, 198b14–16). As, again, commentators have often observed, Plato fulfills his own wish in the *Timaeus* where the things in nature that happen or are present for the sake of some good are all the result of the operation of intelligence or *nous*, either the divine *nous* of the Demiurge or that of the world soul. In *On the Soul* I.2–3, Aristotle himself uses similar language to that he uses in A.7 both to describe Anaxagoras' conception of *nous* as practical intelligence (*phronêsis*), which is "the cause of what is well and rightly ordered in nature," and to describe the work of the Demiurge in the *Timaeus*, who also designs things in nature practically in accord with his calculation of what is "better" (404b1–6, 407b9–11).

Given this close connection, we can see from Aristotle's critique of Anaxagoras in *Metaphysics* A.7 exactly what his main objection to Plato's approach to teleology in the *Timaeus* is going to be. In Plato's approach there our eye structures, for instance, do enable us to regularly achieve certain goals in our standard environment (45b–e). But those eye structures, in Plato's account, are not literally in place for the sake of those goals. Rather, they are in place because a sufficiently powerful

[5] Contrast Frede 1980.

intelligent designer *desired* that goals of this type should be achieved by us on account of their perceived goodness and *chose* these eye structures for us to best fulfill that desire. So it is not the actual goal-state itself, in Plato's account, or the fact that the means to it secure it, that explains the presence of the means. It is, rather, the goal as an object of perceived goodness and of desire, or, in plain language, it is the *desire to realize a goal* of a certain type, perceived as good by a suitable rational agent, that explains the fact that the means that secure it are there. In the language of *On Generation and Corruption* I.7, the goal in question is the cause of the means to it, in Plato's account, only metaphorically, as a part or feature of the true, more inclusive efficient cause, namely the desire to realize a goal of that type. The intended goal only explains the means in this way, as an aspect of their efficient cause. As such, the goal in question is not the actual final resulting stable state, which is for Aristotle the only true final cause of the means and the change to it. Aristotle can himself speak of a desired goal in such cases as an unmoved mover, as he does in *GC* I.7, 324a30–b4 and in *DA* III.10, 433b15–17. But this does not belie the fact that it is only a mover at all metaphorically. So, as Aristotle says in *Metaphysics* A.7, in this approach, the approach of Anaxagoras and also, as it turns out, of Plato in the *Timaeus*, goals as objects of intelligence and desire explain only as parts or features of efficient causes, not strictly as genuine final causes. Thus, the sorts of goals that do figure as causes in the *Timaeus* are not, for Aristotle, final causes at all, at least not of the sort that are operative for him in nature.

It is worth noting further that Aristotle's view here fits closely with his standard account of practical action (*praxis*) and of animal self-locomotion, where the intended goal, as the object or content of awareness and desire, explains strictly as a feature of the efficient cause, not as the actual resulting end of an action. As Aristotle says in *On the Movement of Animals* 6:

> It is the object of desire and of understanding that first initiates movement – not every object of understanding but only [one which is] a goal of action. Thus, it is [practical] goods of this sort [i.e. goods desired as goals] that originate movement, not everything fine . . . For the animal is moved and advances from desire [for] or choice [of a means to a goal] when it is affected by perception or an appearance [of that type of goal as good] . . . So it is in this way that animals are impelled to move and act, the ultimate [efficient] cause of their movement being desire; and this arises either from perception or from the appearance and understanding [of a type of good as goal]. (700b24–701a36, cf. *DA* III.10, 433b11–12, *EN* VI.3, 1139a31–33)

So in *Metaphysics* A.7 Aristotle does not in fact simply "strangely ignore" the approach to natural teleology found in Plato's *Timaeus*.[6] Rather, he indicates clearly to us what he thinks is misguided about that type of approach and why he, therefore, does not count Plato as a genuine proponent of natural teleology. This point deserves special emphasis in any attempt to understand Aristotle's natural teleology in the *Physics* and its originality by comparison with Plato's, since it is often not properly appreciated in the literature. It is sometimes supposed, for instance, that in the *Timaeus*, and in the anticipations of its approach that we find in the *Phaedo*, Plato does hold, in agreement with Aristotle, that when in nature certain means are regularly fit for securing certain goals it is the goal itself that is the cause of the means to it as that because of which the means are present.[7] At one point in the *Phaedo* Socrates does seem to speak in this way when he says that where things are ordered by intelligence (*nous*) the cause (*aitia*) of this order is that things are best that way (98a–b). But Socrates quickly clarifies this with the remark that when something is done "with intelligence" it is done, strictly, because of "the *choice* of what is best" on the part of the intelligent agent (99a–b). This, of course, is also the official view of the *Timaeus*, where the means are for the sake of an end in Plato's teleology only because they are such in the mind and plan and deliberate activity of an intelligent agent (30a).[8]

III Implications for *Physics* II

Some have argued, however, against this, that Aristotle's natural teleology as portrayed in *Physics* II is less different from Plato's than we might suppose from our results to this point, on the following grounds. Although for Aristotle, unlike Plato, nature is not the product of deliberate craft, divine or otherwise, nevertheless in *Physics* II, it is claimed, Aristotle views the operations of nature and of craft as so closely parallel that he virtually erases the difference between them, just as Plato does in the *Timaeus*. We can see this most clearly, it is argued, in *Physics* II.8, to which we should now turn. There Aristotle most fully exploits the analogy between nature and craft. He says, strikingly:

> If a house, for instance, were one of the things that come to be by nature, it would come to be in just the way as it now does by art (*technê*). And if the things that come to be by nature were not only to come to be by

[6] Ross 1936: 38. [7] See e.g. Lennox 2001a. [8] Cf. Johansen 2004: 106–110.

> nature, but also by art, they would come to be [by art] just as they naturally do. (199a12–15)

In context, we need to remember, the immediate point that Aristotle wants to make and to support here, in emphasizing the similarity of natural and of artistic or craft production, is just that in both "all of the earlier stages are for the sake of the end" (199a8–9). Aristotle's argument is that, given this similarity, if it is admitted that there are goals for the sake of which such stages occur in the case of art, this should be admitted, by analogical argument, also in the case of nature. This still leaves room for drawing a clear distinction between nature and art, as in *On Generation and Corruption* I.7, on the grounds that in the art the stages are strictly only for the sake of an end in the plan and deliberate activity of a craftsman, so that, in this case, goals explain only as features of efficient causes, while in nature this is not so. But many have wanted to argue that the similarity between art and nature for Aristotle runs much deeper than this.[9] This comes out most clearly, it is suggested, at the end of *Physics* II.8, where Aristotle echoes his point from the earlier passage that we have just considered, with a special addition. He says:

> It is strange to maintain that there is no coming to be for the sake of anything [in nature] unless we see that the productive agent [there] has deliberated. Yet art too does not deliberate. If the art of shipbuilding were in the wood, it [the art] would make the ship in like manner as something is made in nature. (199b26–30)

Here Aristotle attempts to discount an alleged difference between production in art and in nature that, as he indicates, some had used to argue that there is no final causality in nature. The alleged difference depends on the assumption that in art the productive agent is a conscious deliberator, and the argument is that it is only as the intended object of a deliberator that we can make any sense of goals at all. So, in this objection, there is no genuine final causality in nature since there is no deliberator there. This is a line taken, of course, by many philosophers of biology today, and even by some interpreters of Aristotle who hold that for him, as for Kant, there is only *as if* teleology in nature because it is *as if* there were a deliberator there.[10] Aristotle's own response to the objection, however, is that conscious deliberation is not essential even in art. Elsewhere, in *Metaphysics* A.1, Aristotle makes a similar point. He says that in the case of production in the crafts: "Artisans (χειροτέχνας) make things just like

[9] See e.g. Sedley 2007, 2010. [10] See e.g. Nussbaum 1978, Wieland 1975.

some inanimate things do, for instance as fire burns. But while the inanimate things do each of these things by some nature [that they possess] the artisans act by [acquired] habit" (981b2–5, cf. *MA* 10, 703a29–36).

So even in the crafts, for Aristotle, production is not necessarily based on explicit deliberation. It can be due simply to the exercise of an acquired habit that can operate in the same way as fire burns. In *Metaphysics* A.1 Aristotle does reserve the term art (*technê*) for the special skill of director-craftsmen (*architektones*) who can explain *why* skilled artisans correctly make things as they know how to do. But more commonly he ascribes *technê*, in its more standard sense of skill in making, to all types of skilled producers necessary in society, which would include shoemakers, builders, and even *architektones*, who clearly do not and practically speaking could not consciously deliberate about every single thing that they do in exercising their art (*Pol.* IV.4, 1291a1ff., VIII.2, 1337b8ff.).[11]

Here in A.1 Aristotle closely aligns production by art, in certain respects, with the activities of inanimate natural objects such as fire. In so doing he certainly distances himself from the *Timaeus*, where Plato strongly opposes this and argues that the work of art, which is for him necessary for the non-accidental securing of goals in nature, always essentially involves conscious intelligent deliberation (46d–48b).

But still, in *Physics* II.8, as we have just seen, Aristotle does say in emphasizing the similarity of art and of nature that if the art of shipbuilding were in the wood from which a ship is built, the art itself would make the ship. This initially somewhat mystifying remark may be compared with an earlier passage, in *Physics* II.3, where Aristotle says:

> It is always necessary to seek the cause that is topmost (τὸ ἀκρότατον), just as in other matters. For example, a [particular] man builds because he is a builder, but the builder builds by virtue of [possessing] the art of building. This cause [the art of building], then, is prior, and so is it in all cases. (195b21–25)

[11] Alternatively, as we shall see in further detail below, Sedley 2010 connects Aristotle's claim that "art too [like nature] does not deliberate" with a doctrine in *Metaphysics* Z.7 that "the art of [for instance] medicine . . . is the form of health" (1032b13). Aristotle's reasoning, Sedley claims, is that since forms cannot deliberate, only people, then if an art is identical with the form of its product, art does not, indeed cannot, deliberate. But, in Aristotle's official account of an art, "an art (*technê*) is essentially the same thing as a state of capacity to make in accord with a correct course of reasoning [i.e. of deliberation]" (*EN* VI.4, 1140a9–10). The form of health by itself is obviously not identical with any such capacity to make; although of course, as we have already noted, suitable *knowledge* of this form and of how to instantiate it is required for medical art for Aristotle, as he shows us in detail in Z.7, 1032a25–b30. This is doubtless why he indicates there that the form is the art only "in a way" (1032b11). See also Broadie 1990 and Leunissen 2010 for further discussion of this issue.

Here Aristotle claims that the way, in one particular sense, to specify the cause of a building is to say that this cause is the *art* of building, just because it is by virtue of possessing this art that the skilled builder builds. This seems to fit closely with his later claim that if the art of shipbuilding were in the wood, the art itself would be the cause of a ship. But this connection does not yet fully show us how to demystify that later claim. For that, some argue, as noted earlier, we need to turn to still another related passage, in *Metaphysics* Z.7, where Aristotle also compares production in nature and in art, but this time with more detail and in a more fully articulate way. With respect to the products of nature, which he considers first in Z.7, Aristotle says:

> That by which (ὑφ' οὗ) they [natural things] are produced, the so-called nature in the sense of the form (τὸ εἶδος), is their nature. This [nature] is formally the same (ἡ ὁμοειδής) [as the nature in the product], although this [productive nature] is in another individual, since it is a[nother] human being that generates a human being. (1032a24–25)

Here Aristotle ascribes a certain productivity, or causality, in natural production, to *form*. He describes the form in the parent as that "by which," in some sense, natural human offspring are produced. He then proceeds, in the text, to urge that there is a close parallel in production by art to this type of causality in nature.

> [In the same way as in nature a human being comes from a human being so, in production by art,] in a certain fashion (τρόπον τινά) health comes from health and house from house. That is, that [health or house] which has matter comes from that [health or house] without matter (τῆς ἄνευ ὕλης). For the art of healing and the art of building [from which, respectively, health and a house come] are the form of health and of the house. When I speak of the substance without the matter [i.e. the form] I mean the essence. (1032b11–14)

This passage specifies for us now, so it is argued, the respect in which an art is the cause of its products as that "from which" the product comes. An art is the cause of its products in this sense – as that "from which" – in that the *form* of the product is the cause of the product. So, as far as its causality goes, the art is just the form of the product (cf. *GA* II.1 735a2–3). So, in fact, in art just as in nature the *form* of the product is the true agent or efficient cause of the product. When Aristotle says, then, in *Physics* II.8 that if the art of shipbuilding were in the wood from which the ship is built, it would, as efficient cause, make the ship, he means that, just as in nature, if the *form* of the ship were *per impossibile* in the wood from which the ship is built, it would, as efficient cause, make the ship.

This now gives us, it is claimed, the means by which to get a clear picture of Aristotle's natural teleology that shows us how close in fact it is to Plato's. We start from Plato's approach in the *Timaeus*, where nature is the product of divine craft. There, form (Platonic form, of course) serves, in the mind of the divine craftsman, as the intended goal or object of desire that he aims to imitate and instantiate as far as possible in constructing the world. If we now subtract from this picture the causal work done by the desires, calculations, and choices of the divine craftsman made with a view to the realization of Platonic form, and we assign this causal work rather in each case to the form alone, but now the immanent Aristotelian form of the natural object, which itself works through various instruments to achieve the goal, then what we are left with is Aristotle's teleology. Here we see, then, what it is that grounds Aristotle's teleology, and we also see how close Aristotle is to Plato in spite of their differences. Or so it is argued.[12]

IV Questions for this account

Here then is one conception, compelling in various ways, and influential, of the main core of Aristotle's teleology in *Physics* II with an indication of its historical origin in Plato and of the textual basis for it. But how well does this reading square with other things that Aristotle says? Let us first ask how it fits with what we find in *Metaphysics* A.7. There, as we have seen, Aristotle argues in effect that even Plato in the *Timaeus* has no proper conception of a final cause. The reason for this is that in Plato's account the goal of some natural process or structure – as an object of desire – explains not as the actual final cause of the process or structure but only as a part or feature of the efficient cause. That is, the process or structure is not there in Plato's story strictly because it serves to realize the actual goal, but rather because it serves to realize the desire for a certain type of unrealized goal on the part of a sufficiently powerful divine craftsman, and is chosen and put in place as such.

Suppose, then, that we eliminate the divine craftsman from this account and substitute for it Aristotelian form, understood as a non-conscious, non-deliberative productive agent which prompts and directs via certain instruments the development of, for instance, the eye structures in the

[12] Some of the main elements of this line of interpretation, which is now fully developed by Sedley (2007, 2010), are laid out already by Cherniss (1944), who was himself following LeBlond. Cherniss used it as a part of his attempt to convict Aristotle of bad faith in claiming in *Metaphysics* A.7 that Plato had no final cause, since Aristotle's natural teleology, according to Cherniss, very closely copies that of Plato. See Cherniss 1944: 454ff., esp. at n.411.

human embryo so that they are fit for achieving certain goals. Here too the goal, understood as the unrealized object of this impetus or direction from form, explains as a feature of the efficient cause of our eye structures, not as their true final cause. That is, our eye structures are not present in this type of account because they secure the actual final goal-state itself; they are there because they satisfy the sufficiently powerful impetus from form to secure this type of not yet present goal. Thus, the goal, so understood, as the unrealized object of this impetus or direction from form, is as much an efficient cause, or a part or feature of an efficient cause, as it is in Anaxagoras or in the *Timaeus.* This type of account, then, is as much ruled out by Aristotle's strictures in *Metaphysics* A.7 as is that of Anaxagoras or Plato.

It is a further problem for this approach that in Aristotle's own account of animal generation and development our human form is not actually present to direct anything when our eye structures, for instance, are first being developed in the human embryo. Only the potential for this form is present (*GA* II.3, 736a27–737a18). It is argued by some that human form and function is present even in the earliest embryo, in a very low degree of actuality.[13] But that the locomotive, perceptual, and rational capacities that are necessarily included in human form are actually present to any degree at all in the earliest embryo seems a very difficult idea, and one ruled out by *Generation of Animals* II.3.[14]

Matters are not improved, moreover, if we say, as some do, not that our form is the agent of the development of our eye structures in the embryo, but rather that the "irreducible potential" for this form is the directing agent.[15] For here too the eye structures do not strictly come to be because of their actual goal or end; rather, they come to be because they satisfy the non-conscious impetus for the unrealized end or the "primitive directiveness upon the end," as some put it, which the irreducible potential for form supplies. So once again, the unrealized end, as the object of this non-conscious impetus or "directiveness," explains not as a final cause but rather as a feature of an efficient cause.

This same difficulty also faces those accounts in which the agent of our development for Aristotle is some so-called "program" in the embryo for a certain type of form and end, a program that causally directs development and is, in what it directs, sensitive to changes in the environment of the embryo.[16] Here too the structures developed are not present for the sake of

[13] Code 1987. [14] Cf. Lewis 1988. [15] Gotthelf 2012.
[16] See e.g. Bradie and Miller 1999 and M. R. Johnson 2005.

their actual end, but rather because they are brought about by the context-sensitive "program" for this as yet unrealized type of end. As *Metaphysics* A.7 and *On Generation and Corruption* I.7 show us, this sort of account of what grounds Aristotle's natural teleology does not explicate this teleology – it eliminates it. It is argued by some that the understanding of final causes as features of efficient causes is the approach that Aristotle *should* have taken in *Physics* II, though not the approach that he actually did take. They see that, in the case of human action, for Aristotle the goal explains as a feature of the efficient cause and argue that Aristotle should have carried this over to his natural teleology.[17]

Another serious difficulty for this approach arises from the fact that, in this type of account, a distinct, intrinsic, efficient causal role, in embryological development for instance, is assigned by Aristotle to form or, alternatively, to the "directiveness" supplied by the irreducible potential for form. A form itself, as we have seen Aristotle explicitly say in *Metaphysics* Z.7, is not a full substantial thing, nor is the so-called irreducible potential for form given that it, like form or soul, is only a capacity of such a thing and thus not a full substantial thing itself, as Aristotle points out in *DA* II.1, 412a16–22. The same holds of the so-called program for form which some introduce. However, to saddle Aristotle with entities such as human form as intrinsic efficient causal agents is to treat a form as a full substantial thing and to turn him into a soul–body dualist, a stance that he clearly rejects in *On the Soul*. As we know from *DA* II.1, the form of a living natural entity for Aristotle is its soul (412a19–21). But Aristotle refuses in *On the Soul* to ascribe intrinsic efficient causal agency to the soul itself. As he says in I.4:

> We say that the soul [i.e. the human form] is pained and rejoices, is bold and afraid, and further that it is angry and perceives and reasons . . . Yet to say the soul is angry is just as if one should say that the soul weaves or builds. Doubtless it is better not to say that the soul pities or learns or reasons [or weaves or builds] but that a human being does this by virtue of [possessing] its soul [e.g. by virtue of being able to desire and to choose to pursue goals of various *types*] . . . The intellect (*nous*) might seem [to be an exception to this and] to come to be in us as a [self-subsistent] substance and so to be imperishable [since the intellect itself reasons] . . . But reasoning, and loving and hating, are not attributes of the intellect but only of the one who has it insofar as he has it . . . These were not ever attributes of the

[17] See Irwin 1988: ch. 5. Irwin blames Aristotle's failure on the fact that he uses "pure dialectic" as his method of inquiry in the *Physics* and not the "strong dialectic" that Irwin finds in the *Metaphysics*. But, as we have seen, Aristotle maintains the same firm distinction between efficient and final causes that Irwin finds in the *Physics* also in the *Metaphysics*, in A.7. He does the same in A.3, 983a30–32, and in Δ.2 and in H.4.

intellect [itself] but rather of the common thing [i.e. the ensouled or living body]. (408b1–29, cf. I.I, 403a3ff.)

Here Aristotle denies that the soul, that is, the human form, builds or weaves or reasons or is in any way on its own an agent, or a patient, of craft activity or other change. It is only "the common thing," the ensouled body or person, that has such attributes. No entity such as the soul or human form, or indeed the art or form of health, has any such attributes. Soul or form is only a power to do things, not an agent that does anything. Aristotle uses this point here to undermine an argument for soul–body dualism. The argument is that the soul, at least the intellectual soul, is a substance because it can do things, such as reasoning, on its own. Aristotle rejects this argument on the grounds that its premise – that the soul, the human form, is on its own an agent – is false. If human form or soul were on its own an agent, Aristotle supposes, dualism would follow.

There are passages, for instance in *DA* II.4, where Aristotle does say that the soul is in one way an origin of change (ἀρχὴ κινήσεως), or a mover, using language like that which he standardly uses to describe an efficient cause (415b9–12, 22ff.; cf. *PA* I.I, 647a27–28). But he has already warned us in I.4 that while he, or we, may say this sort of thing, it is not to be taken literally. The soul, or human form, is never literally an agent or efficient cause. In *DA* II.4 itself, to show us that and in what way the soul is a "source of change," Aristotle cites the fact that the soul is a source of qualitative change (*alloiôsis*), since the soul is a source of perception: "For perception is held to be a type of qualitative change and nothing which lacks a soul has perception" (415b24–25). So the soul is a source of change not because the soul is the efficient cause of perception. For Aristotle, as we know well, it is the external object of perception, not the soul itself, which is his efficient cause of perception (*DA* II.5 417b20–21). The soul is a cause or source of perception and thus of change, as he says, only in that something must be an ensouled thing of a certain sort in order to have perception.

Our passage in *DA* I.4 is important also for a proper understanding of the text we considered earlier, in *Physics* II.8, where Aristotle says that if the art of shipbuilding were in the wood, it, the art, would make the ship. As the reference to building in *DA* I.4 shows, for the art of shipbuilding to be in the wood some shipbuilder and his body would have to be in the wood, and all of his tools as well, and the only shipbuilding that could be done would be done by the exercise by this shipbuilder of his art (cf. I.3 407b15–26). This shipbuilding might be done non-deliberatively, to a greater or lesser extent, by habit, but that would not affect this point.

So when Aristotle says in *Physics* II.3 that the causality of the art is prior to that of the craftsman, he cannot mean to imply that there is any literal efficient causal agency of the art itself that is responsible for the product. He must mean rather that it is the possession of the art, achieved through habituation and other means, which explains why the active craftsman is the true agent and not *vice versa*. That is, the priority in question here, of the art over the craftsman, is not ontological priority or, thus, any priority of the art as an actual agent or efficient cause. In *Physics* II.3 itself, Aristotle points out that when we are specifying causes, efficient or otherwise, we must consider whether we are talking about the particular cause of something or about the cause understood in some general way, and we must specify also whether we are talking about the actual cause or only some potential cause. In the case of a building, Aristotle says, the *actual particular efficient cause* on any given occasion is "the builder engaging in building" (195b6). So the *art* of building or the *form* of a building is never the actual particular efficient cause in any case of building.

Given this, we cannot use the fact that in *Metaphysics* Z.7 Aristotle says that the art of housebuilding is the form of the house to argue that the actual particular agent in housebuilding is ever the form of the house. For the art of building is not the cause of a house in the sense that it is the actual particular agent or efficient cause. Thus, we cannot infer from Aristotle's remark that the art is the form, taken together with the parallel between art and nature, that in nature the form, e.g. of a human being, is the actual particular agent in human generation and in embryological development. As we have seen, the only actual particular agent of such development is the bodily thing that does have the form or soul and does exercise the capacity which that confers. That is, the only actual particular originating efficient cause of human generation for Aristotle is the father fathering, through appropriate instruments, of course. In *Parts of Animals* I.1 Aristotle says:

> The productive agent [i.e. the efficient cause, of human generation] is prior [to the product] not only in explanation (*logos*) but also in time. For a human being [is the productive agent that] generates a human being, so that it is because *that* human being [the father] is *such* that the generation of *this* human being is *so*. (640a24–26)

V Approaches to an alternative conception of Aristotle's teleology

This conception, then, that we have been considering of Aristotle's natural teleology, in which the agency of the immaterial Demiurge in the *Timaeus* in securing goals in nature is replaced by the agency of Aristotelian form (or by the agency of an equally non-substantial irreducible potential for form or program for form), not only deprives Aristotle of final causes by his own lights by turning them into efficient causes or aspects of efficient causes: It is also inconsistent with Aristotle's hylomorphism. We need then to try to find an alternative conception which avoids these pitfalls. In a word, this alternative conception of Aristotle's natural teleology must not be inconsistent with Aristotle's view of the soul–body relation, and, most importantly, it must be a conception in which the means to actual goals in nature literally are, without qualification, there for the sake of those actual goals and not because they were desired or chosen, or non-consciously urged or directed to be present by some productive agent in order to achieve some as yet unrealized type of goal. It is not sufficiently well appreciated, I believe, how difficult it is to articulate such a conception.

To help bring out further the difficulty I see here let us turn now briefly to the contemporary scene and consider perhaps the most prominent account in the recent literature of purpose or function, namely the so-called etiological account, one that has been used by many to interpret Aristotle's teleology.[18] In this etiological account, our heart, for instance, has the *function* or *goal* of circulating our blood if, and only if, two conditions are satisfied. The first condition is that our heart does effectively circulate the blood, and the second is that our heart is present in us *because* it circulates the blood. Here the notion of cause, or of "because," covers, and was intended to cover paradigmatically, causes introduced in evolutionary explanation. In such an explanation, very roughly, we have hearts *because* they circulate blood, in the sense that the presence of efficient blood circulation by the heart conferred a survival advantage on our ancestors who thereby lived to produce us with a genetic makeup that gives us hearts. One might complain that what is explained here is not so much the fact that we have hearts as that we exist at all with hearts. One could argue that the real explanation of why we as humans have the type of heart we do has to do with our human genetic makeup.[19] But, however that may be, the supposed cause of our having hearts in this case is not an Aristotelian final cause. It is, rather, a complex efficient cause, or efficient/material cause, having to do with our evolutionary history, one in which the fact that we

[18] See e.g. Godfrey-Smith 1994, 2014; Millikan 1989; and Wright 1973, 1976. For applications to Aristotle, see Depew 1997; Gotthelf 2012; and Lennox 2001a.

[19] Cf. Nagel 1977.

have hearts that circulate blood is given a certain historical *cum* genetic explanation.

To see this more clearly, let us spell out the causal sequence in question somewhat more fully. There is, first, genetic variation in a population which gives rise to a certain type of heart in certain members of the population. Then, the action of this heart in these members of the population gives rise to a certain pattern of blood circulation for them by comparison with other members of the population. This pattern of blood circulation then gives rise to an improved survival and reproduction rate for these members of the population, and for their offspring, including us, to whom they transmit such hearts. We, then, have hearts *because* of this. Notice, however, that in this story it is not even mentioned that the function or goal or final cause of the heart is blood circulation, but rather only that we have hearts that do circulate blood in a certain manner, together with a certain efficient/material causal story as to why that fact came to obtain. It is this that reveals that our heart has the function or goal of blood circulation in the etiological account.

To see the import of this point for our present discussion, consider more abstractly now the contemporary etiological analysis of function when adapted as an analysis of Aristotelian final causality. In such an analysis:

B is the final cause of A if and only if:

(1) A accomplishes B and
(2) A is present *because* it accomplishes B.

The first question to ask about such an analysis from the Aristotelian point of view is: What kind of causality is introduced by the "because" in clause (2) of the *analysans*? For Aristotle there are four kinds of causality, so we need to determine which kind or kinds are intended or permitted by (2). It is clear enough that the one mode of causality that cannot be intended or permitted in (2) is final causality, since that would make the analysis trivially circular. If the causality in (2) were final causality, then clause (2) would in effect say or be satisfied by: A is present for the sake of B, or, in other words, B is the final cause of A. But final causality is just the item of which we are trying to give an analysis. So the causality covered in clause (2) must be other than final causality. (Matters are not improved if we write clause (2) as: A *came to be* present because of B, where that means *or* allows: A *came to be* present for the sake of B. Here again we have a circular analysis.) This shows, from the Aristotelian perspective, that in this type of analysis we are in effect explicating or defining final causality reductively, in terms of other types of causality. That, of course, is just the sort of thing one might expect from a contemporary philosophical analysis where the aim, often, is to define some presumed problematic notion – here final causality – in terms of less problematic notions. This analysis permits us, in effect, to understand the presumed unclear notion of final causality in terms of the clearer notion of efficient or efficient/material causality. By this means we sanitize, or naturalize as some would say, the notion of final causality. Many philosophers of biology today would argue this is the best that we can hope for if we want to find respectable teleological explanation in real biology.[20] But this aim is not something with which Aristotle would agree. For him, as we have seen, final causality is not, and is not an aspect of, efficient or efficient/material causality.

So how do we locate a conception of final causality in Aristotle that avoids this reductive character found in the contemporary analysis and also, I have been arguing, in the views of Anaxagoras and of Plato? We do best, I think, to start in a simple way, from Aristotle's claim in *Physics* II.8 that *nature* brings things about for the sake of something, and that in nature the final cause or end of a thing is due to its nature or is *natural* to it (198b10ff., 199a30–32). Among other things, to take one of Aristotle's own examples from II.8 (198b23ff.), this means that the fact that our broad back teeth are fit for breaking up food is itself something that holds by nature (cf. *Phys.* II.1, 192b35–193a2). That, indeed, is the main thing that Aristotle emphasizes in his critique in II.8 of Empedocles, who claims, according to Aristotle, that the fact that our broad back teeth are fit for breaking up food only holds by chance (198b27). To say, against this, that this fact holds by nature implies, given Aristotle's view of nature and of natural science, that this fact is either a primitive principle of natural science which explains other things or that it is a derivative theorem or conclusion explained ultimately by such primitive principles.

In *Physics* II.8 itself, that our broad back teeth are fit for, and do effectively perform, the beneficial function of breaking up food *by nature* is presumed to be confirmed not by reference to any analysis or definition of the notion of function or of final cause, but, in the first instance, by simple empirical means. Aristotle takes it to be clear from our ordinary experience that in humans, "always or usually," broad back teeth are fit for breaking up food. (He says "or usually" to take account of the fact that there are defective humans (199a33–b7).) This, he supposes, is an adequate ground for taking this fact to be a *natural* regularity (198b34–199a8). Some

[20] See e.g. Godfrey-Smith 1994, 2014.

critics of Aristotle's stance here have worried that he allows himself too easy a victory over Empedocles in supposing that in humans broad back teeth are fit for breaking up food. But, as Aristotle presents Empedocles' view, Empedocles agrees that as things stand our broad back teeth are fit for breaking up food. What they disagree about is the status of this fact. Here is how Aristotle presents Empedocles' own view in *Physics* II.8: "Our teeth come up of necessity, our sharp front teeth fitted for tearing, our back teeth broad and useful for breaking up food . . . [But] things are so organized *in a fitting way* (ἐπιτηδείως) by chance (ἀπὸ τοῦ αὐτομάτου)" (198b24–31).

Contrary to what is sometimes assumed in the literature, the main point of difference between Aristotle and Empedocles here does not concern how to explain the fact that we have broad back teeth. In particular, Aristotle and Empedocles do not disagree here over whether there is only a mechanistic explanation, or only a teleological explanation, or both, or neither, of the fact that we have broad back teeth. Rather, their disagreement here concerns the status of the very different fact that our broad back teeth are *fit for* breaking up food. According to this passage, Empedocles agrees with Aristotle that our broad back teeth *are* fit for breaking up food – that is, he agrees that as things stand that is the good and beneficial result that they efficiently secure. Empedocles even argues further, according to Aristotle, that the fact that our broad back teeth do efficiently accomplish this beneficial result confers a crucial survival advantage on us so that, in that sense, our broad back teeth are present in us *because* they perform this task, given appropriate attendant circumstances (198b27–32). This means that Empedocles actually accepts teleology in nature by our contemporary etiological standard. That is, in Empedocles' account our broad back teeth do regularly break up food, and they are present in us – in the sense that we survive with them – *because* they do this. This shows us again that the contemporary etiological account cannot be Aristotle's, since Aristotle clearly does not count Empedocles as a proponent of natural teleology. But, in any case, Aristotle does not allow himself any easy victory over Empedocles on the point that, as observation shows, our broad back teeth are fit for and do efficiently perform the beneficial function of breaking up food, since Empedocles agrees with this.

Where Empedocles differs from Aristotle here is only in his contention that the fact of this easily observed fitness holds by chance. This means, as we have seen, that while our broad back teeth are fit for and do efficiently function to break up food as things stand, and we survive with such teeth because of this, the fact that they are fit for this is not a natural regularity or what we would call a natural law. That is, again in more Aristotelian terms, the fact that our back teeth are fit for this neither is one of the fundamental necessary principles that govern and explain what happens in nature, nor is it a regularity explained by these principles. Aristotle, by contrast, holds that the fact that this functional regularity obtains *is* guaranteed by the fundamental principles of nature in the sense that it either is one of these principles or it is in a suitable way deducible from and explained by them. (Aristotle's critique of Empedocles leaves each of these two options open.) That is the key in *Physics* II.8 to the teleological outlook that Aristotle espouses there by contrast with the outlook of Empedocles, and, we may add, of his contemporary followers today.

Some would discern two contrasted theses of Empedocles at 198b24–30: (1) We have teeth of necessity, but (2) we have teeth fit for eating food by chance.[21] However, there is nothing in the text indicating a contrast between "our teeth come up of necessity" and "our sharp front teeth [come up] fitted for tearing" (198b24–25). The latter phrase is grammatically in apposition to the former and, thus, is explicative of it. So, as Aristotle indicates, for Empedocles, our sharp front teeth come up fitted for tearing by necessity (198b24–25). For Empedocles and the others whom Aristotle is considering here, everything happens by necessity, so if for Empedocles it is a fact that our sharp front teeth come up fitted for tearing, then that fact holds for him of necessity (198b12). But this is not incompatible with the claim that this fact holds by chance and not by nature as Aristotle understands this distinction. If this fact is neither a primitive explanatory principle of natural science, nor a conclusion explained by such principles, then it holds not by nature but by chance, even if it holds – for as long as it does hold – by necessity. Not everything determined or necessitated by the fundamental principles needs conform to a fixed natural law or be explained by such laws. In *Metaphysics* Z.7 what happens by chance is opposed by Aristotle to what happens by nature or by art, not to what happens by necessity (1032a12ff).[22]

So we may say that for Aristotle, according to *Physics* II.8:

B is the final cause of A if and only if:

(1) A is fit for B and
(2) that A is fit for B is a natural regularity.

[21] See e.g. Cooper 1982.

[22] For a related account of the difference between Aristotle and Empedocles, from a different but at least partly complementary perspective, see Sauvé Meyer 1992. For further discussion of the issues, with additional references, see Leunissen 2010.

Aristotle's own notion of final cause, then, is based on these two other notions – that of fitness and that of natural regularity. The notion of fitness is, then, a thinner one than that of final cause, although it is still, of course, an evaluative notion, and ineliminably so for Aristotle.[23] As we have seen, moreover, in Aristotle's approach in *Physics* II.8 the fact that it is a natural regularity that our broad back teeth are, evaluatively, fit for the breaking up of food is taken as something easy to directly confirm, at least in an initial way. It is confirmed, Aristotle supposes, by the easily observed fact that in humans broad back teeth are fit for breaking up food "always or usually" (198b35–36). Of course, Aristotle would need to show that this presumed natural regularity does fit into the whole body of natural scientific regularities in an appropriate way – either as principle or as derived theorem – to finally confirm this. Otherwise, his victory over Empedocles is perhaps too easy. But this is all that he needs to do.

To help see this more clearly consider, finally, how different the main problem is that motivates Aristotle's stance on teleology in *Physics* II.8 from a problem that has motivated contemporary analysts. In recent discussion, one main problem has been to show how to distinguish the goal or final cause of some structure or process, when it has one, from a mere regular consequence of that structure or process. For instance, we today want to be able to say why the circulation of the blood is the goal or function of the heart while the production of heartbeat noises is not. But this is not Aristotle's problem. He does understand this distinction. He notes in *Physics* II.2, as something of a joke, that death is the end of human life in a different way from that in which our best condition is the end of human life (194a28–33). But he regards it as very easy, on the whole, to distinguish (evaluatively good) goals from other regular consequences. In *Physics* II.3, when he introduces final causality, Aristotle does so simply by means of what he takes to be uncontroversial examples. He does the same in *Metaphysics* Δ.2 and H.4, and elsewhere. He never evinces the need for an analysis of function or of final cause to enable him to distinguish goals from other regular consequences. His problem, rather, as we have seen, is to show that in many cases where things obviously are fit to achieve certain goals and beneficial results, that they do achieve these results in a fitting way is something that holds by nature and not by chance.

So if we want to understand Aristotle's own special conception of natural teleology in *Physics* II we need to see that it has these two crucial components: (1) Final causality is not reducible to, or analyzable or

[23] On the issue of value in teleology, see Nagel 2012 and references there.

explicable in terms of, any other kind of causality, such as efficient causality; and (2) in the natural world, that things have the sort of fitness they clearly now do is typically something that holds by nature, not by chance. This shows us how very different Aristotle's approach to natural teleology is not only from the alternative of Empedocles but also from that of Plato.[24]

[24] See Bolton 1997, 2011 for further discussion and for a defense of the view that the natural fitness of means to goals is not primitive for Aristotle but is explained by primitive principles.

CHAPTER 8

Substantial generation in Physics 1.5–7

Devin Henry

The focus of this chapter is on the conceptual machinery of Aristotle's theory of generation in *Physics* 1.5–7. According to the standard interpretation, *Physics* 1.7 imposes two requirements on all processes of change. First, all change requires an underlying subject from which the change proceeds. During the course of the change this subject comes to be F from being not-F. Second, the subject must remain through the change to become a constituent of the finished product. For example, Socrates survives the healing process and simply exchanges sickness for health, and bronze survives the production of the statue and simply changes from being unshaped to being shaped. The central question of this chapter is whether or not Aristotle thinks that both of these conditions hold in cases of substantial generation. Does he think that when a new substance comes into being there is a pre-existing subject that survives that change and becomes a constituent of what comes to be? The orthodox view is that he does, and that his argument for this is to be found in *Physics* 1.7. This, it is held, forms part of Aristotle's argument for hylomorphism, the claim that every particular thing is a composite of matter and form. My aim in this chapter is to re-evaluate this entrenched view of the *Physics*. I shall argue that *Physics* 1.7 is silent on the question of whether or not substantial generation requires a persistent subject, while evidence external to the *Physics* strongly suggests it does not. In this reading, while Aristotle thinks substantial generation requires a subject from which the change proceeds, he does not claim that this subject survives as a constituent of the substance that comes into being.

To begin it will be useful to call attention to the distinction Aristotle makes in the works on natural science between qualified coming to be (γίγνεσθαι τι) and coming to be in the unqualified sense (γίγνεσθαι ἁπλῶς).[1] All change for Aristotle is a form of coming to be (or passing

[1] See also Jones 1974: 478–480.

away) in one of these two senses.[2] As a first pass at this distinction we can say that when "coming to be" (γίνεται, γίγνεσθαι) is used in the qualified sense, the verb is always transitive, in the sense that it remains incomplete without specifying what the subject comes to be. For example, we say that a human comes to be healthy, or that he comes to be bigger. When used in the unqualified sense, the verb is intransitive, in the sense that it is complete all on its own. In this case we say that a human comes to be (full stop), not that he comes to be so-and-so. This linguistic distinction has a metaphysical basis. For the difference also turns on which category of being the change concerns.[3] When a thing comes to be in the qualified sense, the change always occurs with respect to one of the non-substantial categories. Alteration (ἀλλοίωσις) is coming to be in the category of quality (e.g. Socrates comes to be healthy), growth is coming to be in the category of quantity (e.g. Socrates comes to be bigger), and locomotion is coming to be in the category of place (e.g. Socrates comes to be in the market). All of these count as instances of coming to be in the qualified sense because the substance (Socrates) already exists and merely becomes qualified in some particular way. Coming to be in the unqualified sense, by contrast, is a change in the category of substance; it is substantial generation. In this case, a new substance comes into being that was not there before. In what follows I shall reserve the term "generation" for coming to be in the unqualified (ἁπλῶς) and strict (κυρίως) sense, and use the more general term "change" (μεταβολή, κίνησις) for the broader category that includes coming to be in the qualified sense (alteration, growth, locomotion).

At the outset of the *Physics* Aristotle tells us that when the objects of any inquiry have principles, causes, or elements (ἀρχαὶ ἢ αἴτια ἢ στοιχεῖα), knowledge and understanding are acquired by grasping these: "For we do not think we know a thing until we are familiar with its primary causes or first principles and have carried out our analysis as far as its elements. Clearly, then, also in the science of nature our first task will be to determine its principles" (184a10–16). Aristotle's positive inquiry into these principles is set out over *Physics* I.5–7.[4] According to the results of that inquiry, every change involves a subject, S, that undergoes the change (τὸ ὑποκείμενον) and a pair of contraries, F and G, that represent the two extremes of the

[2] For the main distinction, see *Phys.* I.7, 190a31–33; *On Generation and Corruption* I.3, 318a34–35. My focus in this chapter is mainly on coming to be, although much of what I have to say applies equally to passing away.

[3] With the following, see *GC* I.3, 317a27–28; I.4, 319a11–16; and *Metaph.* XII.2, 1069b9–14.

[4] Compare *GC* II.9, and *Metaphysics* VII.7 and XII.4. Of these four places only *Metaphysics* XII.4 offers the complete list of principles of coming to be (form, privation, subject/matter, primary moving cause).

change. Within this tripartite schema, S is always a body of some kind (e.g. human, wood, seed), while the contraries are identified with the form (the possessed state, ἕξις: e.g. musical) and its privation (the absence of that state, στέρησις: e.g. unmusical). When the change proceeds from the privation to the form, it is coming to be. When it proceeds in the opposite direction, from the form to the privation, it is passing away. Together *Physics* 1.5–7 argue that these three items (form, privation, and subject) constitute the principles of coming to be for all natural things: *Physics* 1.5 argues that the contraries are principles; *Physics* 1.6 shows that there must be some third thing besides these; *Physics* 1.7 argues that this third thing is the subject that underlies the change.[5]

Aristotle's main position is developed at length in *Physics* 1.7. The core of that theory can be extracted from two key passages: 190a13–21 and 190b10–23. For our purposes we can focus on the celebrated passage at 190a13–21:

> Once these distinctions have been made one can grasp the following from all the cases of coming to be, if one looks at them as we suggested (ἐάν τις ἐπιβλέψῃ ὥσπερ λέγομεν), namely, that there must always be an underlying subject (which is what comes to be), and that this, though always numerically one, is not one in form. By "in form" I mean the same as "in account"; for what it is to be a human is not the same as what it is to be musical. And one survives, while the other does not: for what is not an opposite survives (for the human survives), but not-musical or unmusical does not survive, nor does the compound of the two, namely, the unmusical human.

Aristotle makes two main claims here. First, coming to be requires an underlying subject, which is the thing that comes to be (τὸ γιγνόμενον).[6] This subject is numerically one, but not one in form; rather, it changes from F to G as a result of coming to be. Second, in the process of change one part survives (the subject), while the other does not (the privation or the subject-cum-privation). I shall call these the Subject Requirement and the Survival Requirement, respectively:

Subject Requirement. Coming to be requires an underlying subject from which the change proceeds.

Survival Requirement. When something comes to be, the subject endures and remains as a constituent of the finished product.

[5] For a general discussion of these chapters, see Bostock 2006: ch. 1 and Kelsey 2008.

[6] Aristotle uses τὸ γιγνόμενον to refer both to the subject from which a change proceeds (the *terminus a quo*) and to the thing that results from that change (the *terminus ad quem*). In the present context Aristotle means the former, while at 190b10–23 (quoted below) τὸ γιγνόμενον picks out the latter. In *Generation of Animals* Aristotle uses τὸ γιγνόμενον almost exclusively for the embryo that comes to be an F.

Most scholars take *Physics* 1.7 to be developing the concept of underlying subject in close connection with the idea of being a continuant of change, so that the Subject Requirement and the Survival Requirement come out as equivalent. Bostock (2006: 9), for example, argues that the expression τὸ ὑποκείμενον is being made to do "double duty" in *Physics* 1.7, both for the starting-point of the change (the *terminus a quo*) and for what persists through the change. On this reading, to say coming to be requires a subject that underlies the change just means there must be something that persists through the change and survives as a constituent of the finished product.[7] In the reading I shall defend, the concept of subject and continuant come apart: The concept of subject (τὸ ὑποκείμενον) is playing only the one role of starting-point of change; it is a separate claim to say that this subject also persists through the change.[8]

It is obvious that cases of non-substantial change meet both requirements. For example, when Socrates comes to be healthy from being sick there is an underlying subject from which the change begins (Socrates). And that subject persists through the change and remains as a constituent of the finished product (healthy-Socrates). However, while *Physics* 1.7 argues that substantial generation must also satisfy the Subject Requirement (190b1–5), there is a question about whether or not Aristotle means to commit himself to the view that in cases of substantial generation this subject survives as an element of the substance that results from that change. Before turning to that I want to start by clarifying the Subject Requirement.

The Subject Requirement says that in (all) coming to be there must be a subject from which the change begins. Like the contraries, the concept of an underlying subject of change was not Aristotle's invention. At 189b5–6 Aristotle suggests that some of his predecessors already invoked the idea of a ὑποκείμενον as something different from the contraries, although they disagreed over what that underlying subject was. But how does Aristotle understand the concept? *Physics* 1.7 offers plenty of examples of subjects, including a human (190a17, b14, b20), a seed or embryo (190b5; cf. *GA* 1.18, 724b14–18; 1.20, 728b34–35; and 1.23, 731a2–4), bronze, gold, stone (190a25, 190b16–17), and wood (191a7). We can shed some light on this motley crew of subjects by turning to *Metaph.* VII.13, 1038b2–7:

7 See also Broadie 1982: 47; Gill 1989: 6, 90, 106; and Witt 1989: 66.

8 Compare Charlton 1970: 77 and Bostock 2006: 7–8. Charlton denies that substantial generation involves a persistent subject. While I shall eventually side with this reading, here I only mean to make the point that the Subject Requirement and the Survival Requirement are independent of one another. That point does not require taking a stand on the question of whether or not Aristotle is committed to the Survival Requirement in cases of substantial generation.

> Just as the underlying subject, the essence, and the composite of these are called substance, so too is the universal. We have already spoken about two of these, namely the essence and the underlying subject, and the latter we said underlies in two senses: either as a determinate individual, which is the way that an animal underlies its attributes, or as the matter underlies the complete actuality.[9]

Here Aristotle contrasts matter with the determinate individual (e.g. the individual human or horse) as two kinds of underlying subject. The latter is the more familiar sense of subject from the *Categories*. These primary subjects are independently existing "thises" (τόδε τι) that remain numerically one and the same throughout their existence, and stand as ultimate subjects of predication (*Cat.* 5). The other sense of subject is the proximate matter that underlies the form of the composite individual. In the case of living things, for example, the material subject is the living body whose organs have been specially adapted to realize the capacities of soul that make up its form (*DA* II.1, 412a17–19).[10]

What I want to suggest is that these two senses of underlying subject – subject as individual determinate substance and subject as matter – track Aristotle's distinction between qualified and unqualified coming to be, respectively. The underlying subject of qualified coming to be is always some determinate individual; for example, human is the subject in the change from an unmusical human to a musical human. By contrast, the subject of unqualified coming to be is always matter of some kind; for example, bronze is the subject in the production of a bronze statue. While Aristotle does not avail himself of this distinction in *Physics* I.7, it is clearly implicit in his account. For example, when he is enumerating the principles of naturally generated substances at the end of *Physics* I.7 he attempts to elucidate the concept of "the underlying nature" (ἡ ὑποκειμένη φύσις) by setting up a contrast between matter, on the one hand, and the individual substance, on the other (191a7–12), and stresses that the former underlies the change, but not as a unified "this" (οὐχ οὕτω μία οὖσα οὐδὲ οὕτως ὂν ὡς τὸ τόδε τι; 191a12–13). Again at 190b13–14 (see below) he distinguishes bronze or stone from the human being as two kinds of

[9] See also *Metaph.* IX.7, 1049a34–b1.

[10] Gill 1989: 31 argues that the aim of *Metaphysics* VII.3 is to refine the conditions on subjecthood so that matter actually turns out not to be a subject. In her reading, the refined concept only picks out the individual, i.e. what is separate and a "this" (τόδε τι). But that is not what *Metaphysics* VII.3 concludes. Aristotle says that being separate and a "this" are characteristics of substance, not subject, and that for this reason matter fails to be a substance. Aristotle never retracts the original claim that matter is an underlying subject. And *Metaphysics* VII.13 presents that claim as something already established.

underlying subject. So there is at least some recognition in *Physics* 1.7 that when τὸ ὑποκείμενον is used for the material subject (e.g. bronze, gold, stone, wood) it is to be contrasted with those subjects that correspond to individual substances (e.g. human, horse).[11]

According to the Subject Requirement, then, all coming to be requires an underlying subject from which the change proceeds. When a thing comes to be in the qualified sense this will be a determinate individual (e.g. human). In cases of substantial generation the subject from which the change proceeds is always matter of some kind. The question is whether in *Physics* 1.7 Aristotle thinks the subject must persist in all cases of change, including substantial generation, or whether this is true only for changes in the non-substantial categories.

Before turning to that question let us complete the basic model of coming to be by introducing the other core passage from *Physics* 1.7:

> From what has been said, it is clear that everything that comes into being (τὸ γιγνόμενον) is complex.[12] There is, on the one hand, what comes into being (τι γιγνόμενον) and, on the other, something that comes to be that (τι ὃ τοῦτο γίγνεται). And the latter in two senses: either the subject or the opposite. By "the opposite" I mean the unmusical and by "the subject" I mean the human. Likewise I call the absence of shape, form, or organization "the opposite" and the bronze or stone or gold "the subject." Clearly, then, if there are causes and principles of naturally existing things from which (ἐξ ὧν) they primarily are and have come to be (I mean what each is said to be according to its essence and not incidentally), then everything comes to be from both the subject and the form (ἔκ τε τοῦ ὑποκειμένου καὶ τῆς μορφῆς). For in a certain way the musical human is composed of human and musical, for you can analyze it into the account of each. It is clear, then, that things that come to be (τὰ γιγνόμενα) have come into being from these. (190b10–23)

Consider the generation of an animal. The thing that comes into being as a result of the change is the fully developed animal that can be analyzed into a body, which is the underlying subject, and a soul, which is predicated of

[11] Aristotle's remarks at the beginning of *GC* 1.5 also suggest that these two senses of underlying subject mark the distinction between qualified and unqualified coming to be. There he says that matter in its "most proper and strict sense" is the underlying subject of generation and destruction, although "in a certain way" it is also the subject of other changes (320a2–5). What he means, I take it, is that when the musical human comes to be musical, "man" can be treated as the matter in so far as it stands in an analogous relation to the (un)musical human as bronze stands to the statue or wood to the bed (cf. *Phys.* 1.7, 190b20–23). Thus, in a way, matter is the subject even of non-substantial changes (including alteration, increase, and locomotion). But matter in the "proper and strict sense" is the subject only of unqualified coming to be and passing away.

[12] I take τὸ γιγνόμενον here to refer to what results from the change (the *terminus ad quem*). For an alternative reading, see Gill 1989: 102–108.

the body as its subject (*DA* II.1, 412a16–19). The soul is the form picked out by the definition specifying what comes to be (τι γιγνόμενον),[13] while the thing that comes to be that (τι ὃ τοῦτο γίγνεται) is either the embryo (*Physics* I.7, 190b1–5, 9–10) or the privation of the form (*Physics* I.5, 188b20; cf. Plato *Phaedo* 70b10–72e2).

The remainder of the passage is meant to answer the central question of *Physics* I: What are the principles from which (ἐξ οὗ) naturally existing things primarily are and come to be? Aristotle's answer can be understood by drawing on a distinction he makes in *Metaphysics* XII.3 this time between two kinds of principle: "Since not only the elements that are present in a thing (τὰ ἐνυπάρχοντα) are its causes but also something external (ἐκτὸς) to it, viz. the moving cause, clearly while 'principle' (ἀρχή) and 'element' (στοιχεῖον) are different both are causes, and 'principle' is divided into these two kinds" (1070b21–25). The two kinds of principle here are (1) those that are present in a thing as its constituent elements and (2) those that are not constituents of the product but are external to it. Of the three principles listed in *Physics* I.7, the privation counts as a principle only in the sense that natural generation originates from it. But it is not a principle in the sense of an element of the thing that comes to be. For it does not remain in the finished product as a constituent (οὐκ ἐνυπάρχοντος γίγνεταί τι; *Phys.* I.8, 191b16), but is destroyed in the change. Only subject and form are principles in the sense of being constituent elements of naturally generated substances. This is what Aristotle means when he says that everything that comes into being is composed from subject and form (γίγνεται πᾶν ἔκ τε τοῦ ὑποκειμένου καὶ τῆς μορφῆς; 190b20).[14]

Traditionally scholars have taken this as an argument for hylomorphism. But that is not quite what Aristotle says here. While matter is certainly

[13] Compare *Metaph.* VII.7, 1032a13–25 and XII.3, 1069b36–1070a2. In both places Aristotle uses the expression τι γιγνόμενον to pick out the form that tells us what it is that comes into being. While it is true that the *Metaphysics* VII.7 passage offers "human" and "plant" as its examples of τι γιγνόμενον, which look like composite substances, Aristotle insists that the same term can designate both the form and the composite (*Metaph.* VIII.3, 1043a29–37). That he has the form of the product in mind here is confirmed by *GC* II.9, where he explicitly identifies the three principles of coming to be with the matter, form, and primary efficient cause. Of the first two he says: "This [i.e. the capacity to be and not be] is a cause in the sense of matter for those things that are such as to come to be, while the cause in the sense of their end is their shape or form, which is also the account stating the substance of each of them" (335b4–7). Once again contrast Gill 1989: 121.

[14] *Physics* I.7 does not list the primary moving cause among the principles of naturally generated substances. However, in *Physics* II we learn that form and matter are, in fact, both inner principles of change for naturally generated substance. The external efficient cause mentioned in the *Metaphysics* passage is the (form of the) generating parent (*Phys.* II.7, 198a25–28; *Metaph.* VII.7, 1032a20–25; *GA* II.1, 735a2–4; and *PA* I.1, 640a19–26).

one kind of constitutive subject, we have seen that *Physics* 1 conceives of τὸ ὑποκείμενον in the broader sense that includes both matter (e.g. bronze) and the determinate individual (e.g. human). This is explicit even in the current passage, for "man" is there identified as the subject in the composite "the musical human." So if there is an argument for hylomorphism in *Physics* 1.7, it is not the explicit aim of this passage.

As we have seen, most commentators assume that Aristotle's concept of underlying subject already involves the idea of being a continuant of change, so that the Subject Requirement and the Survival Requirement collapse into a single condition imposed on all coming to be.[15] In this section I re-examine that orthodoxy. We can carve up interpretative space into two broad readings. According to the orthodox view, Aristotle holds that all change requires a subject that persists through the change and remains as a constituent in the finished product, including substantial generation. We can call this the Inclusive Reading, since it takes the Survival Requirement to be inclusive of all forms of coming to be.[16] By contrast the Exclusive Reading takes the Survival Requirement to apply exclusively to coming to be in the qualified sense (change in the non-substantial categories).[17] In this reading, while Aristotle thinks substantial generation requires a subject from which the change proceeds, he does not claim that this subject must survive as a constituent of the substance that comes into being. What I want to argue is that the Exclusive Reading is more attractive as an interpretation of *Physics* 1.7 than the more traditional

[15] I suspect this reading gains a lot of its naturalness from the fact that most gloss τὸ ὑποκείμενον as "that which underlies the change." To avoid begging the question I have adopted the more neutral expression "underlying subject," without any assumptions about what this subject underlies (the form/privation or the change itself).

[16] This reading is by far the most common. It is defended in some form or another by Broadie 1982: 46–47; Gill 1989: 6, 98–108; Code 1995: 415–417; Bostock 2006: 8–9; and Kelsey 2010: 112. Broadie (2004) seems to have changed her position on the matter. Broadie 1982: 47 endorses the Inclusive Reading and takes the concept of underlying subject to entail the idea of persistence. However, Broadie 2004: 129–130 takes the weaker view that *Physics* 1.7 is actually silent on the issue, while *GC* 1.3 positively rejects it in favour of the Exclusive Reading. (In light of this I take her remarks in her 2004 essay in favor of the Inclusive Reading to be offering reasons why one might be tempted by it without actually endorsing it.) Bostock defends the Inclusive Reading, but he is careful to restrict his defense to *Physics* 1 (2006: 8). He does not speculate as to whether or not Aristotle endorsed the Survival Requirement outside *Physics* 1, specifically, whether or not *GC* 1.1–5 actually rejects it. Finally, Kelsey parses *Phys.* 1 .7, 190a13–21 slightly differently, so that the Survival Requirement is not actually at issue. In his reading, Aristotle argues (1) that coming to be always requires an underlying subject, and (2) that this subject must be "two in form." He then takes the claim that one part survives and one part does not survive as evidence for (2). However, Kelsey counts himself among those who accept the Inclusive Reading (personal communication).

[17] Charlton 1970: 77, 135; Jones 1974; and Broadie 2004.

Inclusive Reading. There are two reasons for this. First, although most commentators prefer the Inclusive Reading, the evidence from *Physics* 1.7 is not conclusive enough to force that interpretation on us. As we shall see, it is possible to read that chapter in a way that is compatible with the Exclusive Reading. Second, there is evidence external to *Physics* 1 that strongly suggests Aristotle does not think substantial generation involves an enduring subject. This creates a strong presumption in favor of the Exclusive Reading as an interpretation of Aristotle's more general views about coming to be in the unqualified sense. In light of this I shall propose that *Physics* 1.7 is best read as remaining silent on the question of whether or not substantial generation must meet the Survival Requirement. This conservative reading has the obvious virtue of avoiding a conflict among Aristotle's different discussions of change.

Traditionally, *Physics* 1.7 has been seen as the *locus classicus* for the Inclusive Reading. The most explicit piece of evidence for this reading is 190a13–21 (translated above), where the Survival Requirement is first mentioned. Aristotle introduces that passage by saying: "one can grasp the following from all the cases of coming to be" (ἁπάντων τῶν γιγνομένων; cf. 189b30–31).[18] Most take this as definitive evidence that coming to be in the unqualified sense is subject to the Survival Requirement. Bostock argues that 190b10–23 also implies that some part of the subject remains even in cases of substantial generation:

> The only ground Aristotle could have for saying that whatever comes into being is composite (*sunthetos*) is that we can distinguish in it two "elements," one the persisting element (what underlies) and the other the acquired element (the form). If the element said to underlie did not persist in the end product, there would be no ground whatever for saying that the end product was composite, and Aristotle explicitly claims that all products of becoming are composite. (Bostock 2006: 8)

Finally, scholars typically point to the example of a statue coming to be from bronze at 190a23–26 as a case of substantial generation. And there Aristotle definitely says that the bronze survives.[19]

The trouble starts for the Inclusive Reading when we turn to *On Generation and Corruption*. The question of whether or not the subject persists through the change lies at the heart of *GC*'s distinction between qualified and unqualified coming to be. In *GC* 1.1 Aristotle tells us that the monists, who constructed all things out of a single element, were

[18] Bostock 2006: 8; cf. Broadie 2004: 129–130.

[19] For what is in my view an unsuccessful argument against the latter point, see Jones 1974: 483–488.

forced to reduce substantial generation to alteration precisely because they held that all change requires a persisting subject: "For they must affirm that the underlying subject (τὸ ὑποκείμενον) always remains one and the same: and we call this sort of thing 'being altered'" (314b3–4). On Aristotle's view, when a thing undergoes alteration (or growth or locomotion), the subject survives the change and simply exchanges one property for another. In substantial generation the subject does not survive the change, but is completely transformed in the process (cf. 332a8–10). There are at least three passages where Aristotle makes his position explicit:

> T1. Coming to be and passing away in the unqualified sense are not effected by combination and disassociation but when a thing changes from this into that as a whole (μεταβάλλῃ ἐκ τοῦδε εἰς τόδε ὅλον). But they suppose that all such change is alteration, whereas in fact there is a difference. For one component of the underlying subject corresponds to the form and the other the matter. There is coming to be and passing away when there is a change in these, but when the change is in its incidental properties it will be alteration. (*GC* 1.2, 317a20–26)
>
> T2. Seeing that we distinguish between the underlying subject and the property whose nature it is to be predicated of that subject, and that change occurs in each of these, it is alteration when the perceptible subject remains but changes in its own properties (the properties in question being either opposites or intermediates); for example, the body, although enduring as the same body, is now healthy and now ill, and the bronze is now spherical and now angular, and yet remains the same bronze. But when the thing is transformed as a whole (ὅλον μεταβάλλῃ) and nothing perceptible remains as its subject (e.g. when an embryo as a whole is changed into blood[20] or water into air or air as a whole into water) such an event is the coming to be of the one sort of thing and the passing away of the other, especially when the change proceeds from an imperceptible something to something perceptible (either to touch or to all the senses), as when water comes to be out of or passes away into air (for air is pretty imperceptible). (*GC* 1.4, 319b8–21)
>
> T3. Our account of growth must preserve the features of the subjects that are growing and diminishing (τὰ ὑπάρχοντα τῷ αὐξανομένῳ καὶ φθίνοντι). And these are three: First, any and every part of the growing magnitude is made bigger (e.g. if flesh grows, every particle of flesh gets bigger); second, it gets bigger by the accession of something; and third, it does so in such a way that the growing thing is preserved (σωζομένου) and endures (ὑπομένοντος)

[20] The example is peculiar and does not match anything in Aristotle's own account of animal generation. It would make better sense if "embryo" and "blood" were transposed, or if "blood" was replaced by (e.g.) "animal." As far as I know, there is no manuscript evidence for that. Nevertheless, Aristotle's point is clear.

> the change. For, whereas a thing does not endure (οὐχ ὑπομένει) in the process of unqualified coming to be and passing away, in both growth and alteration the growing subject itself and the thing that undergoes alteration does endure (ὑπομένει), although in alteration the quality (τὸ πάθος) and in growth the size (τὸ μέγεθος) does not remain the same. (*GC* I.5, 321a17–26)

In each of these passages Aristotle uses the Survival Requirement as a means for distinguishing growth and alteration from generation. When the underlying subject persists through the change and simply exchanges one accident for another, it is growth/alteration. Generation occurs when the subject from which the change proceeds does not endure but is wholly transformed in the process of becoming a new substance, which Aristotle glosses in T1 as changing in both matter and form together.[21]

In T2 Aristotle again distinguishes between a subject and its accidents (πάθος here corresponds to the "incidental affections" mentioned at the end of T1, 317a26). And again alteration is said to occur when the subject remains one and the same and changes in one of its accidents, while generation occurs when there is complete transformation (ὅλον μεταβάλλῃ) so that nothing remains of the original subject.[22] T3 is even more explicit about this. Here Aristotle is discussing growth (coming to be in the category of quantity) rather than alteration (coming to be in the category of quality). He mentions three characteristics of genuine cases of growing. The last of these is that the subject is preserved (σωζομένου) and endures (ὑπομένοντος) the change just as in cases of alteration, "whereas it does not endure (οὐχ ὑπομένει) in the process of unqualified coming to be and passing away."

Some commentators have tried to read the *GC* doctrine in a way that brings it in line with the Inclusive Reading. In T1 Aristotle tells us that the subject from which substantial generation proceeds is itself a complex of matter and form. For example, the subject from which animal generation proceeds is the embryo, which is a compound of

[21] For a similar reading of *GC* I, see Broadie 2004. See also *Metaph.* VII.7, 1033a20–23. There Aristotle says that the matter from which a thing comes to be does not persist (οὐχ ὑπομένοντος) but is transformed (μεταβάλλοντος) in the process of coming to be.

[22] Defenders of the Inclusive Reading who take the continuant of substantial generation to be prime matter emphasize Aristotle's reference to the destruction of a perceptible subject. To say that nothing perceptible remains (this reading holds) leaves it open that something imperceptible survives the change, namely, prime matter. In my view the focus on a distinction between a perceptible and an imperceptible subject is a red herring. If Aristotle meant that some imperceptible subject always endures, we would expect him to make that crucial point here. And yet nowhere in *GC* does he ever mention an imperceptible continuant. For more discussion of this point, see Gill 1989: 42–53 and Broadie 2004: 126ff.

menstrual blood from the female and form from the male. According to Gill, when Aristotle says that the underlying subject "does not remain" (μὴ ὑπομένοντος), he means it does not remain what it was at the start of the change. For it loses the form that determined its original identity in the process of being transformed into something new.[23] But the subject does remain *qua* matter; the matter of the underlying subject does not undergo a change. While this way of understanding the *GC* account would save the Inclusive Reading, that is not what Aristotle says in the above passages. T2 and T3 say nothing like that. In T1 Aristotle says that one component of the subject corresponds to form and the other matter, and that generation occurs "when there is a change in these" (not "in one of these"). The most natural antecedent for the plural τούτοις here is the form and the matter (not the form alone). That, I take it, is what Aristotle means by saying that in generation and destruction the subject changes "as a whole." If Aristotle had meant to say that the subject of generation and destruction changes with respect to its form but not with respect to its matter, we should have expected him to make that crucial distinction here – especially since he has made it a point of emphasis that those who posit an enduring subject are committed to the view that unqualified coming to be and passing way are just forms of alteration (*GC* 1.1, 314b3–4).

According to *GC* 1.1–5, then, what distinguishes substantial generation from changes in the non-substantial categories is precisely the fact that in the latter case the underlying subject persists and changes only in its incidental properties, whereas in the former case the subject does not remain but is completely transformed in the process of becoming a new substance.

Another place to look for a defense of the Inclusive Reading is Aristotle's mature theory of animal generation in *Generation of Animals*. If Aristotle had strong reasons for insisting that substantial generation must meet the Survival Requirement, then we would expect this to form an important part of that theory, especially since he treats animal generation as substantial generation par excellence. Unfortunately, there is no evidence that Aristotle was concerned in *GA* to show that the subject from which the process begins persists through the change to become a constituent of the mature animal. By contrast, much of the inquiry in *GA* is arguably motivated by a concern with the Subject

[23] Gill develops this reading over the course of chapters 2 and 3 of her book. For a precise statement, see Gill 1989: 62.

Requirement. For example, *GA* I.18–23 is about the nature of *sperma*, which is characterized as "that sort of thing from which naturally generated organisms first come into being" (724a18–20). Again, in *GA* II.1 Aristotle investigates how the parts of the animal are formed. That inquiry takes as its starting-point the principle that "everything that comes to be by nature or by craft comes into being by the agency of something existing in actuality from what is already potentially that sort of thing (ἐκ τοῦ δυνάμει τοιούτου)" (*GA* II.1, 734b20–22; see also 734b34–36). This reflects Aristotle's general solution to the Eleatic challenge, defended in *Physics* I.8 and *GC* I.3, that coming to be in the unqualified sense proceeds from what exists in potentiality, and that this potential F is the proper referent of τὸ μὴ ὄν ἁπλῶς: "In one way things come to be from what is not in the unqualified sense, yet in another way they always come to be from what is. For there must be something existing beforehand (προυπάρχειν) that is potentially but not actually an F. And this is spoken of both as being and as not-being" (317b14–18). Aristotle repeats this a few lines later: "If something comes to be <in the unqualified sense>, then clearly there will be some substance that exists potentially but not in complete actuality (ἐντελεχείᾳ) from which coming to be proceeds and into which the thing that is being destroyed necessarily changes" (*GC* I.3, 317b23–25).

Freeland (1987) draws attention to Aristotle's claim that blood serves as both the matter for the parts of the mature animal and the matter out of which the embryo is originally constructed (*PA* II.3, 650a32–b13; *GA* I.18, 725a21–28; I.19, 726b1–15; and II.4, 740b34–35). She takes this as evidence that blood is the persisting substratum of biological generation, which satisfies the *Physics* I.7 requirement that all change requires an enduring subject. If Freeland is right, then Aristotle must have abandoned the *GC* distinction between growth and generation in the biological works. For *GC* is categorical that in the case of growth the subject endures through the process, but when a thing comes to be in the unqualified sense the subject does not endure but it is wholly transformed in the process (see T3 above).

Yet Aristotle did not develop his theory about blood as the matter of the parts with the Survival Requirement in mind, and so by saying that "it is the same matter by which the <existing animal> grows (αὐξάνεται) and out of which it is first constructed" (*GA* II.4, 740b35–36), he does not mean to commit himself to any views about a persistent substratum. So we can't take this claim as a defense of the Survival Requirement. More importantly, this claim does not entail that blood is a persistent substratum. Freeland simply helps herself to this by assuming that the material from which generation begins must persist to become the proximate matter out of which the adult animal is constituted.[24] But that does not follow. When Aristotle claims that the matter out of which the animal grows and the matter out of which it is first constituted are the same (*GA* II.4, 740b35–36), he need only be saying that growth and generation begin from the same material, i.e. that blood plays the role of (material) starting-point in each case. This does not commit him to the further claim that blood from which generation begins persists through the change to become the proximate matter out of which the adult animal is constituted. And there are good reasons to think that Aristotle wants to resist this conclusion. First, as we have seen, Aristotle treats growth and generation as two distinct kinds of change, and insists that only the former involves a persistent subject: "For, whereas a thing does not endure (οὐχ ὑπομένει) in the process of unqualified coming to be and passing away, in both growth and alteration the growing subject itself and the thing that undergoes alteration does endure (ὑπομένει), though in alteration the quality (τὸ πάθος) and in growth the size (τὸ μέγεθος) does not remain the same." Second, according to Aristotle's mature theory of animal generation, blood could not satisfy the Survival Requirement. To meet the Survival Requirement, the blood from which generation begins must remain numerically the same throughout the change ("this, though always numerically one, is not one in form"). Yet Aristotle holds that the blood that is used to form the embryo is not even specifically the same blood that is later used to grow and maintain the parts of the completed offspring (*GA* II.6, 744b32–745a4).[25]

[24] See Freeland 1987: 406.

[25] Much of Freeland's argument depends on the claim that menstrual blood is the proper subject of animal generation in the *Physics* sense, which I think is incorrect. Aristotle associates the subject that serves as the starting-point of substantial generation with the potential F (e.g. *Phys.* I.8, 191b27–29 and *GC* I.3, 317b23–25). And according to *Metaphysics* IX.7, a thing counts as the potential F when it is in a state where nothing further must be added, removed, or changed before it can straightaway be transformed into an F (1049a10–11) and, in cases of natural generation, possesses the generative principle within itself so that it develops into an F through itself, if nothing external impedes it (1049a13–18). While Aristotle thinks that menstrual blood is the matter out of which the embryo is formed, and while he does say in *GA* that it is already potentially an animal of the same kind as the mother (*GA* II.3, 737a22–25; II.4, 738b3–4 and 740b18–20; II.5, 741b7–8; cf. 729a32–33), there is good reason to think that it is not the subject (τὸ ὑποκείμενον) of generation in the *Physics*' sense. It fails the conditions for being the starting-point of the change laid down in *Metaphysics* IX.7. For it is not yet in a state where it will straightaway change into a new organism "through its own principle" (διὰ τῆς αὑτοῦ ἀρχῆς). It still needs to undergo a further change before it is in that state, namely, it needs to be fertilised by the semen and then transformed into an embryo (cf. *GA* I.20, 728a26–30). When we apply the *Metaphysics* IX.7 test to the *GA*, it turns out that it is only when the embryo has been formed and has developed to the point where it has a rudimentary heart that can process nutriment that it is in a state where

Defenders of the Exclusive Reading can use the evidence from *GC* I.1–5 and Aristotle's mature theory of animal generation as a reason to reconsider the standard account of *Physics* I.7. Consider, first, the passage at 190a13–21. It is possible to read the opening sentence of that passage as restricting its scope to those cases of coming to be that Aristotle has just been discussing (ἐάν τις ἐπιβλέψῃ ὥσπερ λέγομεν), namely those where a thing comes to be in the qualified sense (189b32–190a12).[26] (Remember that in *GC* Aristotle says in at least three places that when a thing comes to be in the unqualified sense the subject does not survive.) In support of this is the fact that *Physics* I.7 does not present any argument for the (surely controversial and not obvious) claim that even in substantial generation something persists through the change. Aristotle does argue that when a thing comes to be in the unqualified sense there must be some underlying subject from which the change proceeds (190b1–5). But he does not also say that this subject must persist through the change to remain as a constituent of the finished product.

Bostock claims that 190b10–23 implies that the subject persists through the change even in cases of substantial generation. But this is an interpretative move on his part. At 190b10–23 Aristotle says that it is clear "from what has been said" that what comes into being is always complex. Now, what has just been said is that all cases of coming to be involve a subject from which the change proceeds (190b9–10). Bostock simply assumes that the concept of "underlying subject" already includes the idea of persistence. Thus, he reads 190b10 as a straightforward inference from the claim that the subject persists to the claim that what comes into being is therefore complex. Moreover, he claims to be unable to see any other reason for drawing such a conclusion: "If the element said to underlie did not persist in the end product there would be no ground whatever for saying that the end product was composite, and Aristotle explicitly claims that all products of becoming are composite." [27] But this is not the only way to read the passage. Aristotle had already made the point that what comes to be is complex back at 190a3–4, prior to introducing the idea that something might survive the change (190a17–21). So that point could hardly be said to

it will immediately develop into a new animal when nothing external impedes it (*GA* II.1, 735a12–26 and *GA* II.4, 735b33–740a13). This makes the primitive embryo the proper subject of animal generation. (This is confirmed by the fact that throughout *GA* Aristotle refers to the embryo, and not the menstrual blood, as τὸ γιγνόμενον, which in the *Physics* is synonymous with the underlying subject of change.) For a similar reading, see Gill 1989: 228–233.

[26] This is the sense of the Oxford translation. Cf. Bostock 2006: 8. [27] Bostock 2006: 9.

depend on that idea. Instead, Aristotle's grounds for claiming that what comes to be is complex seem to be linguistic: It is implicit in the way we speak, for example, "when we say the non-musical human becomes a musical human" (190a3–4). The bulk of the chapter is an attempt to unpack this implication with more precision, using his own distinction between subject and contrary. In light of this, "from what has been said" should be taken to refer to everything that precedes 190b10 (rather than just the immediately preceding lines, as Bostock assumes), so that the whole passage can simply be read as a reiteration of the chapter's central point.

Finally, while Aristotle does say at 190a23–26 that, in cases where a statue comes to be from bronze, the bronze survives the change, defenders of the Inclusive Reading take it for granted that he is using the example to make a point about substantial generation.[28] But there are good reasons for thinking he is not. First, Aristotle does not extend the argument of the chapter to substantial generation until 190b1. Prior to that, he seems to be speaking only about changes in the non-substantial categories. Second, and more importantly, there is no evidence that Aristotle actually thinks of this as a case of genuine substantial generation. The only clear examples of substantial generation in *Physics* 1.7 are cases of biological generation (e.g. when an animal or plant comes to be from *sperma*, 190b4). And in T2 (*GC* 1.4, 319b8–21) the case of a statue coming to be from bronze is explicitly treated as an alteration of the bronze, which Aristotle contrasts with coming to be in the unqualified sense.[29] At the very least *Physics* 190a23–26 should be read with caution. For Aristotle elsewhere says that we should not speak of a statue coming to be from bronze without qualification, because that from which the change begins does not persist (οὐχ ὑπομένοντος) but gets transformed (μεταβάλλοντος) in the process of becoming a statue (*Metaph.* VII.7, 1033a20–23).[30]

The evidence from *Physics* 1.7 is thus not strong enough to force the Inclusive Reading on us. In light of this I think Broadie (2004: 129–130)

[28] For instance, Code 1976.

[29] Code 1976: 357–358, 365 claims that in *GC* Aristotle thinks we can describe the same change as either an alteration or substantial generation. This cannot be right. The central aim of *GC* 1.1–5 is to establish that substantial generation exists. And the way Aristotle does this is by showing that it is distinct from and not reducible to changes in the non-substantial categories.

[30] For an interpretation of this passage that is consistent with the Inclusive Reading, see Code 1976: 363ff.

has it right when she says that the model of substantial generation on offer in *Physics* 1.7 is at best undeveloped with respect to the Survival Requirement. While Aristotle is concerned to show that substantial generation must proceed from an existing subject (190b1–5) – a claim that forms the basis of his response to the Eleatic challenge in *Physics* 1.8 (191a30–31) – he is not focused on the further issue of whether or not this subject also persists through the change. If this is right, then *Physics* 1.7 is not incompatible with *GC*, as it would be on the Inclusive Reading. For Aristotle did not mean to commit himself in the *Physics* to any particular view concerning the persistence of the subject of generation. When it came to substantial generation Aristotle's primary concern was simply to establish that substantial generation exists as a genuine type of change (cf. *GC* 1.1, 314a4–6; 1.2, 315a26–28). This only required defending the Subject Requirement (*Phys.* 1.7, 190b1–4). On the one hand, Aristotle must show that there are genuine cases of substantial generation where something new comes into being without committing himself to the possibility of generation *ex nihilo*.[31] He meets this challenge by arguing that even in the case of substantial generation the change proceeds from some pre-existing subject (*Phys.* 1.7, 190b1–4), and that this subject is the potential F (*GC* 1.3, 317b14–25). This solution to Parmenides' challenge does not depend on the further claim that this subject also persists through the change to become a constituent of the finished product.[32] On the other hand, Aristotle needs to show that when a new substance comes into being it is not reducible to a change in one of the non-substantial categories (*GC* 1.1, 314a4–6). And this, we have seen, required him to reject the Survival Requirement.

[31] Aristotle agrees with Parmenides that generation *ex nihilo* is impossible.

[32] Some argue that Aristotle must take this subject to be persisting through the change in order to avoid the idea that substantial generation involves sheer replacement (Gill 1989: 7; cf. Code 1995). If there were no persisting subject, the argument goes, then substantial generation would be a case where the pre-existing subject perishes into nothing and is replaced by a product that emerges out of nothing, which violates Eleatic principles that Aristotle accepts. I find this argument for the Survival Requirement unmotivated by the text. Even if there is a philosophical problem here (to me this sounds like a pseudo-problem), there is no suggestion that Aristotle was concerned with it in the works on natural science, let alone that he developed the Survival Requirement in response to it. Nor does positing a persistent subject solve that problem (cf. Ebrey unpublished). It is also worth noting that in *GC* 1.4 Aristotle actually characterizes substantial generation in a way that sounds like sheer replacement. In T2 he says that with substantial generation the thing "is transformed as a whole" so that "nothing perceptible remains as its subject." "Such an event," he says, "is the coming to be of one sort of thing and the passing away of the other."

If the arguments of this chapter are reasonable, then we should be cautious about attributing to Aristotle in *Physics* 1.5–7 the view that substantial generation involves a persistent subject. Not only is there a shortage of conclusive evidence in those chapters for the Inclusive Reading, the bulk of the evidence outside *Physics* 1 shows that Aristotle had little reason to endorse it and every reason to reject it.[33]

[33] *Phys.* 1.9, 192a13–14 remains problematic for this reading. There Aristotle uses "what persists" as shorthand for matter. But this may just be a simplification of the *Physics* 1.7 doctrine that ignores the difference between qualified and unqualified generation. As we have seen, when he is careful to distinguish the two, he does so by reference to the survival requirement (cf. *Metaph.* 1.3, 983b7–19).

CHAPTER 9

A dynamic ontology: On how Aristotle arrived at the conclusion that eternal change accomplishes ousia

Diana Quarantotto

Nature is difficult. As Aristotle reminds us at the opening of *Physics* 1, the principles of nature are obscure to us, at least at the beginning, although they are clearer in their own right. So, Aristotle's preferred methodology in approaching the study of nature is to start with what is more familiar to us and to move from there to what makes more sense in its own right. Aristotle's use of the craft analogy is a well-known instance of this methodology. He begins with a miscellany of craft-related examples to elucidate the distinction between the four causes before moving to its application to nature, where the four causes are much trickier to disentangle.

This chapter sets out to show that Aristotle applies this methodology also to a fundamental aspect of the distinction between being and change. Broadly, Aristotle starts his physical inquiry by relying on a clear-cut distinction between being and change and by giving priority to being over change. Going on, approaching more and more nature's inner functioning and deep structure, and considering things from a different, more fundamental and more revealing perspective, Aristotle arrives at the conclusion that the substantial being (*ousia*) of natural entities is accomplished through a particular kind of change (eternal change), and hence that being in its primary sense is itself dynamic.[1] This is a major innovation by Aristotle within the Greek tradition of natural philosophy and of its treatment of the relation between being and change. Aristotle's *Physics* provides a long and important segment of the investigatory path that brought Aristotle to this result.

[1] While reconstructing Aristotle's research on this issue, I shall clarify and define progressively what I mean by '*ousia* accomplished through eternal change', and what the interpretation proposed here is an alternative to.

The topic proposed here is not easy to tackle or even to frame. As we shall see, the thesis that being is accomplished through change is formulated along a research path and from a point of view that, albeit fundamental, are both rare and not fully articulated by Aristotle. However, there is enough evidence in Aristotle's writings to go on, and the matter is too important because of its implications just to be left aside. Basically, the issue is intimately connected to how the prime mover can be both the first principle of being and the first principle of eternal change for natural entities. Of course, my remarks here do not amount to a full treatment of the topic. Take them as early steps along a road that has been little travelled,[2] but which may make a big difference.

I The initial evidence: The road less travelled by

Our starting-point will be *GC* II.10, 336b26–337a7. This is one of the rare passages where Aristotle offers an overview of the natural world, looking at it, as he says elsewhere, 'from above' (*GA* II.1, 731b23). This means looking at it from the point from which 'nature and the heavens are suspended' (*Metaph.* XII.7, 1072b14), i.e. from the perspective of the prime mover (PM), and to follow the path that proceeds from this principle towards the different levels of the hierarchy of the natural world.[3]

> [The continuity of generation and corruption] happens with a good reason. For we say that nature, in all things, always desires what is better, and that being is better than not-being (it has been said elsewhere how many senses there are in which we say 'being'), but it is impossible for this [i.e. being] to belong to all things, since they are far removed from the principle. Therefore god filled up the whole in the way that was left, by making generation unceasing. For this was the way to connect being together as much as possible, since coming-to-be eternally and generation are the nearest things there are to *ousia*. The cause of this, as has frequently been said, is circular locomotion, since this alone is continuous. That is why also the other things that change into each other in respect of their affections and powers, as do the simple bodies, imitate circular locomotion. For when air is generated from water and fire from air, and in turn water from fire, we say that the generation has come round in a circle, because it has returned again to the beginning. So also rectilinear locomotion, by imitating circular locomotion, is continuous.

[2] To my knowledge, the only scholar who has addressed this area is Michael Frede (2000: 12–17). He claims that the relation between being a sensible substance and change is much more intimate than one may think, i.e. that the *ousia* of natural entities is identical with a certain pattern of change.

[3] On the distinction between the path towards the principle and the path from the principle, see *NE* I.2, 1095a30–b4.

The immediate goal of the passage is to provide a teleological explanation of the continuity and eternity of natural generation. To do this Aristotle sets natural generation in the wider net of the relations that link it to celestial bodies and to God. Aristotle describes the world as a hierarchy ordered according to the teleological principle that being is better than not-being and according to the degree with which different kinds of natural entities manage to accomplish being. Being is conceived as eternal being and as *ousia*.

Aristotle claims that, although every natural entity strives for being, this cannot belong to every one of them. Among natural entities, being belongs to celestial bodies, which accomplish it through their circular locomotion, whereas it does not belong to sublunary entities. Yet sublunary natural entities, due to their striving for being, accomplish eternal generation, which is described as the nearest thing to *ousia*. This idea of closeness and of approximation to being and *ousia* is also expressed by means of the concept of *mimêsis*: eternal generation is a *mimêsis* of the process whereby celestial bodies accomplish being/*ousia*. Lastly, a divine principle is described as what first causes both celestial locomotion and sublunary generation to be eternal. Arguably, the likeliest candidate for this role is the first unmoved mover,[4] who is indeed both the first cause of being and the first cause of the eternal movements of natural bodies.[5]

The topic of this passage is also tackled in *GA* II.1, 731b18–732a3 and *DA* II.4, 415a23–b7. However, in this *On Generation and Corruption* passage there are at least two particular points that offer an unusual perspective on the matter.

First, Aristotle employs a single notion of 'being' (eternal being conceived as *ousia*) and provides a gradualist perspective on it. He adopts a point of view whereby, strictly speaking, only God and eternal natural entities have being/*ousia*, whereas perishable natural ones just manage to get as close as they can to this goal. Aristotle's purpose is to explain the way that being stems from a single divine source and reaches towards all natural substances.

Secondly, eternal change is presented as the means whereby eternal natural entities achieve being/*ousia* and perishable ones approximate to it. Here eternal change is considered not simply as what brings about *ousia*,

[4] Joachim 1922: ad loc.; Kahn 1985: 185; Williams 1982: ad loc.

[5] In this passage, Aristotle uses locutions that credit God with a providential action and that are, I think, rightly regarded as figurative (Kahn 1985: 185; Sedley 2007: 168). However, I shall claim that, this metaphorical language notwithstanding, the passage provides a true Aristotelian view on God's causal role.

viewed as a mere result of change, but, more intrinsically, as what constitutes it. Indeed, while being is conceived rather literally as 'full being' and applied primarily to God, coming-to-be eternally and generation are said to be 'the nearest things there are to *ousia*', what 'connects being together' (*suneiroito to einai*) and the means whereby God 'filled up the whole' (*suneplêrôse to holon*). In other words, sublunary generation is described as a process that blends parts of being and fills up gaps in order to accomplish full being.

II Some objections from the road more travelled by

The perspective offered by *On Generation and Corruption* II.10 clashes with some common ideas both about the relationship between being and change and about the relationship between the being of the prime mover and that of natural entities.[6] A possible formulation of these common ideas might be the following:[7]

The 'being-est' of beings is the PM, which is eternally unchanging. The PM is the first cause of being for everything. Moreover, it is also the first cause of the eternal change whereby natural entities imitate its perfect being. However, the being of celestial bodies lies not in their eternally *changing* but in their *eternality*. Similarly, the elements and natural corruptible substances 'aspire' to perfect being by constituting an *eternal*-process, not an eternal-*process*.

This interpretation ignores both the two points of the *On Generation and Corruption* II.10 passage that I have just highlighted: the gradualist perspective on being/*ousia*[8] and the idea that the *ousia* of natural entities is accomplished through eternal change. According to this interpretation, the reason natural entities are involved in an eternal change is to emulate God's perfect being, and not to accomplish their own being: having an *ousia* and striving for eternal being are independent or loosely connected matters. Secondly, this interpretation ascribes to Aristotle a universal application of the clear-cut distinction between being and change: it conceives of the

[6] Note, however, that the proposed interpretation of what Aristotle says in *On Generation and Corruption* II.10 is compatible with its other parallel passages (i.e. *GA* II.1, 731b18–732a3 and *DA* II.4, 415a23–b7). The only difference between them is in focus and emphasis.

[7] I picked up on these ideas reading the comments by the editor of this volume and one of the anonymous readers, whom I thank for this. The unanimity of these comments encourages me to think that this interpretation is *de facto* standard, although I have not found it explicitly formulated in the literature.

[8] Instead this interpretation recognizes a gradualist perspective on divine/perfect being.

eternity of natural entities as a mode of their being, which is distinct from the eternal change that accomplishes it.

Now, I shall suggest that this interpretation, despite its sounding Aristotelian, is simplistic and that the root cause of this is that it takes concepts that properly belong to the starting phase of Aristotle's physical inquiry (the path towards the principle) and tries to apply them to a later and more advanced phase, to which they do not pertain (the path from the principle).

Let us consider the first point. A gradualist perspective on being/*ousia* is rare in Aristotle's writings. However, this is not sufficient reason to dismiss it. Not only is it clearly documented by the *On Generation and Corruption* II.10 passage, but it is also not unique to it, as it is attested in other passages from Aristotle's writings,[9] and, as we shall see, it was endorsed also by some of Aristotle's predecessors. Moreover, there are reasons that account both for its rarity and for its importance within Aristotle's thought.

A gradualist perspective on being/*ousia* is rare along the research path that predominates in Aristotle's writings. This is the path that proceeds from what is better known to us and that corresponds to the starting phase of Aristotle's inquiry, which is focused on a local and domain-specific level of explanation. Indeed, while proceeding along this path, and for a long segment of it, we do not even know that there is a single, changeless, and immaterial principle 'from which nature and the heavens are suspended'. Therefore, this initial inquiry might give the impression that the immanent and local causality Aristotle focuses on can be made fully intelligible without any appeal to a universal cause, or that the PM's causal action concerns some extra aspects of the being of natural entities (for instance, their attempt to imitate the PM's perfect being, as the standard interpretation mentioned above claims). But this impression is misleading. For while reaching the first principle and while starting the path that proceeds from it, one gets a different view on the natural world. One learns that the PM's influence on the natural world is omnipresent. Moreover, along the path from the principle, a gradualist perspective on being/*ousia* is perfectly legitimate, and perhaps even necessary. Indeed, this path aims at showing both that and how all beings stem from one single being, i.e. both that and how being, in its primary sense, stems from a single source towards the natural world. The PM is the source of being for all natural entities, and

[9] Examples of gradualist perspectives in Aristotle: *Cael.* I.9, 279a23–30; II.12, 292b1–293a15; *Meteor.* IV.12, 389b29–390a9, esp. 390a4–7; *GC* I.3, 318b35–319a1; *Cat.* 5, 2b22–28; *Metaph.* II.1, 993b24–31. On the gradualist aspects of Aristotle's ontology, see Morrison 1985 and Rashed 2005: lxiii–lxxxv.

not only of their approaching its perfect being. Despite the rarity of his treatment of this issue, Aristotle does not give us reasons to think that the causal action of the PM on the natural world is twofold, or that it is limited to natural entities' approaching its perfection and therefore does not concern their own being – as the standard interpretation claims.[10] In other words, Aristotle does not give us reasons to think that the passages where he uses a gradualist perspective on being/*ousia* and describes natural entities' striving for eternal being concern something different from and further than their own being. Indeed, these are the rare passages where we find clues as to how the PM's causation of being is exerted.[11]

Similar points can also be made about the second aspect of the standard interpretation, i.e. the clear-cut distinction between being and change. As we shall see, a clear-cut distinction between being and change is proper to the path that proceeds from what is more familiar to us towards what makes more sense in its own right. But things change when one looks at the natural world from above. Indeed, the PM is the first cause of being for all natural entities and it is such insofar as it is the cause of their eternal change. Again, Aristotle does not give us reasons to think that the PM's causal action is twofold, i.e. that there is a distinction between the PM's causation of being and the PM's causation of eternal change.[12] There is simply no other way whereby the PM exerts its causal action on the natural world than by determining an eternal change. This is why the first principle is presented as a mover and indeed called the 'first mover'. So, insofar as the PM is concerned, the causation of being is causation of change. I suggest that the PM's causation of being is causation of change because natural entities accomplish being/*ousia* through eternal change.[13]

This idea may sound less awkward if one recalls that the being of the PM is not a state but a process, i.e. a living activity. So, change – or better, a

[10] Of course, this means neither that the PM is the only cause of the being of natural entities nor that their being is identical with God's being.

[11] It is known how scarce Aristotle's statements on this issue are. Even *Metaphysics* XII does not offer much detail. There, the divine substances are called 'primary substances' (1074b9). However, as Frede observes (2000: 50), Aristotle does not tackle the issue of the sense in which they are so and hence, more generally, the question of how these primary substances are causes of being for other substances.

[12] E.g. *Metaph.* XII.8, 1073a23–25, 1074b35–36.

[13] An alternative hypothesis may be the following: the PM causes eternal change directly and the being of natural entities indirectly. I do not see clear signs of this two-step causality in Aristotle's writings and believe that this hypothesis is undermined by all the passages where the PM is presented as the first being and the first source of being in a sense that seems stronger than the abovementioned one (besides *On Generation and Corruption* II.10, see, for instance, *Metaph.* XII.8, 1074b35–36 and *Cael.* I.9, 279a28–30).

particular kind of change (eternal change) – can be viewed as the means whereby natural entities get as close as they can to that particular kind of dynamic being that the PM both is and is the source of.[14] Indeed, this may be why in *Cael.* I.9 279a28–30 Aristotle presents the PM as the source from which 'is suspended *the being and life* for other things, for some in more distinct form, for others in a dim way', and why at the beginning of *Physics* VIII Aristotle describes eternal change as *a sort of life* for all things composed by nature.[15]

Naturally, it is always possible to make distinctions and to claim that there is a difference between an *eternal*-process and an eternal-*process*. However, I wonder whether this distinction is so important in Aristotle's work, especially where the path from the principle is concerned. I suggest, by contrast, that much can be learnt by emphasizing the conceptual shifts that the view from above engenders, among which the idea that being is accomplished through eternal change.

Of course, by claiming that the *ousia* of natural entities is accomplished through eternal change, I do not mean to deny that Aristotle also speaks of *ousia* in other senses. Rather, my proposal is that these different senses, or at least some of them, stem from and belong to different and progressive paths of Aristotle's investigation. Moreover, my proposal is that of emphasizing that only once the path from the principle is followed is Aristotle's inquiry on nature really completed. Indeed, this is, at least in part, the methodological and theoretical meaning of Aristotle's introduction of first philosophy and its distinction from physics, i.e. of Aristotle's conviction that first philosophy is the ultimate fulfilment of the human desire to know, and that the understanding of God, conceived as the principle of all things and as a non-physical entity, is necessary to fully understand the natural world.[16] This is especially important to recall, since each of these two paths provide a different view on the natural world. Indeed, grasping the cause of all things shows fundamental connections and relations that are invisible from the bottom up, and hence requires a modification of our understanding of the things that are caused by the first principle. Thereby

[14] It is true that the PM is of itself entirely changeless. However, this does not imply that a similar distinction between being and change, or even a similar extraneousness of being from change, holds also for natural entities (I tackle this issue in Quarantotto 2014). Note, moreover, that there is another reason the particular kind of relationship, suggested here, between the immutable eternity of the PM and the mutable one of natural entities may sound less awkward: it has a precedent in the Platonic description of time as a 'movable image of eternity' (*Tim.* 37c).

[15] According to *DA* II.4, 415b13, the being of living things is life: it is not a state but a particular kind of process. I tackled this issue in Quarantotto 2010.

[16] Frede 2000 and 2004.

the view from above determines a revision of (at least some of) the conceptual tools and frameworks that originated from, and were used in, the previous, more localised, perspective. The idea that *ousia* is accomplished through and realised in eternal change is among these revisions.[17]

Now, in what follows, I shall provide some independent evidence that Aristotle has taken the point of view from above seriously. I shall suggest that, by adopting this point of view and by claiming, as a result, that the *ousia* of natural entities is accomplished through eternal change, Aristotle inserts a novelty into the age-old debate on the relationship between being and change, and on the status of the being of natural entities. Moreover, this further evidence will help us understand the meaning of the *On Generation and Corruption* passage and reconstruct the research path that Aristotle followed to reach the result presented there.

III Some independent evidence on *ousia*, eternity, and eternal change

Within Aristotle's *Physics*, the topic of natural entities' *ousia* is dealt with for the first time in Book II, chapter 1. Book I provides the first steps of an investigation into the principles of natural science and only mentions the problem of *ousia* (*Phys.* I.7, 191a19–20). Instead, in *Physics* II.1, the issue of *ousia* is explicitly addressed (193a9–11, 16, 20, 25). Aristotle considers two possibilities: the *ousia* is either matter or form. The first hypothesis is attributed to some of Aristotle's predecessors and is argued for with their own arguments. According to these thinkers, only matter can be the *ousia* because only matter meets a certain requirement: to be eternal.

> Some identify the nature or substance of a natural object with that immediate constituent of it which taken by itself is without arrangement … As an indication of this Antiphon points out that if you planted a bed and the rotting wood acquired the power of sending up a shoot, it would not be a bed that would come up, but wood *which shows that the arrangement in accordance with the rules of the art is merely an accidental attribute, whereas the substance is the other, which, further, persists continuously through the process.* But if the material of each of these objects has itself the same relation to something

[17] I suspect that one of the causes of this conceptual revision is the shift, which results from taking up the view from above, from a perspective that focuses on finite entities and processes to a perspective that focuses both on eternal entities and processes and on the way in which finite entities and processes are part of something infinite. This is also suggested by a closely related topic, i.e. by the difficulties arising from the application of the same principles to both perishable and eternal sensible substances and by the enormous strain under which the notions of these principles are put (cf. Frede 2000: 17). I shall tackle this last issue in section VI.

> else ... that, they say, would be their nature and substance ... For whatever any one of them supposed to have this character ... this or these he declared to be the whole of substance, all else being its affections, states, or dispositions. *Every such thing they held to be eternal (for it could not pass into anything else), but other things to come into being and cease to be times without number*). (193a9–28)[18]

This passage shows that the idea that *ousia* is something eternal is an ancient and traditional one. Eternity is considered by the pre-Socratics both a distinguishing trait of *ousia* and a fundamental requirement something must fulfil to be an *ousia*. These thinkers are presented by Aristotle as holding that only what is eternal can be an *ousia*, and that therefore what changes, what is generable, and perishable, cannot be an *ousia*. This is why they ruled out the possibility of attributing the role of *ousia* to form.

Aristotle emphasises this point also in other passages. For instance, in *Metaph.* I.3, 983b6–18, he claims:[19]

> Of the first philosophers, most thought the principles which were of the nature of matter were the only principles of all things; that of which all things that are consist, and from which they first come to be, and into which they are finally resolved (*the substance remaining, but changing in its modifications*), this they say is the element and the principle of things, and therefore they think *nothing is either generated or destroyed, since this sort of entity is always conserved* ... So they say *nothing else comes to be or ceases to be; for there must be some entity – either one or more than one – from which all other things come to be, it being conserved.* (trans. Ross 1928)

The idea that *ousia* is something eternal, together with the associated contrast between being and becoming, had a lasting career: it was also endorsed by Plato, who in certain contexts conceives of being as what always is and never comes to be and contrasts it with what always comes to be and never is.[20]

> First then, in my judgment, we must make a distinction and ask: *what is that which always is and has no becoming; and what is that which is always becoming and never is?* That which is apprehended by intelligence and reason *is always in the same state*; but that which is conceived by opinion with the help of sensation and without reason *is always in a process of becoming and perishing and never really is.* (*Tim.* 27d–28a, trans. Jowett 1875)

[18] Translations are by Hardie and Gaye 1930, sometimes with small changes. Emphasis is always mine.

[19] Cf. also *Metaph.* I.3, 984a9–16 (esp. 983b10).

[20] Significantly, this is also the point of view adopted by Aristotle in *On Generation and Corruption* II.10: he identifies being with eternal being and claims that being cannot belong to all natural entities. However – breaking with this tradition – here Aristotle also claims that eternal generation is the nearest thing to *ousia*.

Aristotle accepts this position, but only to a certain extent. In *Phys.* II.1, 193a9–28 and 193b8–11 he accepts the criterion for *ousia* lying behind Antiphon's argument, but uses it to prove what Antiphon and other pre-Socratics denied: i.e. that form is (in a relevant sense) eternal, and so can be an *ousia.*[21]

> Again man is born from man but not bed from bed. That is why people say that the shape is not the nature of a bed, but the wood is – if the bed sprouted, not a bed but wood would come up. But even if the shape is art, then on the same principle the shape of man is his nature. For man is born from man. (193b8–12)

This shows that, according to Aristotle, form is an *ousia* (also) because it is eternal and that its eternity is accomplished by means of circular generation ('man is begotten by man').

In the first chapters of *Physics* II Aristotle uses the idea that 'man is begotten by man' also while dealing with the issue of the kind of separation that is characteristic of natural forms (194b12–13): '[The student of nature] is concerned only with things whose forms are separable indeed, but do not exist apart from matter, for man is begotten by man and by the sun as well.' In this connection, he criticises other kinds of separation, especially the one Plato attributed to forms:

> Thus on the second account of nature, it would be the shape or form (*not separable except in account*) of things which have in themselves a principle of motion. (193b3–5)

> The holders of the theory of forms do the same, though they are not aware of it; *for they separate the objects of natural science*, which are less separable than those of mathematics. (193b35–194a1)

From these passages and remarks it may be deduced that the point of Aristotle's criticism is to show that forms need not be separate in order to be eternal, since 'man is begotten by man'.[22] More generally, drawing the right consequences from the fact that 'man is begotten by man' is sufficient to prove – against the pre-Socratics – that form too is eternal, and – against Plato – that to ensure the eternity of forms it is not necessary to conceive of them as separate from matter and change.

[21] This move is prepared by Aristotle's insistence in *Physics* I on the requirement that any principle, as such, must be permanent and eternal (188a27–28, 192a25–34; *Metaph.* III.4, 1000b24–28; VI.1, 1026a16–17). This makes impossible the interpretation that Aristotle's acceptance of the pre-Socratics' criterion for *ousia* in *Physics* II.1 is simply dialectical.

[22] *Metaph.* I.6, 987a 29–b18; III.2, 997b6–12; VII.8, 1034a2–5.

In sum, *Physics* II.1 records that Aristotle's predecessors consider eternity a necessary requirement for something to be an *ousia*, that Aristotle himself shares this idea and that he nevertheless believes this requirement can be met without also endorsing (at least not completely) the contrast between being and becoming; in fact, Aristotle believes that a certain kind of becoming (the one picked out by the 'man is begotten by man' formula) is what enables forms to be eternal and hence to be *ousia*.

Now, this is very close to what Aristotle claims in *On Generation and Corruption* II.10: eternal generation is the means whereby perishable natural entities accomplish, as far as they can, eternal being and *ousia*.

Hence, *Physics* II provides independent evidence to support the proposed interpretation of *On Generation and Corruption* II.10. Moreover, *Physics* II shows that the idea expressed in *On Generation and Corruption* II.10 – that the *ousia* of natural entities is accomplished through eternal change – is already germinating in the starting phases of Aristotle's investigation into nature.[23] However, this idea still needs to be developed and clarified. We still need to know what change is and what kind of change is able to accomplish *ousia*. And, especially, we still need to understand how and in what sense change (even a special kind of change) accomplishes *ousia*.

IV The first steps of Aristotle's inquiry into change: Things more familiar to us

The idea that change accomplishes *ousia* is an enormous innovation by Aristotle in the traditional debate about the relation between being and becoming. There is even something paradoxical about it, since traditionally – e.g. according to the Heraclitean flux theory – what was contrasted with being and *ousia* was an eternal change.[24] Naturally, in the *Sophist*, Plato rejects the contrast between being and becoming, judging as absurd the hypothesis that *to pantelôs on* is without movement or life and putting forward an ample ontology in which being includes 'all changeless things and all changing things' (248e–249d). But Aristotle goes a step further than this. He claims that being/*ousia* not only includes change and all changing

[23] Note, moreover, that in *Phys.* I.7, 190b17–20 Aristotle speaks of the principles of change and of being without making any clear-cut distinction.

[24] On the flux theory, cf. Plato *Theaet.* 152e2–8, 160d6–8; *Crat.* 402b1–5. Of course, the eternal change that, according to Aristotle, accomplishes *ousia* is different from the eternal change of the flux theory (*Phys.* VIII.3, 253b6–254a1, 8, 265a2–12).

things, but also that the being/*ousia* of natural entities is accomplished through eternal change.

Aristotle introduces and develops this innovation gradually. Here I shall highlight the main steps of the path that, within the *Physics*, brought him from the traditional idea that change is something highly problematic and close to not being to the idea that it is the means whereby being, in its primary sense, is accomplished. I shall claim that, in the starting phases of his investigation (especially in *Physics* I and III), Aristotle employs the ordinary perspective – also adopted by his predecessors – which distinguishes being from change, and gives priority to being over change. This perspective highlights the problematic nature of change and makes it appear close to not-being. Later, in a more advanced phase of his inquiry, Aristotle introduces a different perspective. In *Physics* VIII, the concept of 'change' is extended and put under strain: Aristotle presents a kind of change whose distinguishing features seem antithetical with those whereby he defined change in *Physics* III.1–3: e.g. the former is said to be something complete (*teleion*),[25] whereas the latter is described as incomplete (*ateles*).[26] The change introduced in *Physics* VIII is circular locomotion, which is one of the changes that, according to *On Generation and Corruption* II.10, accomplishes *ousia* and which is introduced, tellingly, at the beginning of *Physics* VIII as 'a sort of life, as it were, to all naturally constituted things'.

Let's start with *Physics* I. In this book, Aristotle deals with the first principles of natural entities and addresses some of his predecessors' theories, including the most radical and influential one: Eleatic monism and immobilism. He criticises both the Eleatic theory that being is one and changeless, and one of its main assumptions: the idea that being has only one meaning (186a22–25). In Aristotle's view, this assumption was the main cause that brought both the Eleatics to their conclusions and other philosophers to unsatisfactory accounts of change (185b31–32, 186b35–187a10).

In *Physics* I.2–3, Aristotle challenges the Eleatic theory by setting it against the thesis that 'being' and 'one' have many different senses and hence that being is not one in the way they think it is (185b25–186a3). Thereby the issue of change is dealt with on the basis of a theory of being and of its meanings. This becomes particularly clear in chapter 8, where the problem of change is directly addressed. Here Aristotle presents the Eleatic dilemma used to deny the possibility and intelligibility of change. Change

[25] *Phys.* VIII.8, 264b27–28; VIII.9, 265a13–27 (esp. 17, 23).
[26] *Phys.* III.2, 201b32; VIII.5, 257b8–9; *Metaph.* IX.6, 1048b28.

is denied because a changing thing should meet two requirements that seem to be incompatible: what changes must come from something that both underlies (i.e. from being) and is not already present (i.e. from non-being) (191a27–31). So, a changing thing appears to be impossible, because it must both be and not be, or it must come to be neither from being nor from not-being. The main tool Aristotle uses to solve this dilemma is his innovative idea that being has many senses: what changes, changes both from what is, but not insofar as it is, and from what is not, but not insofar as it is not (191b13–15, 17–18). So a changing thing can both be and not be without there being any contradiction and without violating the principle of excluded middle.

The inquiry into change continues in *Physics* II, where Aristotle focuses on teleological processes and, as we have seen, hints at the idea that eternal change is the means whereby natural entities accomplish *ousia*. However, it is only in Book III that Aristotle finally provides a definition of change. In *Physics* III.1 Aristotle defines change as the 'actuality [*entelecheia*] of that which is potentially as such' (201a10–11) and then rephrases this definition in various ways. These different formulations record the difficulty of defining change and of grasping its way of being. This is particularly clear with '*energeia ateles*' (201b31–33), which – especially when *energeia* is replaced by *entelecheia* (257b8–9) – is a genuine oxymoron, at least on the level of the expression: it says that change is an incomplete completeness, i.e. that it both has and does not have the *telos* (*en̲telecheia a̲teles*).

In Aristotle's view, the problematic nature of change stems from the fact that 'it can be classed neither in the *dunamis* of things that are nor in the *energeia*' (201b28–29): again, also from this point of view, change seems to violate the principle of excluded middle. Because of this, change is something indeterminate (*aoriston*). This is why some of Aristotle's predecessors defined it in privative terms as difference, inequality, or not being (201b20–26). Aristotle's solution is to characterise change neither as potentiality (*dunamis*) nor as actuality *simpliciter* (*energeia haplê*), but as a sort of actuality (*energeia tis*), i.e. as an incomplete actuality (*energeia ateles*) (201b31–33).[27]

From a terminological point of view, Aristotle's account of change is built around the notion of goal (*telos*/*ergon*): change is defined according to its relation with the *telos*. Change is something insofar as it is an *entelecheia*/*energeia*; change lacks something insofar as it is *ateles*. What brought Aristotle's predecessors to define change in terms of privation (i.e. the

[27] Cf. also *Metaph.* IX.6, 1048b28; *EN* X.3, 1174b2–5.

indeterminate nature of change) is conceived by Aristotle in terms of incompleteness: the lack of the *telos*.

Change is incomplete – says Aristotle – because the potential being it is the *entelecheia* of is itself incomplete (201b31–33). The potential being is incomplete because it lacks the *telos* towards which change is directed, and change inherits this privative nature. Hence, change is characterised by a lack, i.e. by the fact that it does not have the *telos* towards which it is directed. It is a being that comprises not-being within itself.

This is difficult to grasp, but not impossible to be (202a2–3). So Aristotle comments on the result reached in this phase of his investigation into change and its relationship to being.

Now, what is the scope of the account of change provided in *Physics* III.1–3? This is a difficult question, but it is necessary to tackle it, especially if we want to understand how eternal change is able to accomplish *ousia*. Scholars tend to see it as a sign of correctness (or, at least, as a strength) for an interpretation of the definition of change provided in *Physics* III.1–2 if it picks out both finite and infinite changes. Instead, here I shall explore another possibility. In my view, there are reasons to think that the definition presented in *Physics* III.1–2 is modelled *primarily* on finite changes,[28] since they are what is more familiar to us and are therefore the first objects of Aristotle's scientific inquiry into change.[29]

This is suggested, first of all, by the way the investigation into change is set up from its very beginning in *Physics* I. Here Aristotle describes change as something that occurs between opposites (188a19, 188b21–26). This description suits only finite processes. For, as we learn from *Physics* VIII,[30] changes that occur between opposites cannot be continuous and eternal.

The same point of view is taken up in *Physics* III.1–3. This inquiry is introduced both by a reference to the oppositions characteristic of change (201a3–9) and by a distinction between what is only in actuality and what is both in potentiality and in actuality (200b26–27). What is both in potentiality and in actuality is then identified with the *kineton* (201a19–25). This suggests that the change Aristotle is here focusing on is

[28] Cf. *Metaph.* III.4, 999b9–10: 'no change is infinite, but every change has an end'.

[29] This, however, does not imply that the definition given in *Physics* III.1–2 cannot be adapted subsequently to eternal motions. Indeed, it can be adapted since eternal motions can be considered also as combinations of finite parts (cf. section VI). The question whether and in what sense eternal motion is a *kinêsis* is closely connected with the question of whether it is or it is not an *energeia* (in the sense of *Metaphysics* IX.6). On this latter issue, see my (Quarantotto 2014).

[30] *Phys.* VIII.8, 261a28–b3, 2, 252b7–12, 252b28–253a2, 7, 261a29–b3.

temporally limited: a *kineton* is conceived as something that does not always have a certain attribute (i.e. it is not only and always in actuality a certain thing), but, if it does not have it at t_1, it has the potentiality to acquire it and to have it at t_2.[31]

The same conclusion can also be drawn from the fact that, in *Physics* III, change is characterised as incomplete, i.e. in terms of its lacking the *telos*/end-point towards which it proceeds.[32] Indeed, in *Phys.* VIII.9, 265a20–21 Aristotle explicitly links the notion of finiteness to that of incompleteness, by claiming that a finite change is incomplete.

Note, moreover, that the notion of 'infinite' (and the questions whether it is something or not, and what it is) is introduced and discussed only after the conclusion of the inquiry into change provided in *Physics* III.1–3, i.e. in chapters 4–8 of this book. Also, the very possibility and existence of eternal change are demonstrated not before Book VIII. By contrast, in Book VIII, finite change is considered evident and used as a starting-point of the investigation. At the end of this inquiry eternal change is then described with properties that are antithetical to those ascribed both to intermittent change and to change as defined in III.1–3: eternal change does not occur between opposites and is complete (264b27–28).

Another clue in this direction is given by the way in which the definition of *kinêsis* provided in Book III is used in Book VIII. In *Phys.* VIII.1, 251a8–16, from this definition Aristotle infers that 'there must be something burnable *before* it is burnt and something able to burn things *before* it burns them', i.e. that the subject of change that this definition picks out is something that is not always in motion. Now, in *Phys.* VIII.7, 261a31ff. Aristotle makes the following claims: 1) this kind of subject of change pertains to change insofar as it occurs between opposites; 2) not all changes occur between opposites; 3) eternal change does not occur between opposites and this is why it is eternal. In other words, in this passage Aristotle claims that the properties of change and of what changes he

[31] Cf. also *Phys.* VIII.1, 251a8–16; III.1, 201b5–11.

[32] *Phys.* III.2, 201b31–32. Scholars generally maintain that the interpretation that allows the definition of *kinêsis* given in *Physics* III.1–3 to pick out both finite and infinite changes is the one according to which change is the actuality of a potentiality for changing. Instead, I believe that this interpretation either does not explain why change is incomplete or, if it does, it does not allow the definition of *kinêsis* to pick out infinite changes. Arguably, the only way for this interpretation to account for the incompleteness of change is to claim that change does not accomplish its final goal (e.g. a house = $goal_1$), but only the one that is the actuality of a potential for changing (e.g. the construction of a house = $goal_2$). Now, if this is why change is incomplete, then the definition does not pick out infinite processes. For, on Aristotle's view, only circular processes are infinite, and circular processes are said to be complete in the sense that they always reach their goal in each of their points, i.e. they always reach the goal that, in this interpretation, *kinêsis* does not reach ($goal_1$).

deduced in *Phys.* VIII.1, 251a8–16 from the definition of change provided in Book III do not belong to eternal change or to what is in eternal motion.

If this is true – if the definition of change given in *Physics* III.1–2 is provided to account, at least primarily, for finite processes – then change thus defined is not a good candidate for being the means whereby *ousia* is accomplished. Because of its relationship with what is not, it cannot be eternal and complete. Put more carefully, because of its occurring between opposites, change must come to a stop once it has reached its *telos*.[33] Using the language of *On Generation and Corruption* II.10, we may say that change thus defined cannot accomplish *ousia* because it does not 'connect being together'. However, this is not surprising, given Aristotle's progressive research. Indeed, in *Physics* VIII Aristotle not only – as just noted – uses the concept of change introduced in *Physics* I and III to tackle some arguments against the possibility of an eternal motion, but he also sets up to demonstrate the existence of an eternal motion. Moreover, this eternal motion is identified with the circular motion of celestial bodies and, unlike the *kinêsis* dealt with in *Physics* III.1–3, it is described as complete (*teleion*).[34] So it seems that Aristotle's inquiry proceeds by first focusing on finite changes (which are more familiar to us) and then going on to consider infinite changes (which are less familiar to us, but arguably make more sense in their own right). Put differently, Aristotle moves from one kind of change, which cannot accomplish *ousia,* to another kind of change, which has exactly the features needed to accomplish *ousia*: eternity and completeness.[35]

Let us see then what circular locomotion is and whether Aristotle's account of it in *Physics* VIII gives us some clues to understanding how and in what sense eternal change accomplishes *ousia.*

V Eternal changes accomplishing *ousia* (circular locomotion): Things less familiar to us

In *Physics* VIII the existence of an eternal change is the object of a demonstration. Aristotle's inquiry begins with the question *ei esti* (is there eternal change?), and then, once it is established that there is eternal change, it continues with the question *ti esti* (what is eternal change?).

[33] *Cael.* I.8, 277a14–23; *Phys.* VIII.7, 261a31–b15, 8, 261b31–36.

[34] *Phys.* VIII.8, 264b27–28, 9, 265a17; *Cael* I.2, 269a18–21. *Phys.* VIII.9, 265a17 suggests that also rectilinear motion can have a degree of completeness, which it possesses insofar as it doubles back (*Phys.* VIII.9, 265a20–22).

[35] Eternal motion is also complete: *Phys.* VIII.8, 264b9–28, 9, 265a13–22; *Metaph.* XII.6, 1071b10–11.

Unlike intermittent change (i.e. change of things that are sometimes in motion and sometimes at rest, and which occurs between opposites), eternal change (i.e. the change of one single thing that is always in motion) is not considered evident, i.e. something we can observe in ordinary experience and take for granted. Moreover, both the existence and structure of eternal change are studied starting respectively from the existence and from the structure of the intermittent change: here, too, Aristotle proceeds from what is more familiar to us towards what makes more sense in its own right.

Insofar as the structure of eternal change is concerned, Aristotle defines it by first showing why motion between opposites cannot be eternal and what 'modification' it must undergo to be so. He claims that motion between opposites cannot be continuous and eternal but only consecutive. This is because what moves from opposite to opposite with a rectilinear motion, once it has reached the *terminus ad quem*, if it is to go on, must move on a straight path at the same time from one point towards the same point. So this kind of change cannot be continuous, but there will be a time interval between changes. In sum, 'motion that doubles back along a finite line is a composite motion consisting of two parts, while if it does not double back, it is *incomplete and perishable*' (265a20–22).

To be continuous and everlasting, a motion must be such that what moves makes its movement at the same time from one point towards the same point (264b10–11).[36] This is not possible for rectilinear motion, but it is possible for circular motion, i.e. it is possible if the beginning and end, which in rectilinear motion are distinct and opposite, come to coincide and become the same point, like bending a straight line so that its two ends meet: the continuity and eternity of circular motion are conceived in terms of the overlapping and coincidence of what is instead distinct and opposite within straight and finite motion (264b27).

This is also why circular motion, unlike rectilinear motion, is complete (264b27–28). Circular motion is complete since it has an end (*telos*), but no limit (*peras*), i.e. since, unlike straight motion, each of its points is both a beginning (*archê*) and an end (*telos*).[37] Each point is an end, since each is where the movement is completed; and each point is a beginning, since each is where the movement 'starts' again and again. So, to have an end without being limited means reaching completion without coming to a stop. Whereas rectilinear motion is incomplete, since it does not have the

[36] The possibility of an infinite rectilinear motion is ruled out in *Phys.* VIII.9, 265a17–20.

[37] *Phys.* VIII.8, 264b9–28, 9, 265a27–265b1; *Cael.* I.9, 279b2–3.

telos towards which it proceeds, circular motion is complete, since it always reaches this *telos* in each of its points.

VI The problem: How can eternal change accomplish *ousia*?

Do the structure and completeness of eternal motion enable it to accomplish *ousia*? How, and in what sense? What does it mean for a process to accomplish *ousia*? Note that *Physics* VIII does not explicitly aim at answering these questions. The book deals more generally with the existence and structure of eternal change and with its cause. However, given that Aristotle has never fully articulated the view from above, this book – together with some passages from *Cael.* – is our only source of information on this topic. Let us then attempt to provide an answer. We shall see that this inquiry will give us clues to understand better how and in what sense eternal change 'connects being together' and how it is the means whereby God 'filled up the whole', as the *On Generation and Corruption* II.10 passage claims.

Remember that it is mainly the relationship of change to not-being that determined the contrast between being and becoming as formulated by Aristotle's predecessors.[38] Moreover, in Aristotle's account, this very same relationship prevents *kinêsis*, as defined in *Physics* III.1–2, from being complete and everlasting, i.e. from having the features whereby the processes that accomplish being/*ousia* are described in *On Generation and Corruption* II.10. Now, insofar as *kinêsis* (in the sense of incomplete and finite change) is concerned, not-being is given by the opposition between beginning and end: one of these points is the privation of the other. So it seems that, in order to be eternal and complete, a process must 'neutralise' the not-being that is present in finite changes: again, borrowing the language from the *On Generation and Corruption* II.10 passage, it must 'connect being together' and 'fill up the whole'. And, indeed, this is what happens with circular motion: in circular motion the opposition between the starting and the end points are neutralised because each point is both a beginning and an end. This, however, does not mean that there is no opposition at all between these points. If that were so, there would be a state, not a process. So we probably need to distinguish between the neutralization and the complete nullification of the opposition. With this distinction in mind we may say that insofar as circular motion is

[38] Cf. also *Phys.* I.9, 192a15, where Aristotle speaks of the 'destructive power' inherent in change.

concerned, opposition is in a sense absent and in another sense present, i.e. it is not completely nullified but disabled by the circularity of the motion.

So here is my proposal: for a process to accomplish *ousia* it needs to neutralise (without nullifying) the not-being, which an incomplete and finite change instead includes; and this is what the coincidence of beginning and end, characteristic of circular motion, amounts to; moreover, this is what enables a process to go on eternally.

This proposal has several advantages. As we shall see below, the idea that, in a circular process, the opposition between the starting point and the end point (and hence the not-being) is neutralised is suggested by some crucial passages of Aristotle's description of circular motion. In particular, this idea offers a way to explain some problematic and, apparently, conflicting aspects of this description. Moreover, the proposed interpretation is a good fit with the picture of the world's structure offered in *On Generation and Corruption* II.10: it enables us to deepen our understanding of the way whereby being/*ousia* stems from its primary divine source to the different levels of the natural world's hierarchy, as it gives celestial motion a mediating role in this causal chain.

Let us start with the textual evidence. The kind of neutralization I am suggesting is particularly clear in Aristotle's description of circular motion in *Phys.* VIII.9, 265a27–b1. Here he claims that, whereas in rectilinear motion the beginning, middle, and end points are defined (*horistai*), in circular motion they are undefined (*aorista*), since each is at once a beginning, a middle, and an end. This shows that there is a difference between these points, but this difference is neutralised by their ability to interchange their roles within the movement.

The hypothesis just sketched may explain why Aristotle gives apparently conflicting descriptions of circular motion, saying sometimes that it occurs between opposites and sometimes that it does not. As we have just seen, Aristotle claims that, insofar as circular locomotion is concerned, what moves from point A moves at the same time and in virtue of the same direction towards point A, without undergoing contrary movements. By contrast, in other passages, circular motion is said to occur between points that are either opposites to or, at least, different from each other. In *Metaph.* VIII.8, 1050b20–22, for instance, we are told that what moves circularly is endowed with a potentiality of being and not-being in respect of place. Similarly, in *Metaph.* XII.7, 1072b3–10, Aristotle claims that what moves circularly, insofar as it moves, can be otherwise with respect to place, unlike the first unmoved mover, which cannot be otherwise *simpliciter*. In *Cael.* I.8, 277a24–26, Aristotle says that circular locomotion, just like rectilinear locomotion, has in a way the opposites. Lastly, the matter that celestial bodies are made out of is called *pothen poi hulê* ('from one place to another'), since it is potential being with respect to place,[39] and this potentiality is called by Aristotle 'potentiality for opposites'.[40]

One way to account for this discrepancy is to attribute it to Aristotle's use of two points of view on circular locomotion. This hypothesis is suggested especially by *Phys.* VIII.8, 264b24–28 and *Cael.* I.8, 277a23–26. Here the double nature of circular locomotion seems to stem from the possibility of viewing it both as an infinite whole and as a combination of finite parts. If one considers just a segment of the process, a body that moves circularly moves between points that are opposites to each other. Instead, if one considers the whole process, i.e. if one connects the end with the beginning (*Phys.* VIII.8, 264b27–28), the opposition disappears: the circularity of the process neutralises, as it were, the opposition between being and not being with respect to place.[41]

The explanation for why Aristotle considers circular locomotion from these two perspectives may be found in the difference between the completeness of circular motion and that of a state or that of the PM's activity.

The completeness of circular motion is different from the completeness of a state (e.g. being pale in actuality).[42] The completeness of a state is – let us say – simple, whereas the completeness of a process is complex, as it involves two elements: according to Aristotle, every change is from one point to another point.[43] If the opposition between these points is kept, the process is incomplete and finite. Instead, if they are neutralised without being nullified, the process is complete and everlasting without turning into a state. Indeed, if the opposition were nullified, there would not be any difference between a complete process and a complete state.

Similarly, the completeness of circular motion is different from the completeness of the PM's activity. Whatever it is, this activity is not a state, but shares the simplicity of a state: the PM is described as a point-like

[39] *Metaph.* IX.8, 1050b20–22; XII.2, 1069b24–26. [40] *Metaph.* IX.8, 1050b8–9.

[41] This point is connected with the 10th *aporia* of *Metaphysics* III: whether perishable and eternal substances have the same principles or not. In the light of the interpretation advanced so far, I suggest that eternal substances have principles that, from a certain point of view, are the same as those of perishable substances (i.e. three principles: two contraries and matter), whereas, from another point of view, they are different (i.e. one principle: since the contrariety is neutralised, the three principles become one). On the hypothesis that celestial bodies, unlike perishable ones, have just one principle, cf. Fazzo 2013.

[42] *Phys.* III.1, 201b5–13. [43] *Phys.* VIII.2, 252b10; *Cael.* II.12, 292b6–7.

entity that lacks any kind of duality (*Cael.* II.12, 292b4–7).[44] Now, unlike the PM's activity, circular motion entails a duality, but this duality is neutralised (without being nullified) by the circularity of the process, and this is what turns it into something complete.

So, circular motion seems to have a double nature. Aristotle says that what moves circularly 'both moves and is at rest' (265b1–2). What moves circularly moves, since, while rotating, its parts occupy opposite points. What moves circularly is at rest, since the opposition between these points is neutralized by the circularity of the process.

The fact that circular motion has a double nature may stem from its being the intermediate element between the PM and the sublunary world, i.e. from its having a mediating role in the 'transmission' of being/*ousia* from the first immaterial point-like unmoved mover to the material complexity of sublunary natural entities. Indeed, circular motion has this intermediate position both spatially and functionally: it is the first natural member of the causal chain that stems from the PM, and so it is both what is moved and what moves (*Metaph.* XII.7, 1072a24). Moreover, it is what ultimately brings, and accounts for, the opposition present in the sublunary world (*Phys.* VIII.6, 260a1–5). But how does it do so? I suggest that it does so because, unlike the PM, circular motion is a change, i.e. something that entails a duality or an opposition and so partakes in the nature of change; however, unlike sublunary changes, circular motion neutralizes this duality and opposition and so partakes in the simplicity of the PM. This is why Aristotle calls it the 'first motion'.[45]

VI Eternal changes accomplishing *ousia* (circular generation): Things less familiar to us

According to *On Generation and Corruption* II.10, sublunary natural entities accomplish being/*ousia* through circular generation. Actually, through circular generation sublunary natural entities manage to get as close as they can to being/*ousia*: they do not return upon themselves so that the same individual comes to be a second time, but what recurs is the same only in species (*GC* II.11, 338b6–19).

Circular generation achieves this result because it is an imitation (*mimêsis*) of circular locomotion, i.e. it is similar to but not identical

[44] On the similarities and differences between the PM's activity and that of natural entities, see my (Quarantotto 2014).

[45] *Phys.* VIII.9, 265a13–27, b10; *Metaph.* XII.7, 1072b8–9.

with circular locomotion.[46] Circular locomotion and circular generation are similar insofar as both are circular, everlasting, and complete processes. Their differences are due to the different matter that celestial and sublunary bodies are respectively composed of.[47] Celestial bodies accomplish being/*ousia* through circular locomotion, since they are composed of a matter that is non-generable and imperishable, and whose nature is to move circularly and eternally.[48] Instead, sublunary matter is not one single element endowed with a circular and eternal locomotion. Rather, it is a set of four different elements, each of which has its own movement that is not circular and eternal, but rectilinear and finite (*Cael.* I.2, 269a25–27). Therefore, insofar as they are endowed with rectilinear and finite movements, sublunary elements are numerically generable and corruptible, and are such that every compound of them shares their same destiny of numerical generability and corruptibility. This means that, differently from celestial locomotion, sublunary generation is a process that, albeit unitary, comprises a multitude of rectilinear and finite movements as its parts.[49]

This suggests a further implication: sublunary generation is prior to all other sublunary changes,[50] as celestial locomotion is prior to sublunary generation.[51] Indeed, according to the *On Generation and Corruption* II.10 passage, circular generation is the sublunary process that is directly involved in the top-down causal chain of the world stemming directly from the PM. Because of this, circular generation is prior to all other sublunary changes not only from an ontological point of view, since it is the process whereby the *ousia* of sublunary natural entities is accomplished, but also structurally, since it is a complete process, i.e. it is the whole of which all other sublunary changes are parts, and hence the whole without which either of them could occur. Indeed, every one of the partial changes comprised by circular generation is a rectilinear, finite, and hence incomplete process (i.e. a process that occurs between opposite: being

46 *GC* II.10, 337a1–7.

47 An interesting passage about hierarchical systems where entities belonging to different levels reach or approach a certain goal through different means is *Cael.* II.12, 292a22–b19; on this, see my (Quarantotto 2014).

48 *Metaph.* XII.2, 1069b24–26; *Cael.* I.2, 269a2–9.

49 This does not imply that celestial motions have no internal complexity (cf. *Cael.* II.12, 292a22–b19).

50 Spontaneous generation would be an exception.

51 This is why Aristotle calls eternal locomotion the 'first *kinêsis*' (cf. note 45 above) and celestial matter the 'first element', the 'first body', and the 'first substance' (*Cael.* III.1, 298b6; *Meteor.* 338b2, 339b16–17, 340b11; I.3, 270b3, 10; II.12, 291b32). Note, moreover, that eternal being, which is accomplished through eternal change, is prior to the other kinds of being (*Metaph.* IX.8, 1050b19; *Int.* 13, 23a21–26; *GA* II.1, 731b23–31).

and not-being in respect of place or quality or size or *ousia*). Therefore, for sublunary generation, to be complete means turning a multitude of rectilinear, finite, and incomplete changes into a circular, infinite, and complete one. And this is what Aristotle may mean when, in *On Generation and Corruption* II.10, he says that 'also rectilinear locomotion, by imitating circular locomotion, is continuous'.

Insofar as the generation of organisms is concerned, this is why Aristotle describes generation and the partaking in the eternal and divine as the goal of the first soul – which accordingly he calls *gennêtikê* (*DA* II.4, 416b23–25) – as the most natural of the activities of organisms (415a26–b1) and as that for the sake of which they do whatever they do by nature (415b1–2). Moreover, the view from above enables us to understand these claims in a way that is different from how they may be understood in the ordinary perspective: generation can be seen as the complete whole of which nutrition and the self-maintenance of each single organism are incomplete parts,[52] i.e. as a unitary process that goes on through the metamorphosis of matter (i.e. through the life-cycles of a series of individual organisms).

Something similar holds for sublunary simple bodies. In this case, too, circular generation is prior to the rectilinear motions performed by each single body. Indeed, the reciprocal transformation of sublunary elements is one of the main reasons the whole sublunary world has not broken down into its parts (*GC* II.10, 337a7–15). In this sense, sublunary elements are not independent from each other: their rectilinear motions are just a phase within a broader and complete process that is their reciprocal transformation.[53]

So, the sublunary world must be kept in continuous change to avoid its own destruction. Continuous/circular change ensures the being of the whole and of its parts through the neutralisation of not-being (*GC* I.3, 318a13–26). This may be one of the reasons why, in *On Generation and Corruption* II.10, Aristotle describes eternal generation as the process whereby God 'filled up the whole' and 'connected being together'. These expressions suggest that the being of the whole is something that has gaps of non-being within itself that must be filled up, or holes that must be neutralized by means of connections between 'parts' of being.

[52] The living activity of individual sublunary organisms is a complete process (cf. *Metaph.* IX.6, 1048b18–36), but it is so in a weaker sense than circular generation, since it has temporal limits and is therefore, from this point of view, rectilinear (*GC* II.11, 338b6–11).

[53] Cf. Bodnár 1997: 81–117.

Conclusion: The view 'from above'

Following the bottom-up path, you will first meet a clear-cut distinction between being and change. Moreover, you will first meet only rectilinear and finite changes. From the bottom-up, you will see parts without knowing that they are parts of something, what they are parts of, and how. But when you come to the top and have a view from above, you will see the entire picture, the building blocks, and the main relationships between them.

You will see that the world is a hierarchy comprising three main levels, each of which causes the being of the lower one without being caused by it. These three main levels are similar to each other, since each is something everlasting, complete, and capable of accomplishing *ousia*: the complete being of the first unmoved mover, the complete locomotion of celestial bodies, and the complete generation of sublunary natural ones. These are the building blocks on which the being and continuity of the world depend. All the other kinds of being and of change, those occurring in the sublunary world, are incomplete parts of sublunary generation.

Moreover, from above you will see that, insofar as natural entities' *ousia* is accomplished through change, Aristotle's ontology is radically dynamic and that the world keeps being insofar as it keeps moving and changing. As already noted, this may sound less awkward if one recalls that the being that is the primary source of being for all other things is not a state but a living activity. Hence, perhaps it is not just because of rhetorical emphasis that, at the beginning of *Physics* VIII.1, Aristotle describes eternal change as a *zoê tis ousa* for all things composed by nature (250b14). But what does it mean? Is it a sort of life or a life of a certain sort (*Cael.* I.9, 279a23–30)? Let's keep investigating.

CHAPTER 10

Aristotle's processes

David Charles

I Introduction

Aristotle describes *changes[1] (*kinêseis*) as continuous (*Phys.* v.4, 229a1, viii.10, 267a20), as quick or slow (v.4, 228b29), as lasting from one time to another (vi.8, 239b1ff.), as going on for ever (viii.7, 261a30ff.), as even or uneven (v.4, 228b18 and 228b27ff.), incomplete or complete (*EN* x.4, 1174a28ff.), interrupted or uninterrupted (*Phys.* viii.4, 255b7, ii.8, 199b16), accelerating or slowing down (vi.7, 238a6ff.), and as forced or in line with nature (iv.8, 215a1ff.), and also talks of different *changes proceeding in different directions (v.6, 229b27ff.).

As these descriptions show, particular *changes have different properties at different times: some are initially slow and then fast. *Changes are (non-derivatively) the subjects of these properties at different times. They last through time and move through space. Some are interrupted, failing to reach their goal: these could have continued longer.

In this chapter, I shall consider some aspects of Aristotle's account of *change, beginning with his attempt in the first three chapters of *Physics* iii (hereafter *Physics* iii.1–3) to say what *change is. There, he suggests that it is 'the actuality (*entelecheia*) of that which is potentially as such' (*Phys.* iii.1, 201a11–12), illustrating this suggestion by examples such as 'when the

I am indebted to Tom Ainsworth, Elena Cagnoli Fiecconi, Jason Carter, Ana Laura Edelman, Paolo Fait, Verity Harte, Mariska Leunissen, Fred Miller, Takashi Ochi, Michael Peramatzis, Diana Quarantotto, Stefan Sienkiewicz, and Breno Zuppolini for their comments on the penultimate draft of this chapter. I also gained from earlier discussion with Lucas Angioni, David Bronstein, Laura Castelli, Kei Chiba, Timothy Clarke, and Ursula Coope. The present chapter engages with, and re-assesses, issues already discussed in Charles 1984.

[1] I shall use *change (with an asterisk) to stand for whatever is denoted by the Greek term *kinêsis*. It is a substantial thesis that Aristotle's *changes are changes. I use this notation so as not to assume from the outset that *kinêseis* are, for example, events that occur at or between times, rather than processes that continue through time.

buildable, in so far as we describe it as such, is in actuality, there is building' (201a16–18). In sections II and III, I shall consider differing interpretations of these telegrammatic remarks, placing them in the context of Aristotle's discussion of *changes elsewhere in the *Physics* and *Metaphysics Theta*. I shall next examine his remarks on the individuation of *changes in *Physics* III.3 (section IV), and finally locate his account of *change in the context of his discussion of substances and activities (*energeiai*) in *Metaphysics Theta* (section V).

II The definition of *change: 'a *change is the actuality of that which is potentially . . . as such' (*Phys.* III.1, 201a11–12)

Several interpretations have been offered of Aristotle's initial account of what it is to be a *change.[2] They differ in their understanding of its two key terms: 'actuality' and 'what is potentially as such'.

'Actuality' is sometimes understood as an actualisation. However, this seems problematic if 'actualisation' means the process of transitioning to an end-state:[3] for this appears to make Aristotle's account immediately and unacceptably circular. Many, as a consequence, translate 'actuality' as actuality, taking this to be different from the process of actualisation.

But what is an actuality? For some, it is the state achieved when the *change is completed. They take this to be the only alternative to the (immediately circular) account in terms of actualisation. So understood, Aristotle's subject of change has the potentiality to be in a given state: the one achieved at the end of the change. However, as it stands, their account is flawed. It fails to capture the idea of the subject proceeding to, and having the potential to proceed to, the relevant end-state. But it is precisely because it has this latter potential that the subject comes to be at the end-state.

Two lines of interpretation address this difficulty. One, favoured by David Ross, Robert Heinaman, and my earlier self, understands the phrase 'what is potentially as such' to refer to something *qua* having potentiality to *change or be *changed.[4] This potentiality is realised when there is a *change. However, this interpretation (which I shall call Interpretation A) is, as it stands, incomplete. What is needed is an appropriate way, suggested in Aristotle's own writings, of understanding the term 'actuality'. I shall seek to fill this gap in section V.

[2] Some are contrasted by Coope 2009.

[3] Although this view is defended by Kostman 1987, I shall not consider it further here.

[4] Ross 1936, Charles 1986, and Heinaman 1994. See also Beere 2009.

A second interpretation, initially proposed by Aryeh Kosman and subsequently developed by Sarah Broadie and (with reservations) Ursula Coope, has recently gained currency. It takes Aristotle's phrase 'as such' to modify not the potentiality of the subject to be in a given state but the type of actuality involved.[5] The relevant actuality is not the end-state achieved but the actuality-as-being-potentially-in-a-given-end-state of what has the potential to be in that end-state. Housebuilding, so understood, is a distinctive kind of actuality of the potential to be a house: the actuality-as-what-is-potentially-a-house of that which has the potential to be a house. When building occurs there is an actuality-as-the-potential-to-be-a-house of the bricks' potential to be a house. Although this proposal (which I shall call Interpretation B) is ingenious, it is, I shall argue, subject to serious objections.[6]

Two considerations are taken by proponents of Interpretation B to establish that, whatever its difficulties, it must be preferable to Interpretation A.

The first runs as follows. In *Phys.* VIII.5, 257b6–9, Aristotle describes a *change 'as the incomplete actuality of what is *changed'. However, since all potentialities are incomplete, he cannot be referring to a special kind of incomplete potentiality, but rather to a special kind of incomplete actuality. Interpretation B takes the potentiality at issue to be the one which (in the case of housebuilding) bricks possess to be a house. Thus, when Aristotle distinguishes complete from incomplete actualities, he is identifying the former with actually being a house, the latter with actually being potentially a house (or, as it is sometimes said, 'being active as a potential house').[7] This passage, in their view, points to a special kind of incomplete actuality: the actuality-as-potentially-a-house.

It is important to note that Aristotle, in 257b6–9, does not explicitly mention the potentiality to be, for example, a house. He refers only to the subject of *change (e.g. the bricks). Nor does he distinguish, as he does elsewhere, the actuality of the bricks *as* what is buildable (namely the building) from the actuality *as* a completed house (compare *Phys.* III.1, 201b10–12). This latter distinction, however, is significant. Once the house is completed, the *change is over. There is no longer an actuality of the subject *qua* what is buildable. The subject is no longer correctly described

[5] Kosman 1969, Waterlow/Broadie 1982, Coope 2009, and Kosman 2013. An interesting variant of this view is suggested by Hussey 1983.

[6] I shall refer to proponents of these interpretations as A- or as B-interpreters. My interest is to see how far each line of interpretation can be developed and defended, not merely to criticise or support what has so far been said in support of either.

[7] Kosman 2013: 68.

in this way. By contrast, the actuality characteristic of a *change has as its subject something not yet itself complete (as Aristotle notes in 201b11–12: see also *DA* III.7, 431a6–7): something that is not yet a house. With this distinction in place, Interpretation A need not take 257b6–9 to refer to a special kind of incomplete potentiality. The relevant potentiality can, on the A-reading, be that of a special kind of subject: something, not yet a house, with the potential to become a house. The bricks in question might have two distinct potentials: one to be a house, the other to *change into a house. *Qua* possessing the latter, they are – according to A-interpreters – a special kind of subject of change.

But why does Aristotle refer to the *change as 'an incomplete *actuality' in 257b6–9? *Changes, as he often notes, are, in some important respect, incomplete (see *Phys.* III.2, 201b32; *Metaph.* IX.6, 1048b28: 'every *change is incomplete'). This is because, in his view, housebuilding is only complete when the house is built. While the building is under way, the *change is incomplete. This is why *changes are hard to pin down in terms of actuality and potentiality (*Phys.* III.2, 201b28–202a2): unlike other *actualities, they are essentially incomplete even though they differ from mere potentialities (which are present before the process has begun). These remarks suggest an appreciation of what is now called the 'imperfective paradox': *Changes, while they are going on, exist but are not yet complete. However, the incompleteness of the actuality, so understood, is a general truth about *changes, which both Interpretations A and B can easily accommodate. The phrase 'incomplete actuality' in 257b6–9 does not support either view. It is simply a way of distinguishing the *change (which is an incomplete actuality) from the state (of being a house) which is, by contrast, a complete actuality. Aristotle's brief remark in 257b6–9 is consistent with both Interpretations.

A second consideration, emphasised by B-interpreters, is that Aristotle could not have accepted as the definition of *change the actuality of what is potentially to *change or be *changed (as proposed by Interpretation A). This definition is, they say, 'circular.' Charity requires us to see Aristotle as proposing the non-circular definition suggested by Interpretation B.

It is important to spell out this argument carefully. There is no circularity in defining a *change as the actuality of the potential to *change or be *changed, assuming that the notion of actuality is not itself that of a *change. For the potential to *change is not the same as the *change itself. (In this, the present suggestion differs from the immediately circular definition of *change in terms of actualisation.) The problem arises only when one combines this definition with one of the potential to *change in terms of *change (its specific actuality) – a step Aristotle appears to take in *Metaphysics Theta* when emphasising the definitional priority of actuality over potentiality.[8] Once these accounts are combined, a *change is defined as the actuality of the potential to *change (or be *changed) and the latter defined in terms of *change. Even if Aristotle's definition of *change is not immediately circular, it is one of a pair of definitions, each of which depends on the other. Aristotle does not succeed in breaking out of a small circle of interdefined terms (potentiality to *change and *change). Since, for B-interpreters, this failure violates Aristotle's basic requirements on acceptable definitions, the A-interpretation of his definition of *change must be incorrect. Further, as they point out, there is no explicit reference in *Physics* III.1 to the capacity for *change (postulated by A-interpreters). While Aristotle speaks of what is buildable, he does not specify what it is for something to be buildable. So, B-interpreters ask, why saddle him with Interpretation A's 'circular' account when there is no textual need to do so?

It is, of course, true that the two accounts – that of *change as the actuality of what is capable of *change, and that of what is capable of *change as what *changes – when combined, depend on each other. They do not, taken together, offer a reductive definition of *change in terms independent of the notion of *change. If Aristotle was aiming to break out of this circle of terms, he failed. However, there is reason to believe that this was not his goal. Or so I shall argue.

In *Metaphysics Theta* 6 Aristotle writes that one should not seek a definition of everything (1048a36ff.), noting that one may not be able to offer definitions of such basic terms as activity (*energeia*) or potentiality (capacity).[9] The basic term 'activity' is applied elsewhere to actualities (1047a30, 1050a21–22). In *Theta* 6 Aristotle simply gives examples of what is capable: what is capable of building and seeing, and matter capable of being a 'worked-up' object, such as a statue (1048a31ff.). The basic notions of potentiality and activity (and, by implication, actuality) are not defined independently of each other. Indeed, Aristotle seems to think that it cannot be done. When discussing his examples, he describes them as analogous to each other. There is no attempt to analyse one group in terms of the other. He simply says: 'as capacity stands to activity, so matter stands to form' (1048b6–8).

[8] See *Metaph. Theta* 8, 1050a4ff. For an illuminating discussion of the relevant types of priority, see Peramatzis 2011.

[9] Indeed, an activity may be understood as the way a thing is when it is not in the way in which we would describe it as capable of being or of becoming so and so.

Aristotle is offering a conceptual map on which to locate varying kinds of potentiality and activity (including actualities). On one side, there are capacities to be, the kind of capacities matter possesses. On the other, there are capacities for *changes: the capacities to build or to be built with. So far from attempting to give a definition of the latter in terms of the former, Aristotle simply presents them as analogous. He is not attempting the ambitious project required by the B-Interpreter: of defining *change in terms independent of *change.

In *Theta* 6 Aristotle develops his account by means of examples: as that which builds to that which is ready to build, that which is awake to that which is asleep, so that which is seeing is to that which has sight (1048a37–b4). In the first case, he contrasts that which is actually building with that which is potentially building. 'Actually building' and 'potentially building' are two different ways of building, as 'actually seeing' and 'potentially seeing' are different ways of seeing. These different ways of seeing or building are compared with actualities and potentialities in the case of substances. There is no attempt at an independent definition of any of them. Nor does he do so in *Metaph. Delta* 7, 1017a35 when listing cases, some involving *changes (e.g. building), some activities (e.g. seeing), still others substances (like being a statue). While all can be thought of in terms of being potentially and actually something, there is no attempt to reduce one set of cases to the other.

Aristotle takes a further step in *Theta* 7. He suggests that when something is potentially a statue, it has the capacity to be a statue and distinguishes two relevant capacities: the capacity to *be* a house and the capacity to *become* a house. He writes (1049a10ff.):

> Something is potentially a house, if nothing in the matter prevents it from becoming a house, and there is nothing which needs to be added or taken away or changed for it to become a house, it is potentially a house. The same account applies in all the cases where the starting-point of the process of coming to be is external to the matter.

The matter has two distinct capacities: one to be a house, first acquired when it has the potential to become a house, the other to become a house.[10] While the latter requires that the matter in question has the potentiality to be a house, the two potentialities are not to be identified. Indeed, when the house is built, the object no longer has the potential to become a house. So understood, a *change is the realisation of the specific capacity to become a

[10] I discussed this passage in more detail in Charles 2010c.

house which the matter possesses before it has become a house and loses once it has become one.

In sum: Interpretation B's second consideration, with its emphasis on the dangers of 'circularity,' seems ill-founded. Aristotle is not seeking in *Theta* or elsewhere a reductive definition of *change in terms independently of *change. Instead, he is trying to show that various differing kinds of potentiality and activity (and actuality) share a similar structure. The presence of this circle of interdefined terms is, from his point of view, no objection to its philosophical validity. So, even if B-Interpreters were to succeed in devising, using Aristotelian machinery, a reductive definition of *change, their ingenuity would be misplaced. There is no reason to think that Aristotle himself was engaged in any such reductive project. (I shall return to this issue in sections III and V.)

This said, Interpretation B is correct to emphasise that there is no explicit reference in *Physics* III.1 to the capacity, for example, to be *changed into a house. While Aristotle talks of the buildable, he does not specify this in terms of the capacity to *change (or be *changed) into a house. But what does this show – given Aristotle's willingness, in *Metaphysics Theta* 7, to distinguish the capacity to become a house and the capacity to be a house? Perhaps *Physics* III.1 contains a preliminary discussion of *changes, aiming only to specify the extension of the term '*change', not to give a complete account of what a *change is. So understood, its 'definitions' are not designed to give a full account of what it is to be a *change.[11] Instead, Aristotle is introducing an account to be refined later in *Physics* III and elsewhere. Indeed, subsequent passages in *Physics* III seem designed to elucidate key phrases such as 'the buildable' (as used in *Physics* III.1). Let us examine this suggestion further.

In *Physics* III.3, Aristotle gives a 'more knowledgeable' account of what *change is (202b26ff.): the actuality of what is capable of doing and of suffering as such. A more knowledgeable account is presumably closer to the first principles of an account of what it is to be a *change. The actuality of a capacity is a *change if the relevant capacities are ones for acting or suffering, ones whose actuality is acting or being acted upon. However, *Physics* III.3 does not suggest the approach favoured by B-interpreters. Aristotle focuses on making and suffering (*poiein* and *pathein*), which he separates from the result: the thing made or what is suffered (*pathos* or

[11] One might regard *Physics* III.1 as giving an account of what the term '*change' signifies to be contrasted with subsequent accounts of what it is to be a *change. For this distinction, see *APo.* II.9, 93b29–94a10 (discussed in Charles 2000: 23–56). A similar distinction may be suggested at *Phys.* II.7, 198a26ff.

poiêma) (202a25). He speaks throughout of *changing, making and suffering, not the result of the *change (the completed house). In this more advanced definition, Aristotle refers to the capacity to act or to be acted upon, not to the potential to become a result. As *Physics* III progresses, he elucidates the kind of potentialities involved: they are ones required for doing and suffering, not those required for the presence of the relevant result (being a house).[12]

B-interpreters, however, understand the capacities to make or to suffer differently, taking the capacity of the builder not to be to build but rather as one for the house (the result) to be.[13] According to this suggestion, the builder has the capacity for the relevant matter (such as bricks and stones) to be in a new state. Aristotle, so interpreted, focuses on the result and understands the builder's skill as his capacity for something else to be in a new state.

This ingenious suggestion has its own disadvantages. Skill is usually (and plausibly) characterised by Aristotle as the agent's ability to make something (as for example, in *EN* VI.4, 1140a7ff.). The builder does not have the bare capacity for something else *to be* in a new state, rather the ability to *bring about* this new state (by building). The former capacity (if it were to exist) would not be unique to builders: their employers would also have it and exercise it by providing builders with the resources to do the building. However, the employers lack the capacity to bring about the new state themselves by building. Indeed, this is why they need a builder.

Aristotle's account of the relevant kind of capacity in *Physics* III.3 makes explicit his earlier thought: the potential of what is *changeable as such is to be understood in terms of what has the potential to make or suffer as such. He does not seek to define *change in terms independent of capacities for *change. Instead, in this more developed account, Aristotle specifies the capacity of the *changeable in terms (familiar from *Categories* 8, 9a14ff.) of the capacities of the agent to make something (*poiein*) and the patient to suffer (*paschein*). This approach is to be expected if Aristotle is following the strategy favoured by Interpretation A.

To conclude: first, while Interpretation B offers a possible reading of some key phrases in *Physics* III.1 and VIII (taken in isolation), it is less convincing as an account of the unfolding argument in *Physics* III.1–3. Second, its central motivation – that of presenting Aristotle as seeking a way to break out of the relevant circle of terms – is not one which he shared. Indeed, he may well have thought it misguided.

[12] See Heinaman 1994. [13] See Coope 2009.

III The definition of *change (2): Further problems for Interpretation B

Interpretation B fares no better as *Physics* III progresses further. Aristotle talks later of infinite *changes that go on forever (*Phys.* III.8, 208a16–21; see also IV.1, 208a20 and VIII.7, 261a30ff.). Since these have no end-point, they cannot be the actuality of the ability to be at an end-point as that ability. In his account, the infinite is 'that of which there is always something outside' (*Phys.* III.6, 207a1–2). In the case of geometrical construction, 'it is always possible to think [of a bigger number] in the direction of the more, since bisections of a magnitude are unlimited' (*Phys.* III.7, 207b10–11). To grasp that the procedure is infinite is to grasp that it will not actually terminate. In a similar way, to grasp that a given *change is infinite is to grasp that it too will not terminate (208a16–21).[14] Nor can these remarks be dismissed as irrelevant to Aristotle's definition of a *change. *Physics* III, which begins with that definition, ends with an account of infinitely long *changes. Such *changes (including those involved in geometrical construction) play an essential part in his account of the 'potential infinite'. Aristotle could not have consistently developed that account had his definition of *change been as the actuality of what is potentially in such-and-such a given end-state. There is simply no such state in the case of infinite *changes.[15] Since the B-interpretation attributes to Aristotle a major inconsistency in *Physics* III as a whole, it seems preferable to understand his account of *changes in the way recommended by Interpretation A: as the actuality of the capacity to *change.[16] *Changes, so understood, can go on for ever.[17]

Does Interpretation B succeed in attributing, in the case of finite *changes, a plausible non-circular definition of *change to Aristotle? Does the full actuality of the ability to be a house, *as the ability to be a house,* capture all and only cases of housebuilding?[18]

[14] Contrast the idealist reading rejected in *Phys.* III.8, 208a18–19: 'something is outside not because someone thinks it but because it is so'.

[15] White 2009: 260–263.

[16] For further discussion of Aristotle's idea of the potential infinite, see Coope 2012.

[17] Eternal circular *changes are said to be complete (*teleion: Phys.* VIII.8, 264b28 and 9, 265a21) because they are continuous and indestructible. However, they are still 'incomplete' in a different way: it will never be true to say of an object, 'this is moving in a circular way and has already completed that individual circular movement' (compare *Metaph. Theta* 6, 1048b25–31). I am indebted at this point to discussion with Diana Quarantotto.

[18] A. Anagnostopoulos (2010) offers several important criticisms of the translation of key phrases adopted by defenders of Interpretation B. I seek to address his concerns about Interpretation A below.

It is not clear that it does. Many think that bricks can manifest the ability to be a house as that ability without actually being built with: for example, when lying about before being used. According to Interpretation B, however, they do not (at this stage) manifest an ability to be a house (as that ability), only an ability to have the ability to be a house.[19]

Did Aristotle himself think in this way? In *Metaph. Theta* 7, 1049a10ff., he notes that bricks first have the potential to be a house when nothing needs to be added or taken away from them for them to be acted on by the builder to make a house. They have this ability when builders can take them to use in building. Nor should this surprise us. Their ability to be a house is used (as that ability) by builders who choose to use them for building or retain them for later such use. Further, brick-makers use the fact that bricks (when finished) actually have the ability to be a house when they stop working on them once they have reached the point at which they can be used (without further addition etc.) in house construction. Aristotle, it seems, took the (commonsense) view that a collection of bricks, when fully made, have (and manifest) the ability to be a house, not merely the ability to have the ability to be a house.[20]

The case of bricks standing in a half-built house abandoned by the builder raises similar problems. They too seem to manifest the ability to be a house *as that ability*. So too a sapling, considered as a half-grown tree: it appears to exercise its ability to be an oak (as that ability) even though being a sapling (the state) is not itself a *change. Interpretation B, it seems, does not (as stated) succeed in picking out all and only cases of *changes. Its proposed non-circular definition does not yield an extensionally correct account of what a *change is.

Some B-interpreters reply as follows: the ability to be a house as that ability is only *constitutively realised* when building is under way. But why, one may ask, is building the only 'constitutive realisation' of the ability to be a house? Why aren't the bricks, before being built with and when in a half-built house also 'constitutively realising' (albeit in a different way) the relevant ability? Unless the notion of 'constitutive realisation' is itself defined in terms of *change itself, it will apply to many irrelevant cases.

Some B-interpreters, at this point, fall back on the idea of activity (*energeia*), suggesting that there is only an activity of the relevant ability when the house is actually being built. However, this proposal (without further development) seems vulnerable to a similar pattern of objection: *either* the notion of the activity of what is potentially a house as potentially

[19] See Kosman 2013: 61–63. [20] Similar remarks apply to an acorn before it has been planted.

a house is itself to be understood on the basis of that of *change *or* it applies to states such as being a sapling which are not *changes at all.

The difficulty is a general one: B-interpreters seek to avoid the impasse noted in section I by introducing a further actuality – the actuality of the bricks as potentially a house. To avoid immediate circularity, they insist that this is not an actualisation. However, if this new actuality is understood in such a way that states or objects can also manifest it, it will be present in phenomena other than *changes. Merely adding this further complex actuality does not yield the non-circular definition of *change that B-interpreters seek. They seem not to have escaped from the impasse noted in section II.

The (charitable) conclusion to be drawn is, I suggest, that Aristotle was not proceeding in the way suggested by Interpretation B. That said, Interpretation A still faces a serious challenge: to locate an Aristotelian notion of the actuality of the capacity to *change (or be *changed) which avoids immediate circularity. In the final section of this chapter, I shall address this issue. However, before doing so, there is a further aspect of Aristotle's discussion of *changes to consider: his account of their individuation.

IV *Changes and their individuation: Another look at *Physics* III.3

Physics III.3 is a much-discussed chapter.[21] In this section, I shall argue that it is less committal, in certain crucial respects, than it is often taken to be. Its goal is to uncover just enough structure for Aristotle's immediate purposes, not to give the complete, worked-out, account of *change individuation that recent authors have assumed.

Although *changes are characterised in terms of capacity and actuality (*entelecheia*) in *Physics* III.1–2, several issues remain unresolved. In *Phys.* III.3, 202a20–21 Aristotle notes that, in the case of acting and suffering, there is one activity (*energeia*) of what is moved and of what is capable of moving but not one account of it (*logos*). How is his claim to be understood? I shall begin with a brief overview of his subsequent discussion.[22]

Aristotle presents a general difficulty (202a21–22) and gives two arguments against the view that there are two distinct *changes (such as teaching and learning) which occur in the patient (202a28–36). Teaching

[21] See Gill 1980; Charles 1984; Coope 2004; and Marmodoro 2007. My present view differs from the one developed in 1984.

[22] For more detail, see Charles 1984 and Coope 2004.

cannot be in the learner. It is not possible for there to be two *changes which end at the same point in the same object. (Let us call the first argument I, the second argument II.) He then offers two arguments against the idea of there being one *change in the cases under review. The first is that there cannot be one activity of two capacities that differ in form (argument III). The next argument (argument IV) is that if the two *changes were identical, all predicates true of one would be true of the other; but this is not the case. The teacher in teaching does not learn what he teaches (202a36–b4).

Aristotle considers these arguments in turn. He rejects argument I: teaching, he says, can occur in the learner (202b6–8). He next rejects argument III: there can, he claims, be one activity of two capacities (202b8–9). Finally he rejects IV, saying that, even if there were one *change, not all predicates true of one would be true of the other (202b12–16). But what is his attitude towards argument II: the contention that there cannot be two *changes that end at the same point in the same object? Does he accept it, holding that there is only one *change? Or does he accept it, in a quite different way, holding there are two *changes? While he is clear that there is in some way only one *change, how is this claim to be understood? Is there one *change solely in virtue of the fact that doing and suffering happen at the same time in the same place? (For this use of 'being the same,' see *Metaph. Gamma* 2, 1003b22.) So understood, there will be one *change in nature because acting and suffering always (or necessarily) co-occur. Or, alternatively, did he think that there is only one *change because there is, in reality, only one nature present with one essential feature?

Aristotle's remarks seem inconclusive. In 202b8ff., he talks of one activity (*energeia*) which is not the same in being (*tôi einai*), terminology used previously to describe things which are (in some way) the same but differ in the time of their occurrence or some other aspect (201a20). The cases under review in *Physics* III.3 are not the same in the account of *what it is to be each thing* (*to ti ên einai*). They are to be contrasted with his favoured example of cloak and garment (which are the same in the relevant type of account: 202b12–13).[23] If things differ with regard to the account of *what it is to be each of them*, they will differ in definition and essence. But if they

[23] It is important to distinguish between difference 'in being' (frequently used when there are different accidental descriptions of the same thing) and difference in 'in being what it is to be the thing' (where the description is an essential one). I know of no case (in Aristotle) where there is something one in number but with two (non-equivalent) essential descriptions. (I take 'being a cloak' and 'being a garment' to be, in this sense, equivalent descriptions.)

differ in these ways, they must be numerically distinct. (How can there be one thing with two distinct essences?) Therefore, if the definitional accounts of the two *changes (doing and suffering: 202b11) are distinct, these *changes must be numerically distinct. (There is, it should be noted, no further *change mentioned in the immediate context of which these are accidental properties.)

Aristotle, if this is correct, holds that there are two *changes which necessarily accompany each other but are not strictly identical. They will be the same in the less demanding manner already mentioned: they necessarily co-occur but are not the same in definitional account. There will be a distinct individual *change for each separate capacity realised: the capacity to do (teach) and the distinct capacity to learn (suffer). The two *changes will further differ in having different subjects and different goals. (The goal of learning will be, let us say, the acquisition of knowledge (the *pathos*: 202a24), that of teaching to ensure that someone else acquires knowledge (the *poiêma*: 202a24).) In having these different goals, the two *changes resemble the road from Athens that ends at Thebes and the road from Thebes that ends in Athens (202b12–14). Further, the two *changes, although they co-occur, may have different properties (as the road to Thebes is uphill, that to Athens downhill). They will be one *change solely in virtue of their necessarily co-occurring.

While this proposal constitutes an adequate answer to Aristotle's puzzle, he immediately considers another:

> However ('not but what', 202b16f.), not even if teaching and learning were one, would it be the case that to teach is to learn . . .[24]

Since this suggestion is contrasted with the previous one, the *change now under consideration must be one in account. While before Aristotle had previously been thinking of two *changes, different in definition, which necessarily accompany each other, he now considers an alternative: the *change in question is one in definition (one in being what it is). However, even so, he notes, to teach will not be the same as to learn. There is one *change (one in definition) describable as a teaching and as a learning, where these descriptions do not mark out distinct essences of different *changes. In a similar fashion, one might say that there is one and only one distance that can be described either as the distance from A to B or B to A. So understood, the descriptions 'the distance from A to B' and 'the distance from B to A' are not essential descriptions of different distances.

[24] For more on this locution, see Denniston 1934.

Instead, they are different ways of describing one and the same distance, which (in Aristotle's terminology) is one in definition but distinguishable in thought (and accidental properties).

Aristotle's next sentence introduces a relevant generalisation:

> But to speak generally, the teaching and the learning are not the same, nor is the acting and the suffering in the strict sense; but that to which both belong, that to which the teaching and the acting belong, the process, will be the same in the strict sense [as something unspecified]. For the activity of this in that and of that under the influence of this differ in account. (202b19–22)

This passage adds two points. In the case of acting and suffering:

- (i) the suffering (the *pathêsis*) is not strictly identical (*kuriôs*) with the acting (the *poiêsis*);
- (ii) that to which the suffering belongs, the *change, is the same in the strict sense [as something unspecified].
- (i) is clearly true in the second account just sketched. Although acting and suffering are the same in definition, 'suffering' and 'acting' are two different ways of describing the same *change. Suffering and acting are not one in the strictest sense because they are distinguishable in thought. However, (i) is also true in Aristotle's earlier account since, so understood, suffering and acting differ in definition as well in thought. The generalisation holds true in both the accounts Aristotle has so discussed.
- (ii) is more difficult to interpret because the final clause is incomplete. Some take its full expansion to be:
- (ii)* 'the *change is the same in the strict sense as the suffering . . . '

In their view, the suffering (of A under the influence of B) is strictly identical (inseparable in thought and in definition) with the *change. Others take this clause to mean:

(ii)** 'the *change is the same in the strict sense as itself . . . '

where the *change is said to be strictly identical with itself, not with either the suffering or the acting.

The first reading seems preferable. The final clause is naturally read as understanding the whole phrase 'is strictly the same as the learning/suffering' rather than just a sub-phrase 'is strictly the same'. Second, it makes a substantial point, recalling Aristotle's earlier remark that the *change is the actuality of what can be moved under the influence of the mover

(202a15–16): the *change is strictly identical with the suffering in the patient. This point is reinforced in the final lines (202b21–22), where a difference in account (*logos*) between suffering and acting allows the former but not the latter to be strictly identical with the *change. By contrast, (ii)** makes only the obvious point that the *change in question is identical with itself, without specifying what that *change is. But why add this self-evident point here?[25]

It is important to note that (ii), understood as (ii)*, will be true in both the proposals he is considering. In the first, the acting differs from the suffering in definitional account; in the latter, in thought alone. In both, the *change (understood as the actuality of what can be moved) may be taken as strictly identical with the suffering, spelling out the remark in 202a15ff.; they will be inseparable in both definition and thought. So understood, Aristotle's generalising remark ('to speak generally') applies to both his first and second proposal. In both, there are two accounts (two *logoi*) which describe what occurs, which is (in some way) one. In the first, these two accounts define two distinct activities that necessarily co-occur. In the second, there is one *change, one in definition, even though it can be thought of in two ways. Both spell out ways in which there can be one activity that is two in account (202a18–21).[26]

Did Aristotle decide between these alternatives in the present passage? Or was he content to point to a feature common to both: that there is (in some way) one activity which can be seen (in both these the two ways indicated) as falling under two different accounts? There is no need for him to go further at this point. On either view, what occurs is in some way one and the same activity of the capacity of the sufferer and of the maker, even though the two models he has canvassed differ in their ways of making this thought determinate.

Aristotle did not need to go further if his aim in the chapter is to defend the following claims:

(1) There is one activity (*energeia*) of both capacities (202a18).

[25] Bishop Butler, it should be noted, only made his famous comment 'everything is what it is and not another thing' when rejecting an alternative theory. It was not put forward as part of a positive definition of any particular type of thing. Aristotle mentions a similar point briefly in *Metaph. Zeta* 17, 1041a19–20, only to lay it quickly aside.

[26] For cases where Aristotle notes several ways of addressing a problem without finally deciding between them, see *GC* 1.3, 319a29–b5. Two more committed strategies have been suggested: (a) take 202b19–21 as resuming from 202b16 with 202b16–19 as an aside (prioritising the first proposal): see Charles 1984: 14f.; (b) take 202b19–21 as applying only to the proposal made in 202b16–19 (prioritising the second proposal). See Coope 2009.

(2) The activity occurs in the patient and is not different from the activity of the agent (202a15–16).
(3) The *change is the activity of what is capable of *changing under the causal influence of the *changer (202a19–22).

All three claims are true in both models, even though they offer different accounts of the key notions of 'one activity' in (1) and 'not different from' in (2). Both can agree that (3) is strictly true of the *change which they distinguish either in thought or in definition from the activity of the agent. Both develop (albeit in different ways) the idea that the *change is, strictly speaking, identical with the activity of the patient brought about by the agent. In both, housebuilding can be understood as the actuality of what is capable of acting and of suffering, as such (202b26–28), although they disagree as to the precise form of sameness involved.

In recent discussions, some have taken Aristotle to favour the first model, others the second (in one of several forms).[27] My present suggestion is that Aristotle did not, in fact, decide between these alternatives in *Physics* III.3. He could leave this issue open since he has set out sufficient conceptual resources to defend *in some way or other* his three initial claims (noted above). Since he aims to address a problem, an *aporia*, by outlining possible solutions, he does not need to determine here the *exact* kind of sameness in question. His present project is a modest one, with the more minimal goal of finding at least one way out of the *aporia.*

There are, of course, several ways in which Aristotle's remarks could have been developed further. However, he does not do so here. In *Metaphysics Gamma* 2, he notes two ways in which things can be the same: either by being necessarily coextensive or by having the same definition. In *Physics* III.3 either suggestion is sufficient for his purposes. (Nor is it clear that he decided between them elsewhere. But the resolution of that question awaits a further study of many difficult texts throughout his writings.)

V. The nature and role of actualities: interpretation A and Aristotle's ontological project in *Metaphysics Theta*

There remains a major problem: can Interpretation A offer an Aristotelian understanding of actuality (*entelecheia*) in a definition of *change

[27] I took the first approach in Charles 1984, while Ross 1936 preferred the second, individuating *changes in terms of the route followed from beginning to end. Others who prefer the second approach individuate the relevant *change as the joint realisation of two capacities (see Coope 2009). These issues are discussed in Marmodoro 2007.

acceptable by Aristotle's own standards? Such a definition must apply in a non-immediately circular fashion to all and only *changes. It should also fit into Aristotle's ontological project in *Metaphysics Theta*.

Aristotle, as we saw, was happy to take potentialities and actualities (or activities) as basic features of his ontology. In *Metaphysics Theta* 6 to 8, he seems to see the relevant potentialities for *change, together with *change itself, as part of a circle of interdefined terms. But how did Aristotle understand actuality?

Some remarks in *Metaphysics Theta* suggest an ontologically less committed way of understanding 'actuality'. In *Theta* 8, 1050a15ff. Aristotle writes:

> The result (*ergon*) in the case of action is a goal, and the activity (*energeia*) is the result. This is why that term (activity) is predicated on the basis of what is the result in the case of actions and extends also to the actuality (*entelecheia*).

The goal-like status of actions and of actuality allows him to refer to both as 'activities'. In the case of flute-playing discussed in the previous lines, the goal of the teacher is that the pupil plays the flute. While the pupil's actions exhibit the fact that the teacher has taught him how to play, what the learner aims at (and what the teacher teaches) is playing the flute. This is what has the goal within it. The pupil displays that he has achieved this goal in playing the flute. In other cases the relevant goal is seeing or thinking: something present in my or your thinking and seeing (1050a35ff.). In others, what counts as success is something beyond the action. The latter is exemplified by the house produced by house-building (1050a26ff.): the result of the action. However, both housebuilding and the resultant house are described as 'more a goal' than the capacity for house-building (1050a27ff.).

Aristotle's remarks suggest the following account. An actuality is something which when present is the realisation of the relevant potentiality. One might (with due circumspection) describe the actuality as the *success condition* for the realisation of that potentiality. In the case of teaching someone to play the flute, the success condition is playing the flute. This is what teachers aim to pass on and what is achieved when the pupils' potential to play the flute is realised (1050a21–23). When the potentiality is to see, this will be realised when there is seeing (1050a23–25). If the potentiality is to be a house, this will be realised when there is a house. Similarly, if the potentiality is that to *change or be *changed into a house, this will be present when there is a *change of this type (1050a25–29).

Success conditions need not be states. They are simply whatever has to be present for the relevant potentiality to be realised. In some cases, they will be activities such as seeing or thinking, in others states (such as being a house), in others a *change. Since the notion of a success condition applies widely in this way, it can be used without circularity in the definition of a *change. Of course, the success condition for the specific potentiality to *change is the *change itself: but this is precisely what is required. There is no danger of immediate circularity, since the notion of actuality is not analysed in terms of *change.

The notion of 'success condition' can be developed further. In the case of substances, the goal (as in the case of activities such as seeing) is present in the activity itself (1048b21). By contrast, in the case of *changes, the goal is reached only at the end of the *change. Although the builder is engaged in housebuilding, the *change is not completed until the house is built. (This is the imperfective aspect of *change noted in section II.) The relevant success condition for the realisation of the capacity to build does not require that builder complete the house. Housebuilding can be going on even in cases where the builder does not live to finish his project. Although the house may be the final goal, some goal is achieved in the very activity of housebuilding itself (1050a25–28). The success condition relevant to capacities for *change or being *changed is the achieving this intermediate goal. *Changes can do this without reaching their final goal.

Actualities in *Theta* 8 (so interpreted) are one part of Aristotle's basic ontology. In the case of substances and *changes, the actuality, understood as the relevant success condition, is what has to be achieved for the capacity to be realised. When the capacity is one for being a substance, the relevant success condition is being that substance. When the capacity is one for *change, the success condition will be the *changing. In the first case, the success condition (being an axe or being a human being: the *form*; see 1050a15ff., 1050b2) will be met when there is a particular substance (such as an axe). In the second, when the relevant success condition is met, one has a particular *change (such as a particular housebuilding). The notion of actuality in these cases is simply what has to be achieved for the capacity to be realised. This general notion can be used without circularity in the definition of *change.

Aristotle is setting out a map of the relevant ontology. On it he can locate various types of continuant (including *changes and substances), all which can have different properties at different times. (A human can be short at some time, tall at another, as a particular *change can be slow at one time, fast at another.) From this perspective, substances, *changes and activities such as seeing are differing types of continuant, each with its own distinctive features.[28] This is why Aristotle presents *changes and substances as having analogous but irreducibly distinct roles (see 1048b9ff.). His aim, in talking of potentiality and actuality, is to compare and contrast these differing types of continuant.

Aristotle's strategy throughout *Metaphysics Theta* is designed to generate this overview. When he says at the outset 'Let us get clear about the basic case of capacity, an actuality which is not most useful for our purposes' (1046b36f.) his aim is to get clear about the basic case (of *changes) and then to examine the use of 'actuality' and 'potentiality' beyond it. There can be genuine cases of capacities (or potentialities) beyond those for *change introduced in 1046a4–19 and, equally, beyond capacities for activities like seeing. There are, as is clear in *Theta* 6 and 7, capacities to be a given object distinct from those for undergoing a *change or activity.[29] While the most useful sense of 'potentiality' is that in which something is said to *be* something potentially, this is not a capacity for a *change (or for an activity like seeing) but rather a capacity to be. Aristotle carries through the project, outlined in the introductory chapter (1045b35ff.), in the remainder of *Theta*, pointing to different kinds of continuant: *changes, substances, and activities such as seeing, with differing features.[30] His ontology, as becomes clear as *Theta* develops, is made up of these varying types of continuant. Interpretation A captures this ambitious but steadfastly non-reductive account of *changes.

What are *changes in Aristotle's account? As a distinctive type of continuant, they are, it seems, unfoldings or processes (changings), with different properties at different times. They can, as already noted, move through space and end before they should. So understood, his *changes are not changes, understood as events. Events do not move through space.[31] If they unfold, they do so by one (sub-)event happening after another. There is no one event that accelerates and decelerates. Events are not the kind of thing to speed up or slow down.[32] There is no one event that is non-

[28] For the general notion of a continuant, see W. E. Johnson 1921.

[29] For this approach, see A. Anagnostopoulos 2010 and Johansen 2012a. It was developed in some detail in Friday morning discussions on *Theta* 1 in Oxford in 2008. For a contrasting view, see Frede 1989 and Beere 2009.

[30] Examples of differences: (a) matter, conceived as the capacity to be (e.g.) a human, plays an essential role in the case of substances (1050a15–16) but not in that of *changes; (b) in the case of substances, the relevant success condition is a form, such as being an axe or being disposed to cut, not the activity of cutting itself. The latter is the success condition for the potentiality to cut.

[31] This point is made by Dretske 1980.

[32] For further discussion, see Stout 1996, and Galton and Mizoguchi 2009.

derivatively fast at one point and slow at another. Events are occurrents, not continuants: they happen at times, or occur between times, but do not themselves move through time. Although events unfold, they are distinct from the process of their unfolding.

To conclude: Aristotle's *changes (*kinêseis*), as continuants, are processes, not events.[33] A proper appreciation of this point has important implications for our understanding of his accounts of action, time, and causation. The end of this chapter is not the end of the still unfolding story of Aristotle's processes.

[33] Most recent commentators talk of *changes as events: see, for example, Penner 1970 and Coope 2007, some (it seems) misled by translating *changes (*kinēseis*) as changes at the outset. For some relevant conceptual distinctions, see Fine 1999 and Haslanger 2003.

CHAPTER 11

Physics v–vi *versus* viii: *Unity of change and disunity in the* Physics

Jacob Rosen

Introduction

You toss an apple straight up into the air and let it fall back into your hand. The apple moves up, and then the apple moves down. There is an upward motion of the apple and then a downward motion of the apple. Do these two motions compose a single whole motion? Is there such a thing as the up-and-down motion of the apple?

Another question. You read *Physics* v–vi, with its general theory about changes and continua. Then you read *Physics* viii, with its cosmological arguments about the eternity of motion and the existence of a first unmoved mover. Have you just read a single text, something we could call a 'continuous treatise on movement' (Ross 1936: 3)?

The first question matters to Aristotle's cosmology. Aristotle believes he can show, in *Physics* viii.8, that the answer is 'No'. He purports to prove that when something moves back and forth along a straight line, its successive motions in opposite directions do not compose a single motion. This result, together with other theorems of his physics, entails that there cannot be an eternal motion along a straight line. From this he infers that the only possible eternal motion is circular motion. Since, independently of this, he thinks he has proven that there needs to be an eternal motion, he can assemble a demonstration for the existence of eternal circular motion. His candidate for such a motion is the motion of the heavenly sphere in which the stars are fixed. Ultimately, then, his argument in *Physics* viii.8 offers a measure of confirmation for certain astronomical theories of his day, and a sort of explanation for the existence and rotation of the outermost heavenly sphere which is posited by these theories.

The second question matters to the history and interpretation of Aristotle's writings. Its answer admits of degrees. It will affect, among other things, the extent to which we use passages in one book to elucidate passages in another, and the extent to which we may combine doctrines from the different books and then offer up the resulting whole as a view held by Aristotle.

In this chapter I will develop some thoughts about the second question, the one about the text, by way of considering how Aristotle reaches his answer to the first question, the one about the apple. Aristotle offers several arguments in *Physics* viii.8 for his thesis that, when something moves back and forth, it does not undergo a single motion. These arguments occur against the background of a sophisticated theory, expounded in *Physics* v–vi, of the basic structure of motions and of other continuous entities such as times and magnitudes. The arguments in *Physics* viii.8 stand in a surprisingly complex relation to that theory. On the one hand, Aristotle evidently relies on the theory in a number of crucial steps.[1] Yet in other steps he seems to contradict or misapply the theory. This situation offers the occasion to examine Aristotle's views about some fundamentals in the metaphysics of motion, while also raising questions about the unity of the text which has come down to us as the *Physics.*

Let me signal in advance one of the questions we will encounter. In one of the arguments of *Physics* viii.8, Aristotle introduces a thesis about continua that we may call the Potentiality Doctrine. According to this doctrine, a continuous entity has no actually existing proper parts and no actually existing middle-points. Rather, it has parts and middle-points only *potentially* or in capacity. The Potentiality Doctrine is not affirmed in *Physics* v–vi. To the contrary, Aristotle often refers in these books to parts and middle-points without ever suggesting that they are only potentially there. It is natural to wonder whether we may attribute to Aristotle a single overall theory of continua in which the Potentiality Doctrine is combined with the theory of Books v–vi. The answer will depend on many considerations, but an important one is this: when we come to Aristotle's *argument for* the Potentiality Doctrine in *Physics* viii, we will see that it rests on an assumption that is contradicted by a theorem in *Physics* vi.5. That is reason to doubt whether the Potentiality Doctrine is going to combine successfully with the theory of *Physics* v–vi. At the least, it speaks

[1] For example, Aristotle's rejection of eternal rectilinear motion in *Physics* viii.8 tacitly relies on the following two theorems from *Physics* vi: (1) no motion traverses an infinite straight line (*Phys.* vi.10, 241a26–b10); and (2) no motion takes an infinite time to traverse a finite distance (*Phys.* vi.2, 233a31–34, and vi.7, 237b24–25). These theorems imply that if something moves forever along a straight line, it must sometimes turn around; *Physics* viii builds on this by arguing that what turns around does not undergo a single motion.

against regarding these three books of the *Physics* as a single, continuous exposition of such a combined theory.

In all, there are many points of agreement and more than one point of tension between Books V–VI and Book VIII. I will begin by presenting a few fundamental points of agreement. Then I will argue that there are (at least) two tensions. First, in *Physics* VIII.8, Aristotle assumes that any single motion must be homogeneous (in a sense to be explained), whereas according to *Physics* V–VI a motion is not homogeneous. Second, in *Physics* VIII.8, Aristotle assumes a beginning of change (in a sense to be explained), whereas in *Physics* VI.5 he proves that there is no such thing. These tensions both undermine Aristotle's justification of the Potentiality Doctrine and affect his strongest arguments against the existence of eternal rectilinear motion. We will thus need to ask where these tensions leave us in assessing the unity and success of Aristotle's project in the second half of his *Physics.*

I Common ground

Let us begin with three doctrines about change that are endorsed by Aristotle both in *Physics* V–VI and in *Physics* VIII. The first is the thesis that there are changes. The second concerns the way in which changes occupy time. The third concerns the conditions under which a given change is 'one' (full stop), or 'one with' a given change.

The most fundamental commonality between the different texts is their commitment to the existence of such objects as changes and motions.[2] This is a substantial theoretical commitment, going beyond the more modest claim that things move and change, or that there are moving and changing things. (Davidson posed the question: 'Things change; but are there such things as changes?'[3] Aristotle's answer, like Davidson's, is 'Yes.') Aristotle writes:

> It is necessary that if a motion is present, then something moves, and that if something moves, then a motion is present.[4] (*Phys.* VI.1, 231b25–26)

I should mention that this biconditional is preceded in the text with an 'if indeed' and followed by a 'then . . . '. It is the antecedent of a conditional

[2] In my translations of Aristotle, I will use the word 'motion' and its cognates to translate κίνησις and its cognates, and use 'change' and its cognates to translate μεταβολή and its cognates. Every motion is a change, but not *vice versa*. The distinction between motion and change is not important for this chapter, and I will not be careful about it.

[3] Davidson 1970: 25. [4] Translations are my own.

claim, not something that Aristotle asserts on its own. Still, it is beyond doubt that he endorses it. A sign of this is his easy way of switching back and forth between verbs predicating of things that they move or change, and nouns denoting motions or changes. He does so both in *Physics* V–VI and in *Physics* VIII.[5]

As it stands, Aristotle's biconditional is fairly weak. It merely says that something moves if and only if there exists a motion. It does not explicitly say whether there is a motion *of each thing* that moves; or whether a motion is present *at every time when* something moves. Nor does it specify whether motions have particular kinds and characters, corresponding to *how* things move. (For example, is it the case that something moves fast if and only if there is a fast motion? Or that something moves to Venice if and only if there is a motion to Venice?) We must return to the last question later, but the first two have natural, obvious answers, and it seems clear that Aristotle accepts these obvious answers. He assumes that a subject S moves if and only if there is a motion *of S*, and that something moves during time T if and only if there is a motion of it *occupying T*. So I will take it that in *Physics* V–VI and VIII, Aristotle endorses the following principle:

> NOUN–VERB TRANSLATION PRINCIPLE: S moves during T iff there is a motion of S that occupies T.

A second point of agreement between *Physics* V–VI and VIII is their understanding of the way in which changes are temporally extended. Aristotle in both texts conceives of changes as spread out in time like sails. By this I mean that, where a change occupies a given time, it has different parts occupying different parts of the time.[6] (*Physics* VI contains an analogous claim involving spatial extension, namely that, where a change belongs to a given body, it has different parts belonging to different parts of the body.[7]) We should appreciate that this is not the only possible philosophical view Aristotle could have taken. In principle he might have conceived of a change as a continuant: something that endures through time in much the way that substances are naturally thought to do. Thus, he might have thought that a change is wholly present in every part of the time for which it exists. I have heard it said that Aristotle sometimes regards

[5] See, for example, *Phys.* VIII.8, 265a7–12.

[6] See, for example, *Phys.* VI.4, 235a18–24; VIII.8, 263a27–29, 264a24–26. As already mentioned, according to VIII.8 a change has these parts only potentially, not actually (whatever exactly that means).

[7] *Phys.* VI.4, 234b21–24 and following. This claim is neither affirmed nor contradicted in *Physics* VIII, to my knowledge.

changes as continuants in this way. I do not know whether that is true, but in any case he does not so regard them in the texts I am discussing.

The third point of agreement between the texts is a shared framework for addressing issues of unity for changes. Recall that Aristotle's question in *Physics* VIII.8 is whether there can be a *single change* of a certain sort (namely, a single change back and forth along a straight line). In order to answer it, he builds upon a discussion of criteria for oneness of change that was presented in *Physics* V.4. (He refers back to this discussion at VIII.8, 262a1.) Aristotle there discussed three main ways of being one: being one *in genus* (*genei*), being one *in kind* (*eidei*), and being one '*simply*' or '*without qualification*' (*haplôs*).

The last, and strongest, way of being one is the topic of Aristotle's concern in *Physics* VIII.8. On the surface, at least, he adopts the same criteria for it as were laid down in *Physics* V.4. According to both texts, a change's being one without qualification depends upon three factors: what changes (i.e. the subject of change), when it changes (i.e. the time of change), and that 'in which' it changes (i.e. the path of change, if I understand rightly).[8] A change is one without qualification if and only if (i) its subject is one, (ii) its time is one and without gaps, and (iii) what it is in is 'one and indivisible' (V.4), or 'undifferentiated in kind' (VIII.8).[9] The third criterion appears to be the same as the criterion for a change's being one in kind.[10] This explains Aristotle's statement that 'necessarily, a motion that is one (namely, without qualification) is also one in kind, although it is not necessary for a motion that is one in kind to be one without qualification' (*Phys.* V.4, 228b9–10).

In the next section I will discuss a problem in Aristotle's treatment of oneness in kind in *Physics* VIII.8. But before I can do that, there is a point that requires clarification.

Interlude: One-place oneness and two-place oneness

Aristotle's discussion is subject to a complication that will be crucial in the next section of this chapter. The complication is that he shifts between what we may call a 'two-place' use and a 'one-place' use of the notion of oneness.

A two-place statement of oneness has the form 'change A is (or is not) one with change B'. For example, Aristotle tells us that every locomotion is

[8] *Phys.* V.4, 227b23–26; VIII.8, 262a1–4. [9] *Phys.* V.4, 227b29–228a3; VIII.8, 262a1–2.
[10] *Phys.* V.4, 227b6–7, 19, 27–28.

one in genus with every other locomotion (*Phys.* v.4, 227b5). He tells us that every whitening is *one in kind with* every other whitening (227b11). And he tells us that one man's restoration to health is not *simply one with* another man's restoration to health (228a1–3). He gives reasonably clear explanations of what he takes such claims to mean.

A one-place statement of oneness has the form 'A is (or is not) one'. It is sometimes, but not always, clear what Aristotle means to say with a sentence having this form. A fairly clear case is when the term substituted for A signifies a *type* of change. In this case the sentence can be understood as equivalent to a certain two-place oneness claim: it means that each change of the type is *one with* each other change of the type. For example, Aristotle says that *learning* is, to a degree, one in kind (227b13), and this seems to mean that every learning is one in kind with every other learning.

Another sort of case is trickier, although the basic intention behind it is still recognisable. This is the case where the term substituted for A purports to refer to an individual change, and the sentence either affirms or denies that the term's referent is one without qualification. Consider an example (in this passage it is clear from context that 'one' means 'one without qualification'):

> The motion is not one but many, if there is rest between them. Consequently, if a motion is separated by stationariness, it is not one or continuous. (*Phys.* v.4, 228b4–6)

What makes these sentences difficult is that the term 'the motion' or 'a motion' seems as though it must refer to a single thing, if it refers at all. And yet the term is used to say such things as 'the motion is not one' or 'the motion is many'. Are such claims capable of being true? Aristotle seems to be struggling somewhat to express himself properly; indeed he mixes grammatically singular and plural forms in a way that tests the rules of syntax. For example, in his phrase 'the motion is not one but many' there is a predicate ('many', *pollai*), which does not agree in number with its subject ('the motion', *hê kinêsis*).

Nevertheless, Aristotle's basic intention seems reasonably clear. The idea is that there are some changes (plural) that we are interested in, and we are interested in whether or not these changes compose a single change. In order for them to do so, there are some conditions which they jointly must satisfy: for example, they must jointly occupy a time that has no gaps in it. A term such as 'the motion', although grammatically singular, can be used to say things plurally about the changes of interest. Thus, for example, 'the motion is separated' is made true not by the fact that any single individual is separated, but by the fact that the motions in question (plural) are separated. (The plural predication here is like what we use when we say of some people, 'they are crowded together', or 'they are arranged in a square'. Such sentences do not say something about each person individually – they do not say that any individual person is crowded together or arranged in a square – nor do they posit a whole item, composed out of the people, which is a crowded or square item. They simply attribute a predicate to some people plurally.)

II Oneness in kind and homogeneity

As we saw above, Aristotle holds that oneness without qualification implies oneness in kind. Oneness in kind is among the criteria for unqualified oneness.

This point is fairly clear when we deal with two-place statements of oneness. Suppose that my face simultaneously heats up and turns red, and we want to know whether the heating is simply one with the reddening. Well, the subject of the heating is one with the subject of the reddening, and the time of the heating is one with the time of the reddening. Two of Aristotle's three criteria are met. However, his third criterion is not met: the heating is not one in kind with the reddening. Hence the heating is not unqualifiedly one with the reddening, because it differs from it in kind.

When we turn to one-place statements of oneness, we encounter a difficulty. Suppose we now ask whether my face's heating and its reddening are, i.e. compose, a single change (we could put this by asking whether my face's *growing flushed* is unqualifiedly one). Or, returning to our old example of the apple, suppose we ask whether the apple's motion from the apex of its arc down to your hand is simply one. (Do its motion from apex to the level of your eyes and its motion from there to your hand come together to compose a single change?) To answer such questions we will need to apply Aristotle's three criteria in new ways. We will need to know what it means to say that a change's subject or its time 'is one': not one *with* something, but just plain one. And we will need to know what it means to say that a change 'is one in kind': again, not one in kind *with* something, but just plain one in kind.

Above I suggested that we can reconstrue such claims in a plural form. We can look at the components of a putative single change and ask of them, plurally, whether *they belong to one subject*, whether *they occupy one time*, and whether *they are one in kind.* But understanding these plural-form questions is not straightforward. We should notice, to begin with, what the sentence 'they occupy one time' does *not* mean. It does not mean that each of the changes in question occupies *the same time as* each other change. For example, assuming that the apple's motion from apex to eyes and its motion from eyes to hand compose a single motion from apex to hand, we will need to say that they occupy one time. But the motion from apex to eyes does not occupy the same time as the motion from eyes to hand. One occurs after the other. The thought behind Aristotle's criterion, presumably, is that the times occupied by these two motions 'add up' to a single, continuous time.

I want now to focus on the criterion involving oneness in kind. In *Physics* VIII.8, I will suggest, Aristotle interprets this criterion precisely along the lines that we have just rejected for the criterion involving time. That is, he appears to hold the view that if a change is unqualifiedly one, then each of its parts must be one in kind with each other part. Then I will argue that this view is inconsistent with the theory of *Physics* V–VI.

II.1 Homogeneity in Physics VIII.8

Let us say that a change is *homogeneous* if and only if each part of the change is one in kind with each other part of the change. In *Physics* VIII.8, Aristotle apparently holds that a change is one without qualification only if it is homogeneous.

We find this view at work in Aristotle's first argument for the thesis that, if something moves back and forth along a straight line, it does not undergo a single motion. The argument begins:

> For it turns back, and that which turns back on a straight line undergoes contrary motions (261b32–34)

And he concludes:

> but contraries differ in kind, and are not one. (262a5)

In this argument, Aristotle notes that a supposed single back-and-forth motion would have two contrary motions as parts; he observes that these contrary motions differ in kind from each other; and he infers that there is no single back-and-forth motion. This appears to be an application of the view that a single motion must be homogeneous.

Aristotle gives further indication of the homogeneity of changes later on in the chapter. The context is the following claim: if something undergoes a motion which it was not always undergoing, then the motion is preceded by the sort of rest which is opposed to it. (Rest at A is opposed to motion from A to B.) Thus, for example, for some period before you walk from your office to the beach, you are at rest in your office. Now, Aristotle recognises that his claim must be restricted. For motions have smaller motions as parts, and Aristotle does not want to say that all these smaller motions are preceded by periods of rest. On your way from the office to the beach you pass a lemonade stand and an ice-cream booth, and there is a part of your walk by which you traverse the distance between them. Aristotle does not think that this part of your walk must be preceded by a period of rest in front of the lemonade stand. So, he restricts his claim in such a way that it does not apply to motions that are parts of larger motions. He states his restriction in the following way:

> I mean those motions that are different in kind, and not if it is some part of the whole motion. (264a25–26)

Aristotle here contrasts motions that are parts of motions with motions that are different in kind. This suggests that the parts of a motion are one in kind with the wholes of which they are part, and with the other partial motions that precede and succeed them.

II.2 Heterogeneity in Physics V–VI

According to the theory of *Physics* V–VI, by contrast, an unqualifiedly single change will not be homogeneous. The reason is this. Every change is a change from something and to something;[11] let us call these its starting-point and its end-point. A change is one in kind with another change only if it has the same end-point as it. (There are further conditions, too: it must also have the same starting-point, and indeed the same path from starting-point to end-point, as the other change.)[12] But the different temporal parts of a single change have different end-points. For example, the apple's whole downward motion, from its apex to your hand, is unqualifiedly one change. But one of its parts is a motion from the apex to a place before your eyes, and another of its parts is a motion from there to the place in your hand. These parts have different end-points, so they are not one in kind with each other. Hence the apple's whole downward motion is not homogeneous.

A thought along these lines finds explicit expression in *Nicomachean Ethics* X.4 (complete with an apparent cross-reference to *Physics* V–VI).[13]

[11] *Phys.* V.4, 225a1; VI.4, 234b11 and 5, 235b6. [12] *Phys.* V.4, 227b14–20.

[13] *NE* X.4, 1174a31–b5. Aristotle uses this line of thought in order to argue that pleasure is not motion.

What matters to us here, and what I would now like to show, is that the line of thought brings out a genuine consequence of the doctrines of *Physics* V–VI.

According to Books V–VI it is clear that, where a motion occupies a given time, the motion has a part in each part of the time. These parts are themselves motions.[14] And in the lesser times occupied by these lesser motions, the moving thing moves (*kineitai*) or traverses (*dierchetai*) lesser magnitudes. Thus, we find statements such as the following:

> It is evident that the magnitude is continuous if the time is, since in half the time it traverses half the distance, and in general a lesser distance in a lesser time. (*Phys.* VI.2, 233a13–16)[15]

Suppose, for example, that there is a motion from A to C, and that B lies between. Then the motion from A to C has, as parts, a motion during which the thing traverses magnitude AB and a motion during which the thing traverses magnitude BC. It is reasonable to think that these latter motions are, respectively, a motion from A to B and a motion from B to C. Aristotle himself does not have occasion to describe them in precisely this terminology, but arguments can be given to show that this is what they are.

The main thing needed in order to show this is a supplementation to the Noun–Verb Translation Principle stated earlier. We need to add a correlation between a characteristic of a moving thing, namely that it *moves from A to B*, and a characteristic of a motion, namely that it is *a motion from A to B*. If we are justified in doing this, then we may attribute to Aristotle the following expanded principle:

> EXPANDED NOUN–VERB TRANSLATION PRINCIPLE: S moves from A to B during T iff there is a motion of S from A to B that occupies T.

There is good evidence for such an expanded principle in *Physics* V–VI. Here, for example, are two passages from V.1:

> Every motion is from something and to something, for that which primarily moves is different from that to which it moves and that from which it moves. (*Phys.* V.1, 224b1–2)

> (a) Since every change is from something to something ... (b) that which changes could change in four ways: either from a subject to a subject, or

[14] See for example *Phys.* VI.4, 235a18–24. At *Phys.* VI.1, 232a8, an assumption for *reductio* has the consequence that 'the motion would not be composed out of motions', and this is treated as an unattractive consequence.

[15] See also *Phys.* VI.2, 232b7–8, a34–b2, 6, 236b34–237a3, and 7, 237b23–24.

> from a subject into not a subject, or not from a subject into a subject, or not from a subject into not a subject . . . (c) Consequently, it is necessary from what has been said that there are three changes: that from a subject to a subject, that from a subject to not a subject, and that from not a subject to a subject. (*Phys.* v.1, 224b35–225a10)

It seems clear in the first of these passages that Aristotle is equating what *a motion is from and to* with what *a thing moves from and to*. The equivalence is especially clear in the second passage. The details of interpretation of this passage, for example the question what exactly Aristotle means by 'subject', need not concern us. What is noteworthy is Aristotle's switch back and forth between saying that *a change is* from and to something, in points (a) and (c), and saying that *a thing changes* from and to something, in point (b). Aristotle's inference from (b) to (c) shows that he regards the two ways of speaking as equivalent. In other passages in Books v and vi as well, Aristotle switches fairly casually between speaking of what a change is from and to, and of what a thing changes from and to.[16]

Given the expanded translation principle, it only remains to convince ourselves that, during the different parts of a motion from A to C, a thing moves from A to B and moves from B to C. Here we must pause to note an ambiguity in the verbal form 'moves', or rather in the corresponding Greek present tense form. On the one hand, the form can be understood as having perfective aspect. So understood, the statement that S *moves* from A to B during a time implies that, at the end of the time, S *has moved* from A to B. (This in turn implies that S is at B at the end of the time, *Phys.* vi.5, 235b7–8.) But, on the other hand, the form could be understood as having imperfective aspect, equivalent to 'is moving'. The statement that S *is moving* from A to B during a time does not imply that S has moved to B by the end of the time. For example, if it takes you a whole day to walk to Thebes, then it is not true that you *walk* (perfective) to Thebes in the morning, but it is true that you *are walking* to Thebes in the morning.

The question is which verbal aspect figures in the translation principle? Is a motion from A to B a motion during which something *moves* (perfective) from A to B, or a motion during which something *is moving* (imperfective) from A to B? If it turned out to be the latter, then perhaps motions would be homogeneous after all. For it is plausible that a thing *is moving from A to B* during every part of a motion from A to B.

[16] See, for example, *Phys.* v.1, 224b7–10, 12–15; vi.4, 234b10–13 and 10, 241a26–b11 (esp. b9); see also vi.5, 235b13–14, 236b2–4.

In my view, it is more natural on the whole to adopt the perfective reading of Aristotle's present tense 'moves'. This reading is also supported by certain considerations of detail.[17] If you accept my view of the matter, then you will join me in drawing inferences from claims made by Aristotle in the perfect tense, such as the following:

> Let AB have moved from B to C primarily . . . If BC is divisible, there will be something before C to which AB has changed, and another in turn before that. (*Phys.* VI.5, 236b11–14)

Here Aristotle says that, before having changed from one point to another, a thing *has changed* to an intermediate point. I infer that the thing *changes* to an intermediate point, and, applying the translation principle, conclude that it undergoes *a change* to an intermediate point.

Fortunately, there is one passage in which Aristotle himself uses the present tense, so that we may apply the translation principle directly, without settling questions of verbal aspect or the relation between present and perfect tense forms. In this passage, Aristotle supposes that something has changed from C to D. He argues that CD is not indivisible. Then he proceeds:

> Necessarily, what is in between is a magnitude and is infinitely divisible. Consequently, it changes to those beforehand. (*Phys.* VI.6, 237a33–34)

Here Aristotle says that, before having changed from C to D, the thing changes to the various points between C and D (i.e. the points at which CD can be divided). He says this using the present tense. Applying the translation principle, it follows that the thing undergoes a change to each of the intermediate points. Each of these changes is, to review, different in kind from the others, and they are all parts of the change from C to D. Consequently, the change from C to D is not homogeneous.

To conclude, some of Aristotle's arguments in *Physics* VIII.8 against the possibility of eternal rectilinear motion are based on the principle that a change must be one in kind in order to be unqualifiedly one. This principle is common to Books V and VIII. But Aristotle interprets the principle to mean that if a change is unqualifiedly one, then each of its parts must be

[17] Here are two considerations. (1) At *Phys.* V.6, 230a4–5, Aristotle says that if something stands still at A, then a motion to A is or coincides with a coming-to-rest. His claim seems plausible only on the assumption that at the end of a motion to A the thing *is* at A (so that it will rest immediately after the motion). This assumption is validated by the perfective, but not by the imperfective, reading of the translation principle. (2) At *Phys.* VI.5, 235b6–8, Aristotle juxtaposes present and perfect tense claims, in a way that suggests (even if not, I grant, strictly entailing) that what a thing *changes to* and what it *has changed to* are the same for any given change.

one in kind with each of its other parts. This contradicts the theory of *Physics* v–vi, according to which a single change has, as parts, changes to different endpoints, with the result that its parts differ in kind from each other.

III The potentiality doctrine and the beginning of change

Aristotle has other arguments in *Physics* viii.8 that are based on another of his criteria for unity of change. In these other arguments, he appeals to the principle that a change, in order to be one, must occupy a single time without gaps. He thinks he can show that this criterion is failed by back-and-forth motion: when something moves back and forth along a straight line, it must rest for some time at the point where it turns around. (Indeed, he thinks you can *see* that it rests there, *Phys.* viii.8, 262a18.)

In the course of developing these arguments, Aristotle introduces a new general doctrine about continua, which I will refer to as the Potentiality Doctrine. This is the claim that a single continuous thing, such as a motion, line, or time, has parts and middle-points only potentially or in capacity, not in actuality. On the other hand, he argues, when something moves back and forth, the point where it turns around is *in actuality* a middle-point of its motion. Since its motion has an actual middle-point, it follows that its motion is not one. Along the way, he also uses the Potentiality Doctrine to answer one of Zeno's paradoxes of motion (263a4–b9).

Aristotle seems conscious in *Physics* viii.8 that the Potentiality Doctrine is a new thesis relative to the theory presented in Books v–vi. At least, he thinks that the doctrine provides a new and better response to Zeno.[18] There were gestures in the direction of this doctrine in earlier books, during Aristotle's discussions of infinity and of time (*Phys.* iii.6, 206a14ff., iv.11, 220a10ff.). But it is nowhere endorsed in Books v–vi. This is somewhat surprising, given that many of Aristotle's proofs in *Physics* v–vi appeal to the existence of parts and of intermediate points.[19] One would expect it to make a difference to these proofs whether or not the parts and middle-points actually exist, as opposed to being mere *possibilia*, and yet he says nothing about the question.

The Potentiality Doctrine is rather difficult to understand, but it is clearly important to Aristotle. It is closely connected with some claims

[18] He claims to improve on the solution given 'in the first discussion of motion' (ἐν τοῖς πρώτοις λόγοις τοῖς περὶ κινήσεως, *Phys.* viii.8, 263a11), apparently referring to *Phys.* vi.2, 233a21ff.

[19] See, for example, *Phys.* vi.2, 232a32ff., 232b27ff., 4, 234b23ff., 6, 236b32ff., and 10, 241a8ff.

about potentiality found in the *Metaphysics*, in particular the doctrine that no substance is composed out of actually existing substances (*Metaphysics* Z, 13 and 16). (The doctrines are related but not the same, since motions, lines, and times are not substances.) Aristotle seems to need it in order to combine his belief in the existence of continua with his denial of the simultaneous actual existence of infinitely many things.[20] Commentators often regard it as one of his central doctrines about continuity.[21]

What, then, should we make of the Potentiality Doctrine's absence from *Physics* V–VI? Here are two opposed and extreme views. On a unitarian view, we might simply infer that Books V–VI are not intended as a complete, self-sufficient treatment of continua. Part of their job is to build up to Book VIII, and they were written with the intention of their being supplemented by further refinements such as the Potentiality Doctrine. On an opposite view, we might infer that Books V–VI were written in isolation, independently from the concerns of the rest of the *Physics*. Issues about causation, potentiality and actuality, and infinity are simply not on their agenda. They are a more or less mathematical text (relating perhaps to a part of mathematics in the way that optics relates to geometry),[22] plonked into the middle of a more natural-philosophical one.

Presumably the truth lies somewhere between these extremes, and I will not try to determine where precisely. But I would like to point out two issues that are relevant to the question. The first issue is whether Aristotle's argumentation in *Physics* V–VI is even *compatible* with the Potentiality Doctrine. Do his proofs go through when we add the premise that none of the parts and middle-points appealed to therein have actual existence? There is no space here to settle this issue, since it depends on both a detailed interpretation of the Potentiality Doctrine and a detailed analysis of Aristotle's proofs. I am optimistic that the central proofs of Books V–VI can be recast so as to be compatible with the Potentiality Doctrine.

The second issue concerns the argument that Aristotle gives in support of the Potentiality Doctrine in *Physics* VIII.8. The argument is unsound. Worse, the flaw in Aristotle's argument for the Potentiality Doctrine

[20] See Coope 2005: 10. Aristotle's main target in his discussion of infinity (*Physics* III.4–8) is the question of infinitely extended magnitudes. But he also speaks of number, and he seems to deny any simultaneous, actually infinite number of things at III.7, 207b11–15.

[21] See, for example, Ross 1936: 68.

[22] The relation in question I have in mind is described by Aristotle as 'being under'. See *APo.* I.13, 78b32ff. for this relation both within mathematics and across the boundary between mathematics and natural science.

appears to rest precisely on a failure to appreciate a significant theorem from *Physics* VI.5.

III.1 The argument of Physics VIII.8

When you threw the apple into the air and let it fall back into your hand, there was a point where the apple turned around. That point was an actual middle-point of the apple's movement up and down. It served as the end-point of the apple's motion up and the starting-point of the apple's motion down. The apple arrived there from your hand and departed from there back to your hand. For Aristotle in *Physics* VIII.8, these claims imply that the apple rested for some time at the point, and hence that it did not undergo a single motion up and back.

The point where the apple turned around is different from a point before your eyes which the apple passed through on its way down. At the point before your eyes, the apple did not rest for any time. It did not arrive there or depart from there. The point did not serve as the end-point of a motion or as the starting-point of a motion. It was not an actual middle-point of the apple's downward motion. A single motion has no actual middle-points.

Aristotle can parlay his result about middle-points of motion into a general doctrine about parts and middle-points of continua. For, first, Aristotle presumably thinks that a motion has an actual middle-point if and only if it has actual parts: if there is a middle-point, then there are parts that meet at the middle-point, and if there is a part, then there is a middle-point where the part meets its remainder. This explains why, having argued that a continuous motion has no actual middle-points, he later asserts that it has no actual halves (VIII.8, 263a28–29). Second, Aristotle seems to think that a motion has actual parts if and only if the *time* occupied by the motion has actual parts, and if and only if the *magnitude* traversed by the motion has actual parts (compare *Phys.* VI.4, 235a15–17ff.). This explains why he is prepared to assert that continuous lines and times, in addition to motions, have only potential parts (VIII.8, 263b3–9).

Let us follow Aristotle through his argument about middle-points of motion. The central passage is *Physics* VIII.8, 262a19–b8, with the argument then being carried on at 262b17–263a3. Aristotle's first premise is this: if a point is (actually) a middle-point of a motion, then it is (actually) the end-point of a motion (namely, of the part of the whole motion of which it is the final boundary) and the starting-point of a motion (namely, of the part of the whole motion of which it is the initial boundary). He expresses this as follows:

> There being three things, beginning, middle, and end, the middle is both in relation to each. (*Phys.* VIII.8, 262a19–20)

Aristotle's next two premises are these. If a point is actually the starting-point of a motion, then at some time the moving thing *has departed* (*apogegone, apelêluthe*) from the point. And, if a point is actually the end-point of a motion, then at some time the moving thing *has come to be* (*gegone*) at the point. Thus he writes, of something which has undergone a motion from A to C:

> It has departed from point A, i.e. the beginning, and has come to be at C, when it finishes and stops. (*Phys.* VIII.8, 262b7–8)

From these premises it follows that, if a point is actually a middle-point of something's motion, then at some time the thing has come to be at the point, and at some time the thing has departed from the point. So far, Aristotle's reasoning is tenable. What comes next, however, is not so good:

> It is impossible for A simultaneously to have come to be at B and to have departed. Hence it does the one in one point of time, and the other in another. Hence there will be a time in the middle, and consequently A will rest at B. (*Phys.* VIII.8, 262a32–b2)

Aristotle adds a further detail to his argument a little later (by which time point B has been replaced by D):

> It is not the case that it simultaneously has come to be at D and has departed from D: for it would simultaneously be there and not be there in the same instant. (*Phys.* VIII.8, 262b26–28)

In these passages, Aristotle assumes that if A has come to be at a point and has departed from the point, then there is such a thing as the instant, or point of time, when A has come to be at the point, and such a thing as the instant or point of time when A has departed from it. At the instant when A has come to be at the point, it *is* at the point, and at the instant when A has departed from the point, it *is not* at the point. Given the Principle of Non-Contradiction, it follows that the instant when A has come to be at the point is different from the instant when A has departed from the point. Between any two instants there is a period of time. So, there is a period of time between these two instants, during which A has come to be at the point and has not yet departed from the point. Throughout this time, A is at the point. Hence, A rests at the point.

From the perspective of *Physics* VI.5, many of the steps in this argument are acceptable.[23] But there is a crucial error, namely, Aristotle's assumption that there is such a thing as *the instant when* A has departed from the point. It is, of course, sometimes true to say of A that it has departed from the starting-point of its motion. But for every instant at which it is true to say this, there is an earlier instant at which it is also true to say this. There is no first, primary time or instant in which A has changed some, in which A has departed from the starting-point of its motion. This is argued at some length in *Physics* VI.5 (236a13–27), and corollaries are derived from it in the remainder of that chapter (236a27–b18).

When we hold this lesson in mind from *Physics* VI.5, we can see that the argument of VIII.8 does not go through. True, there is such a thing as the instant when A has come to be at the middle-point, and yes, at this instant A is at the point. Every instant at which A can be truly said to have departed from the point is indeed different from, not simultaneous with, the instant at which it has come to be there. So, yes, for any given instant at which A has departed from the point, there is a period of time separating it from the instant at which A had come to be there. But no, such a period is *not* a period during which A has come to be at the point and *not yet departed* from the point. For, prior to any instant you choose at which it is true that A has departed, there is always an earlier instant at which it is already true that it has departed. (There is a last instant at which it has not departed, and every instant thereafter is one at which it has departed.)

As far as *Physics* V–VI is concerned, nothing seems to rule out the possibility that the apple you threw and caught never rested anywhere between leaving your hand and arriving back in your hand. What is more, as far as I can see, nothing in those books rules out the possibility that it underwent a single motion up and back. Related to this, Books V–VI seem to be consistent with the proposition that a single motion has actual middle-points. Now, *Physics* V–VI do not purport to deliver the whole truth about motion. Perhaps Aristotle could derive the results he wants by introducing some further independent principle – something consistent with, though not entailed by, the theory of *Physics* V–VI. But that is not what he does in *Physics* VIII.8. His argument in *Physics* VIII.8 rests on a tacit premise that is inconsistent with *Physics* VI.5. Thus, the argument that Aristotle actually gives for the Potentiality Doctrine, along with the

[23] For example, the connections between having departed and not being there, and between having come to be and being there, appear in VI.5 (235b14–16, 235b7ff.). Similarly, the claim that there is a time between any two instants is ubiquitous in V–VI.

associated argument against the existence of eternal rectilinear motion, is incompatible with the theory of *Physics* v–vi.

IV Conclusion

Let me close with three questions.

The first question is, did Aristotle have a single basic understanding of continua, or did he rather have multiple overlapping fragments of theories? In particular, where does the Potentiality Doctrine belong in his understanding of continua? I have argued that his attempt to establish this doctrine is defeated by *Physics* vi.5. Given this, we should not, without further ado, read the Potentiality Doctrine into the theory presented in *Physics* v–vi. Further research is called for here. A minimal condition on attributing a combined theory to Aristotle would be to show that the Potentiality Doctrine (setting aside Aristotle's attempted proof of it) is consistent with the doctrines and proofs given in *Physics* v–vi.

Second, what should we think about the unity or disunity of the second half of the *Physics*? The tensions we have seen go beyond, I think, what we normally find as a result of mistakes or oversights within a single philosophical work. They indicate that Books v, vi, and viii were not all written in one sitting. This impression is reinforced by differences in technical terminology between the books.[24] On the other hand, the texts do not seem wholly independent of each other, and it is plausible that Aristotle himself assembled them together. It would be useful to have a more complete map of the interconnections and any additional tensions between the texts, so as to make an informed hypothesis about the manner of their composition.

Finally, do the tensions we have seen amount to a major problem for Aristotle's physics and cosmology? I have argued that some of Aristotle's arguments against the possibility of eternal rectilinear motion in *Physics* viii.8 are inconsistent with v–vi. Are *any* of his arguments consistent with the earlier books? I count six arguments in all. Two seem to rely on the

[24] In most books of the *Physics*, including v–vi, a point is called a στιγμή (*Phys.* v.3, 227a27, 28, 31, 4, 227b16; vi.1, 231a25, 26, 30, 231b7, 9, 13, 10, 241a3, 7, 10, 12, 13, 19; there is an exception at vi.9, 240b3). In Book viii a point is called a σημεῖον (viii.8, 262a23, 29, 262b2, 4, 7, 12, 25, 263a24, 31, b10, 12, 264a3). Furthermore, in *Physics* v–vi the starting-point and end-point of a change are normally referred to using the prepositions ἐκ and εἰς, respectively (e.g. v.1, 225a1), while in Book viii Aristotle uses ἀπό for the one and ἐπί or πρός for the other (e.g. viii.8, 262a7, 9–10, b10–12, 19, 264a29–31).

mistake I described about the beginning of change (262a19–263a3, 264b1–6). Two seem to rely, more or less directly, on the 'homogeneity' interpretation of oneness in kind (261b32–262a6, 264a21–b1), which I have argued is incompatible with *Physics* V–VI. One is brief, cryptic, and *prima facie* irrelevant to the desired conclusion (264b6–9). This leaves only one argument which, I think, could be made to work (264a7–21). And this argument seems more suggestive than demonstrative, resting on somewhat uncertain linguistic intuitions rather than clear scientific principles. (Roughly, the intuition behind the argument seems to be that there is something wrong with saying of a thing, 'it is moving back and forth between A and B'. Instead, you should say, at some times, 'it is moving to A', and, at other times, 'it is moving to B'.)

In the first book of *On the Heavens*, Aristotle treats it as a previously established fact that circular motion, and only circular motion, can be eternal (*Cael.* I.2, 269b6–9). We have not questioned Aristotle's positive arguments in the *Physics* for the possibility of eternal circular motion. But it is troubling that his most powerful-seeming arguments on the negative side, against the possibility of eternal rectilinear motion, are all defeated by *Physics* V–VI. When we take away these arguments we weaken his case for the priority of circular over rectilinear motion (cf. *Phys.* VIII.9, 265a24–26), and we hamper his effort to make the stars' circular motion intelligible (why is there an eternally rotating sphere rather than, say, an eternally swinging pendulum?). Still, we should not exaggerate the trouble. Aristotle has other arguments for the priority of circular motion (*Phys.* VIII.9, 265a16–17, b11–12; *Cael.* I.2, 269a18–21). And he has other resources for denying the existence of eternal rectilinear motion. For example, he could mobilise his doctrines about natural and unnatural motions in *On the Heavens* I. According to these doctrines, any simple body that moved back and forth in a straight line would sometimes be moving unnaturally, and no body would do that sort of thing forever.

CHAPTER 12

Perfection and the physiology of habituation according to Physics VII.3

Mariska Leunissen

Introduction: Habituation as a factor in moral development

In his ethical treatises, Aristotle singles out three factors that play a role in the moral development of men, namely nature, habit, and reason.[1] Although a complete understanding of Aristotle's moral theory will thus involve an analysis of all three factors, this chapter focuses mostly on Aristotle's discussions of habit and the process of habituation. Specifically, I aim to offer an interpretation of Aristotle's "physical" discussion of the acquisition of character virtue in *Physics* VII.3 by drawing from his physiological treatment of habituation in the ethical treatises.

In the *Nicomachean Ethics*, Aristotle argues – famously – that character virtues do not come to be by nature (for instance, in the way that humans realize their perceptive capacities),[2] but rather that they are dispositions (ἕξεις) that we acquire "through habits" (*EN* II.1, 1103a17: ἐξ ἔθους)[3] and "by first activating them" (*EN* II.1, 1103a31: τὰς δ' ἀρετὰς λαμβάνομεν ἐνεργήσαντες πρότερον). Virtues thus come to be as the result of the repeated performance of virtuous actions, whereby the quality of the action is said to determine the quality of the disposition that comes to be (*EN* II.1, 1103b6–25). In the *Politics*, Aristotle claims that also practices such as bodily conditioning and exercise, imitations in play, and musical instruction during childhood are productive of the character virtues, and are therefore forms of habituation (ἐθισμός).[4] Together, these practices aim to change the appetitive soul-capacities of men in such a way that they become correctly disposed towards undergoing the appropriate

[1] *EN* X.9, 1179b20–21; *Pol.* VII.13, 1332a38–40 and 15, 1334b6–28; and *EE* I.1, 1214a14–25.

[2] See especially *EN* II.1, 1103a17–b2 and 6, 1106a9–10: "we have the capacities by nature, but we do not become good or bad by nature."

[3] See also *EE* II.2, 1220a39–b1 and *MM* I.6.2.

[4] *Pol.* VII.13, 1332b10–11; VII.15, 1334b8–9; VII.17, 1336a18–19; and VIII.4, 1339a7–10.

affections[5] and obedient to the rational part of the soul.[6] Because of this, Aristotle singles out habituation as the most important factor in the moral development of men, as it can override nature and prepares the ground for moral education.[7] However, even though Aristotle is fairly explicit about the scope, aims, and importance of habituation as a factor in moral development, the ethical treatises – perhaps because they are practical in nature – do not provide much information about what exactly happens *physiologically speaking* in the bodies and souls of men during habituation.[8]

This chapter aims to close this gap in our understanding of Aristotle's concept of habituation by examining the scant physiological remarks in the ethical treatises *from the perspective of* Aristotle's only extended discussion of the changes involved in virtue-acquisition in *Physics* VII.3 and by offering an explanation of how these physiological changes ultimately result in the presence of character virtue as a distinct and unified psychological disposition. Let me note from the outset, however, that the discrepancies between Aristotle's ethical and physical accounts of virtue-acquisition are significant, and that it is not my aim to harmonize the two accounts in all respects. Rather, I will use Aristotle's discussion in *Physics* VII.3 as a starting-point for what I take to be a plausible, "Aristotelian" account of the physiology of habituation, and draw from the ethical treatises in order to flesh out this picture from the *Physics* and to further clarify it.

In section I, I briefly examine three claims Aristotle makes regarding virtue-acquisition in *Physics* VII.3, namely (1) that their acquisition is a case of "perfection" (and not of generation or alteration), (2) that character virtues belong to the category of "relation" (and not to that of quality), and (3) that their acquisition "always occurs with" the alteration of the perceptive part of the soul. In section II, I offer suggestions for how this "natural scientific" account of perfection can be applied to Aristotle's characterization of habituation in the ethical treatises and present a reconstruction of

[5] *EN* II.3, 1104b11–13 and X.9, 1179b24–26; *Pol.* VIII.5, 1339a20–25 and 1339a41–b4. See Moss 2011 and 2012 on character virtue as a non-rational (but cognitive) state and on habituation as a non-rational process of practical induction sufficient for the production of virtue.

[6] *EN* I.13, 1102b25–31; *EE* II.1, 1220a10–12 and 2, 1220b6–8.

[7] *EN* X.9, 1179b4–31 and *Pol.* VII.13, 1332a38–b8.

[8] Cf. Tracy 1969: 23: "[Aristotle] presents the physiological and psychological aspect of moral activity, therefore, only in so far as that is necessary to the practical moralist"; Reeve 1992: 61: "What we cannot do is explain why it is that changing desires, our own or other people's, typically involves a lot of repetition and really hard work . . . There is an explanation of these facts, no doubt, but 'we must hear it from the natural scientists' (1147b8–9), not from philosophers"; and Hursthouse (2006: 112): "But how do we get from the early tuning of toddlers' passions to the enjoyment of fine actions? . . . 'Habituation,' Aristotle tells us, but, as everyone notes, he tells us little about what this involves."

the various types of change involved in habituation mentioned there. And finally, in section III, I will use this evidence from the ethical treatises to make sense of Aristotle's second and third claim in *Physics* VII.3 and provide an account of the changes that occur in the perceptive soul of men during habituation.

I The acquisition of character virtue in *Physics* VII.3

In *Physics* VII.3 Aristotle provides one of the lengthiest discussions of the physiology of virtue-acquisition in the Corpus.[9] However, the main purpose of the chapter is to defend a rather restricted notion of alteration, which Aristotle needs in order to save his earlier argument against kinetic infinity. This argument, which is developed in *Physics* VII.1–2, builds on the premise that every change requires contact between the origin of change and the entity undergoing that change, but in *Physics* VII.3 Aristotle brings up two examples of change – both within the category of alteration – that appear to violate this requirement of contact: namely, the change that items undergo when taking on or casting off figures or shapes, and the change that items undergo when taking on or casting off dispositions (*hexeis*). Aristotle's strategy in dealing with these two alleged counterexamples is, first, to redefine *all* alteration as change in perceptual qualities *only* (245b3–5), thereby restricting alteration to changes in the "affective qualities and affections, such as hotness and coldness" (which constitute the third type of quality as defined in *Cat.* 8, 9a14–10a10);[10] and, next, to show that the two alleged counterexamples are not really cases of alteration at all. The bulk of the chapter deals with the second example, i.e., the acquisition and loss of dispositions of both body and soul (see 246a10–248a9; the example of figures and shapes is discussed in 245b6–246a9), with dispositions being immediately identified with virtues and vices (246a11–12). The subsequent discussion then focuses on the acquisition and loss of these virtues and vices (first those of body, next those of soul), and it is in this context that Aristotle discusses the acquisition of character virtues. In what follows, I discuss Aristotle's account of the acquisition of virtues of

[9] In writing this section I have benefitted much from Maso, Natali, and Seel 2012. I have also relied on their edition of the Greek text of *Physics* VII.3, which has been transmitted in two different versions – version α, which is the version printed by Ross 1950, and version β, which is the version known by Simplicius, Philoponus, and Themistius. My discussion focuses on version α, but also draws from version β where relevant.

[10] *Categories* 8 discusses three more types of quality: (1) states and conditions; (2) natural (in)capacities; and (3) shapes and forms.

the body (in 246b4–20) and that of the acquisition of character virtue (in 246b21–247a19) together, as the two accounts are very similar.[11]

As stated above, Aristotle begins his demonstration of the proposition that the acquisition and loss of dispositions are not cases of alteration by identifying these dispositions with virtues and vices, perhaps because, in the case of living beings, *every* disposition makes its possessor either well or ill disposed (*Metaph.* v.20, 1022b10–12).[12] Based on this identification, Aristotle argues that, since virtues and vices are not alterations but *perfections*, dispositions of the body and soul – and their acquisitions – are not alterations (246a12–b3). Subsequently, he argues that virtues and vices belong to the category of *relation*, thereby reaffirming again that their acquisition or loss cannot be a change in the category of quality (246b3–4). Aristotle then works out these two arguments with regard to virtues and vices of body and of character, while – also in both cases – adding a third argument about how, although their acquisition is not itself an alteration, virtue-acquisition always occurs "with" an alteration of something else. Let me discuss these three arguments in turn.

The first argument identifies virtues and their acquisition with "perfections," while identifying vices and their acquisition with "departures" (246a13–17):

> But rather virtue is a kind of perfection (τελείωσίς τις) – for when it acquires its own virtue (τὴν αὑτοῦ ἀρετήν), then we say that each thing is perfect, for then it is most in accordance with [its] nature (τότε γὰρ ἔστι μάλιστα [τὸ] κατὰ φύσιν), just as a circle is perfect, when it has become a circle in the highest degree and when it is best – and vice is a destruction of this and a departure [from this].

According to this account, which applies equally to virtues and vices of the soul (246b21–247a3), acquiring a virtue is reaching a form of perfection according to which its possessor now exhibits its own nature to the highest degree or in the best way, while acquiring a vice is degenerating from one's nature. Aristotle uses the perfection of houses as an example (246a18–b1): Evidently it would be "absurd" to call the coping and tiling of houses alterations rather than perfections, and so the acquisition of virtues (or

[11] The section concerning the intellectual virtues (247b1–248a6) falls outside the scope of this chapter; on this, see De Haas 2012: 99–108 and Natali 2012: 109–117.

[12] For the sake of this argument, Aristotle must assume that *all* dispositions – and not just those of body and soul – are either virtues or vices. However, elsewhere he also refers to virtues as conditions (*diatheseis* – a term that is perhaps exchangeable with *hexis* in discussions of virtue; see Bowin 2011: 147n.21), and possibly to dispositions of liveless things that are neither virtues nor vices (see e.g. *DA* II.7, 418b18–20 and III.8, 432a5–6). On this, see further Coope 2012: 59–61.

vices) by living beings should also be called perfections (or departures) and not alterations.

What *exactly* is entailed by such a process of perfection is unclear, but the house-example seems to associate perfection with a teleological process of generation. Thus, perfection may either refer to (1) the *final stage* of such a process, as suggested by "coping": for, strictly speaking, houses do not exist without the last (row of) brick(s) being in place yet, so that coping could not be a case of alteration simply because the subject of alteration does not yet exist. Or (2), as suggested by the comparison with "tiling" (and by the example of the circle), perfection may refer to a change of something that has already reached its final stage of generation (e.g., the house and circle already exist, but perhaps imperfectly so), but that is now made to *express its own nature to the highest degree.* The latter is not a case of alteration, since the subject does not come to possess any new attributes. This second reading, which I think is the more plausible one, also fits best with Aristotle's characterization of perfection in version β (246b27–29):

> For virtue is a kind of perfection – for each thing is then most perfect (μάλιστα τέλειόν), when it reaches its proper virtue (τῆς οἰκείας ἀρετῆς) and is most in accordance with nature, just as a circle is then most in accordance with nature, when it is most a circle.

In this version, it is clearly not the generation of the circle that is at stake, but rather the *degree* to which it exhibits its own nature: by being a circle *most*, it is said to be most perfect and to reach its "proper virtue." Similarly, in Aristotle's definition of perfection in *Metaphysics* v (16, 1021b14–1022a3), a being – such as a physician or a fluteplayer – is called perfect when "with respect to its kind it lacks nothing of its proper virtue" (b16–17: ὅταν κατὰ τὸ εἶδος τῆς οἰκείας ἀρετῆς μηθὲν ἐλλείπωσιν) and when it "lacks nothing with regard to goodness and cannot be excelled and has nothing that is found outside of it" (b31–32: τῷ κατὰ τὸ εὖ μηδὲν ἐλλείπειν μηδ' ἔχειν ὑπερβολὴν μηδὲ ἔξω τι λαβεῖν). Again, what is at stake in this definition is not whether something has already realized its own nature or not (the examples refer to people who are already practitioners of their crafts, not students in training), but rather whether it has realized it well or badly, with perfection picking out the condition of that which could not have realized its nature better – it lacks *nothing* of its proper virtue and *cannot* be excelled. Hence, something is perfect when it does not only have everything that is necessary for it to be what it is (that is, for it to be a normal functioning member of its class), but when it also possesses those things in the best possible condition, such that their interplay and combination make it unsurpassable in excellence.[13]

Aristotle's second argument places the virtues and vices of body and character in the category of relation (246b3–4 and 247a1–4; and not in the category of quality, as is suggested in *Cat.* 8, 8b27–29),[14] as they all satisfy the following two criteria (246b8–9):[15] first, "each . . . is in [the category] of relation (ἑκάστη . . . ἐστι τῷ πρός τι)," and second, "[each] puts what has it in a good or bad condition (εὖ ἢ κακῶς διατίθησι) regarding its proper affections," where "proper" is understood as picking out those affections that naturally produce or destroy that which is being affected. And, as there are no changes of relations properly speaking (instead, the items that are part of the relation change, and thereby change the relation between them, but this is not a change of the relation itself), the acquisition or loss of these virtues or vices are not alterations (246b10–14).

Unfortunately, Aristotle does not specify what components are constitutive of the kind of relation that qualifies as character virtue. Perhaps the analogy with the relational nature of health and fitness (246b4–6) is helpful here: For, according to Aristotle, the latter are placed "*in* the mixture and due proportion of hot and cold [elements] (ἐν κράσει καὶ συμμετρίᾳ θερμῶν καὶ ψυχρῶν)," which exist either in relation to themselves, internally in the body, or externally in relation to the mixture of the environment.[16] So, apparently, bodily virtues and vices are either *nothing more* than elements that are in some way related to something (as is suggested in version β, 246b21–22: "just as health is a kind of symmetry of hot and cold [elements], either with what is internal or in relation to the environment"), or they are *ontologically dependent* on those relations without being identical to them.[17] In either case, it seems that, if character virtues are relational *in a similar or analogous way* to the way bodily virtues are relational (as is indicated by ὁμοίως in 246b20), then perhaps character virtues, too, should

[13] Cf. Coope 2012: 69–71 and Strohl 2011: 284–285.

[14] The *Categories* lists the *genus* of dispositions and virtues in the category of relation, but the *particular* dispositions or virtues as qualities: see *Cat.* 7, 6b15–16 and 8, 11a20–36.

[15] The Greek leaves open the possibility that only one of the two criteria needs to be satisfied in order to classify as a relation, but other evidence suggests that health at least requires both: see Freudenthal 1995: 13.

[16] See also *Phys.* IV.3, 210b22–27:

> What Zeno is puzzled about, that if there is a place, it must be in something, is not difficult to solve. For nothing prevents the first place from being in something else, however not as in that place, but *as health is in the hot elements as a state* (ὥσπερ ἡ μὲν ὑγίεια ἐν τοῖς θερμοῖς ὡς ἕξις), and [as] heat [is] in the body as an affection.

[17] Bodnár 2012: 75–77.

be "placed in" or even identified with a mixture and due proportion – not of bodily elements, but rather of "elements" of the soul.

The third argument for why the acquisition and loss of virtues or vices are not alterations claims that this miscategorization is due to the fact that their generation and perishing necessarily always involve *the alteration of something else*. For bodily virtues and vices are said to come to be when the bodily elements in which these states are placed are altered (246b14–20). The account concerning character virtues is more complex, so let me quote it in full (247a6–19):

> But it is necessary that they come to be when the perceptive part [of the soul] is altered (ἀλλοιουμένου τοῦ αἰσθητικοῦ μέρους). And it will be altered by perceptibles. For every character virtue concerns pleasures and pains of the body, and these [occur] in acting or in remembering or in expecting. The ones that occur in action are in accordance with perception, such that they are being moved by something perceptible; the ones that occur in memory and in expectation [come] from this: for either the ones who remember what they have experienced or the ones who expect what they will [experience] feel pleasure. Therefore it is necessary that every such pleasure comes to be through perceptibles. Since when pleasure and pain come to be [in the soul] also vice and virtue come to be (for they [i.e., vice and virtue] concern them [i.e., pleasure and pain]), and pleasures and pains [are] alterations of the perceptive part, it is clear that when one undergoes some alteration one necessarily also loses or acquires those [vices and virtues]. Therefore there is a coming to be of them with alteration, but they themselves are not alterations.

In short, Aristotle argues that when the perceptive part of the soul – in which, presumably, the character virtues are present as relations – is affected by pleasures and pains (which are themselves produced by sense perception) and is thereby altered,[18] character virtues or vices are acquired or lost.

As this overview brings out, Aristotle's remarks about the nature of virtue-acquistion in *Physics* VII.3 are very condensed and leave it unclear as to what extent they apply to his treatment of habituation in the ethical treatises, where – at least *prima facie* – virtue-acquisition appears to be a kind of qualitative change and the language of perfection does not play any substantial role. In what follows, I will analyze Aristotle's treatment of

[18] This characterization of pleasures as alterations of the perceptive soul seems to clash with Aristotle's account of pleasures as perfections of activities in *EN* X.4, 174b31–33: on this, see Wardy 1990: 224–225, who suggests that the former is a simplification of a more complex theory of pleasure that is not relevant in *Physics* VII.3.

perfection and the physiology of habituation in the ethical treatises, and show that, for the most part, it is quite congenial to what has been deemed to be an "excessively physiological"[19] account in *Physics* VII.3, and that these materials together can be used to offer a richer picture of what happens physiologically when men habituate themselves.

II Perfection and the physiology of habituation in the ethical treatises

In the ethical treatises, the view that the acquisition of character virtues through habituation is a kind of *perfection* that results in men who are not only biologically complete and functional, but who also have reached their proper virtue and are therefore *most* in accordance with their human nature, is never developed in much detail.[20] However, as I will argue, it is nonetheless present and fitting with the account in *Physics* VII.3, and, furthermore, its conceptualization in the ethical treatises provides additional insights into the physiological process of habituation. Since the *Nicomachean Ethics* presupposes the student of ethics to have had the kind of upbringing that is outlined in the *Politics,* I start my account with the latter.

In the *Politics*, the view that habituation is a form of perfecting men surfaces in Aristotle's discussion of the tasks of lawgivers in the ideal city. One of the most important of these tasks is the proper habituation and education of children who would otherwise remain imperfect both in their deliberative capacity and in the development of their character virtues (*Pol.* I.13, 1260a13–14 and 1260a31–33; a31: ὁ παῖς ἀτελής) and would not be able to contribute to the virtuousness of the city. Accordingly, Aristotle claims that the educational program should follow the natural division of the development of boys rather than the traditional division into seven-year periods, for "it is the purpose of every craft and [every] education to complete what is lacking in nature" (*Pol.* VII.17, 1336b40–1337a3; a2–3: τὸ προσλεῖπον . . . τῆς φύσεως ἀναπληροῦν).[21] This characterization of education suggests that Aristotle sees the role of political science as a form of helping men realize their human nature ("completing," I take it, is a form

[19] See Ross 1936: 676; Viano 2012: 89–93; and Wardy 1990: 223–227.

[20] The conceptualization of habituation as a form of perfection is, however, worked out in detail in Aquinas, who adopted it from Aristotle: see Irwin 2006: 332–335.

[21] Cf. *EE* VII.2, 1237a2–3 and Aristotle's eugenic program in *Pol.* VII.16 (especially at 1335a11–17 and b29–31), which is designed to help avoid the natural occurrence of imperfections in offspring.

of "perfecting": see *Protrepticus*, Iambl. IX 49.28–50.12),[22] since this cannot be achieved by nature alone.[23] The aim of habituation, then, is to complete nature by instilling the proper virtues in men and thereby producing a kind of "second nature" in them that replaces the imperfect and unstable one they have from birth (see, e.g., *Rh.* I.11, 1370a5–9, *Mem.* 2, 452a27–30, and *EN* VII.10, 1152a29–33). Aristotle does not specify the physiological changes produced in men during this process of perfection brought about by habituation, but his discussion of the educational program of boys in the ideal city includes several components that imply some kind of change: conditioning, imitations through play, and musical education. Let me discuss these in turn.

First, since habituation should track the order of the natural development of the bodies and soul-parts of boys (see *Pol.* VII.15, 1334b10–29), Aristotle stresses the importance of bodily exercise and conditioning straight from birth. By exposing newborns up until the age of 5 to particular environments, diets, and motions, lawgivers primarily condition their bodies for health and (military) strength, and prevent laziness or deformity (*Pol.* VII.17, 1336a25–39 with *Pol.* VIII.2, 1337b5–15), thereby instilling the bodily virtues – a process that is also emphasized during the ages of 7 to 17. However, these practices also change the physiology of the blood to promote natural courage (consuming milk, for instance, produces "a military disposition": see *Pol.* VII.15, 1336a3–8), and possibly also alter the appetitive capacities of the soul by conditioning boys to certain behaviors and affections (e.g., so that they no longer experience cold temperatures as painful: see *Pol.* VII.15, 1336a12–18) – a practice that is also common among some animals.[24]

A second process involved in the early habituation of boys within the household is that of *imitation* through play. Children should not only be exposed to morally good examples (whether in the form of behavior, speech, or artistic representation) while being protected from bad ones, but they also need to be encouraged to imitate those good examples. For instance, children should perform those actions in play that are imitations

[22] For nature does not imitate craft, but it (i.e. craft) nature, and it exists to help also complete those things nature leaves out (τὰ παραλειπόμενα τῆς φύσεως ἀναπληροῦν). For some things nature itself seems capable of bringing to perfection (ἐπιτελεῖν) by itself without actually needing any help, but others it [brings to perfection] with difficulty or is completely unable to do.

Transl. Johnson and Hutchinson 2005: 259; slightly modified.

[23] Reeve 2012: 104. Cf. *Phys.* II.3, 194a21–27 and 8, 199a15–17.

[24] *Pol.* VII.13, 1332b3–4. On animals habituating their young, see *HA* VI.12, 567a5–7, IX.5, 611a19–21; and IX.7, 612b29–31; on humans habituating animals, see *HA* VI.19, 573b25–27; VI.21, 575a33–b2; and IX.3, 610b33–611a2.

(μιμήσεις) of the serious actions they will perform later in life (*Pol.* VII.17, 1336a33–34). Because imitations are naturally pleasant to humans, imitation is a natural mode of learning for them (*Poet.* I.4, 1448b5–24; cf. *Rh.* I.11, 1371b5–10), and it is especially suited for the habituation of children, who easily translate what they see and hear into action (*Pol.* VII.17, 1336b2–8). For through imitative play children are encouraged to reinvoke their perceptions of the actual goods presented to them via imagination – a practice that, perhaps, sharpens their discriminative ability to perceive the actual good as good[25] – and to take pleasure in the performance of those good actions: Imitation thus shapes the preferences of boys by making them enjoy – and love (see: *Pol.* VII.17, 1336b33) – the performance of virtuous actions.

The third process involved in habituation is (presumably)[26] part of the education of boys from age 7 to 14: In addition to performing gymnastics and being educated in letters and drawing, students are to receive *musical education.* Aristotle argues that education in music prepares boys for the proper enjoyment of leisure later in life, once they have become morally perfected men,[27] but he also suggests – and later endorses that this is indeed the case (*Pol.* VIII.5, 1340a1–12) – that music is productive of character virtues (*Pol.* VIII.5, 1339a20–25):

> Or should we rather think that music contributes something to virtue, on the grounds that – just as gymnastics prepares our body to be of a certain quality – so too, music is capable of making our character of a certain quality (τὸ ἦθος ποιόν τι ποιεῖν), by habituating us to be able to enjoy things in the right way (ἐθίζουσαν δύνασθαι χαίρειν ὀρθῶς).

Apparently, education in music – which involves both being trained as audience and, up to a point, also as a performer (*Pol.* VIII.6, *passim*) – is capable of producing positive alterations in human character traits (cf. *Pol.* VIII.5, 1339a41–42: εἰ δύναται τὰ ἤθη βελτίω ποιεῖν) and thereby helps produce character virtue. And, Aristotle claims, music not only makes men able to *enjoy* things in the right way, it also makes them able to *judge* in the right way, where the objects of enjoyments and judgments are not just music itself, but also "appropriate characters and good actions" (*Pol.* VIII.5, 1339a41–b4 and 1340a14–18; cf. *Pol.* VIII.6, 1340b38–39). In this way, habituation through music not only alters the – presumably – natural

[25] Perception is a causal factor involved in action (*EN* VI.2, 1139a17–18), and virtue requires one to be able to perceive the actual good as being good for oneself (*EN* III.4, 1113a29–33 and 5, 1114a31–b20). On practical perception, see further Moss 2011: 251–254 and Reeve 2012: 40–50.

[26] Reeve 2012: 254.

[27] See *Pol.* VIII.3, 1337b25–1338a32 and 5, 1339a32–33, with 1339b15–25.

character traits in the direction of natural virtue, but also the appetitive and the (practical) perceptive capacities of the soul.

Aristotle explains these changes produced by music physiologically in terms of the special representational nature of music: "by hearing imitations ... all people become similarly affected (*Pol.* VIII.5, 1340a12–14: ἀκροώμενοι τῶν μιμήσεων γίγνονται πάντες συμπαθεῖς)" and "change with respect to their soul (1340a22–23: μεταβάλλομεν ... τὴν ψυχήν)," implying that the affection that is expressed in music – when experienced – will produce the corresponding affection (or disposition: 1340a41) in the listener.[28] This in turn happens because sounds possess likenesses to character traits (1340a29–30: ὁμοίωμα τοῖς ἤθεσιν), with melodies representing different kinds of natural character traits and rhythms representing the character of action (1340a38–b10), and humans have a natural affinity to both of these (1340b16–18; cf. *Poet.* 4, 1448b20–21). Or, as the author of the *Problems* explains it, sounds consist of two sets of motions, one moving the sense organs, the other moving us in accordance with the character they resemble (*Pr.* XIX.27, 919b26–37; cf. *Pr.* XIX.29, 920a3–7 and XIX.48, 922b22–23). Thus, by exposing them to *the right kinds of music*,[29] boys regularly realize their natural character traits in the best way possible and are habituated to experience the kind of emotions and pleasures and pains that one ought to feel when faced with the real objects and situations (*Pol.* VIII.5, 1340a23–28). And boys can do so without actually – or yet – having to perform virtuous actions, the qualities of which would have the same effect, as Aristotle claims in the *Nicomachean Ethics*.

In the *Nicomachean Ethics*, Aristotle uses the term "perfection" explicitly and in the relevant way in his earliest discussion of what the character virtues are and how they are acquired in *Nichomachean Ethics* II.1.[30] There, after rejecting the view that character virtues come to be by nature, he states that humans are by nature fit to receive the character virtues, but that "[we] are perfected through habits" (*EN* II.1, 1103a23–26; 25–26: τελειουμένοις δὲ

[28] Especially in melody and rhythms, Aristotle says, there is "the greatest likeness to the true natures of anger and gentleness, and in addition of courage and temperance and of all their opposites and of other character traits as well" (*Pol.* VIII.5, 1340a18–22).

[29] Not all music is appropriate for educational purposes: one should only use melodies that "pertain most to character (ταῖς ἠθικωτάταις)" and do so in a good way (*Pol.* VIII.7, 1341b38–1342a3), such as the Dorian melody (*Pol.* VIII.7, 1342b13–17).

[30] Although Aristotle does not specify what he means by perfection, he clearly understands it as a perfection *of one's nature* (*EN* VII.12, 1153a11–12: τῶν εἰς τὴν τελέωσιν ἀγομένων τῆς φύσεως), as acquiring one's *proper virtues* (*EN* II.5, 1105b29–31, 6, 1106a15–24; and, perhaps, X.5, 1176a27), and as resulting in unqualifiedly good men (*EE* VII.2, 1237a29–30).

διὰ τοῦ ἔθους).[31] Aristotle offers three arguments for the claim that perfection occurs through habit and not through nature, and because they are informative for his conceptualization of the physiology of habituation, I discuss each of them in turn.

First, Aristotle claims that nothing that behaves in one way by nature can be habituated to act contrary to its nature (*EN* II.1, 1103a19–23). Stones, for instance, have a natural tendency to move downwards, and no amount of being thrown upwards changes this tendency. However, when it comes to character, habituation can produce contrary dispositions (see *EE* II.5, 1222a36–42, *Pol.* VII.13, 1332a40–b3, and *EN* II.9, 1109b1–7): The natural character traits we have from birth (which are "natural capacities") might push us into one direction, but by "pulling ourselves into the opposite direction" we can "straighten ourselves out," "change [those traits] towards the better," and attain the mean disposition of character that constitutes virtue. This kind of change is possible because, physiologically speaking, humans do not simply behave in the way they do by nature, but "by desire and choice," and because the character traits they have from birth – although natural in origin – are "rational capacities," which (as opposed to irrational capacities, which can be found in any natural entity but can only be realized in one direction) belong uniquely to ensouled beings and can be realized in opposite directions (*Metaph.* IX.2, 1046a36–b7 and IX.5, 1047b35–1048a5). Specifically, ensouled beings such as humans can themselves determine – via desire or choice – whether to realize their rational capacities in accordance with their own nature (e.g., actualizing medical science for the sake of producing health, which is the proper realization of medical science) or to produce their privations (e.g., sickness, the accidental realization of medical science: *Metaph.* IX.2, 1046b7–24 and IX.5, 1048a10–11). On this account, the *natural virtues* (but not yet the character virtues, since these are not straightforwardly realizations of natural capacities) are the proper realizations of the natural character traits humans have from birth,[32] which have been stabilized in their best condition due to men frequently producing a second actuality of these capacities in the direction of virtue. The change from natural character (a dualizing

[31] Acquiring character virtue with practical wisdom is "merely" the secondary of two ways in which humans reach perfection: "the best and most complete virtue" identified with the human good in Aristotle's function-argument (*EN* I.7, 1098a16–18) is theoretical wisdom, achieved through a life of contemplation (*EN* X.7–8).

[32] Cf. *EN* VI.13, 1144b4–9 and perhaps *EE* III.7, 1234a24–30: Character traits tend by nature towards natural virtue, which suggests that virtue is the *per se* realization of the natural capacities for character.

capacity) to natural virtue (a good disposition), then, is a form of "special alteration," which involves the preservation (σωτηρία) of an already existing capacity.[33]

That character virtue comes to be in a non-natural way is also the conclusion of Aristotle's second argument, which points to a difference in temporal priority between capacities that develop naturally and those that develop non-naturally (*EN* II.1, 1103a26–34). For while capacities that develop by nature – such as perception – are first possessed as capacities and are later displayed as activities, virtue is acquired like crafts, where the capacity for doing something is acquired by first repeatedly doing it.[34] Physiologically speaking, this distinction tracks again the difference between irrational and rational capacities (*Metaph.* IX.5, 1047b33–34):

> Of all the capacities, some are innate, such as the perceptive ones, others are by habit, such as the [capacity] to play the flute, others are by learning, such as the [capacity] of the crafts, and for some it is necessary to possess them by having previously activated them (προενεργήσαντας) – namely as many as are by habit and by learning – but it is not necessary for those that are not like this and that imply being acted upon.

Natural capacities, such as perception, are irrational and are acquired via the substantial change that takes place during embryology (i.e. the capacity to perceive comes to be at the same time as the perceptive organs). This means that we possess such capacities as first actuality from birth (*DA* II.5, 417b16–18; cf. *GA* IV.1, 766a5–10), while their use or activity – as "second actuality" – is a "special" type of alteration. And since these capacities are irrational, their activity does not depend on desire or choice, but simply on the presence of objects of perception that act on the sense organs (*Metaph.* IX.5, 1047b35–1048a24). However, acquiring character virtue is, as Aristotle puts it, just like acquiring the rational capacity to play the flute, which happens by "earlier engaging in their activities" (*EN* II.1, 1103a31: ἐνεργήσαντες πρότερον) and by doing this frequently (*Rh.* I.10, 1369b6–7). In other words, the virtuous activity that constitutes the second actuality produces the virtuous disposition that constitutes the first actuality.

The kind of "earlier activation" that is required for the production of virtues is spelled out more clearly in Aristotle's third argument against the naturalness of moral development (*EN* II.1, 1103b6–25): Since activities are the causes of and means for the production and destruction of virtues (as

[33] And not of proper alteration, which involves the destruction of one affection while another takes its place: *DA* II.5, 417b2–19. Cf. Bowin 2011: 138–161 and Polansky 2007: 234–240.

[34] Cf. *Pol.* VIII.1, 1337a18–21.

they are also in crafts), teachers are required to ensure that men perform the activities that are good rather than bad. For if there were no such causal relation between activity and virtue, and if men were simply born as good or bad performers of those activities, teaching would make no difference and men would have virtues or vices from birth. Aristotle summarizes this causal relationship as follows (1103b21–23):

> In one word, dispositions come to be from like activities, wherefore it is necessary to display a certain quality of activities (τὰς ἐνεργείας ποιὰς). For the dispositions follow the differences between these.[35]

In short, it is the *quality* of the activity that somehow produces a qualitatively identical disposition in the soul. In what follows, Aristotle clarifies what qualities one's actions should have: Actions should be in accordance with correct reason (*EN* II.2, 1103b31–32) and should hold the mean between excess and deficiency in the pursuit of pleasure and pain, and in the experience of emotions (*EN* II.2, 1104a11–27; II.3, 1104b12–16; 1104b25–26; and 1105a6–7).[36] Hence, by performing virtuous actions men experience the qualitative states of feelings and emotions – presumably as mediated by their natural character traits – that are constitutive of virtue, which, if repeated often enough, will produce a qualitatively identical disposition. The physiological account of how one becomes virtuous by listening to the appropriate kinds of music, as suggested in the *Politics*, and the account of how one becomes virtuous by performing virtuous actions are thus very similar. For both cause the natural character traits, the appetitive capacities, and the perceptive capacities to be *repeatedly* affected and thereby altered in the way that is appropriate for virtue, such that feeling pleasure, pain, and emotions correctly becomes second nature to that person, and he becomes prone to responding to perceptual inputs in the way that is characteristic of the person with a virtuous disposition.[37] Because of this, there need not necessarily be any qualitative difference *in the appetitive and perceptive capacities of the soul* in the performance of a virtuous action by someone who is habituating himself and by the virtuous person, although the latter – having also acquired

[35] Cf. *EN* II.2, 1103b29–31: "it is also necessary to investigate the issues concerning actions, namely *how* they should be done: for they are determinative – κύριαι – also of the qualities – ποιὰς – of the dispositions that come to be."

[36] Note that actions with these qualities not only produce virtues but also constitute the proper activities of the virtues once acquired (*EN* II.2, 1104a27–b3).

[37] Cf. *Cat.* 10, 13a23–31 and *EE* II.2, 1220a39–b3, where Aristotle characterizes habituation as being changed frequently in a certain way, such that the soul is eventually capable of acting in that way.

practical wisdom[38] – always performs virtuous actions knowingly (even if this knowledge does not count for much: *EN* II.4, 1105b1–2), deliberately chooses to perform them because of themselves, and performs them from a firm and unchanging disposition (*EN* II.4, 1105a31–33).[39]

This evidence from the ethical treatises concerning habituation is not exhaustive; however, I believe that it is sufficient to flesh out Aristotle's claims about virtue-acquisition in *Physics* VII.3 and thereby to present an Aristotelian account of the physiology of habituation.

III A physiological account of habituation built on *Physics* VII.3

While the ethical treatises do not often characterize habituation in terms of perfection, they are explicit about the fact that habituation is a process through which men realize their human nature to the fullest and are completed in a way that cannot be achieved by nature alone, but that requires the help of political science. At least in this respect, the treatment of habituation in the ethical treatises is consistent with the one in *Physics* VII.3. However, Aristotle makes two further claims about the acquisition of character virtue in the *Physics* that require clarification: namely, that virtue-acquisition is not a case of alteration, because virtues are relations, not qualities, and that, although virtue-acquisition is not itself a case of alteration, virtue comes to be "when the perceptive part [of the soul] is altered."

In the ethical treatises, Aristotle never refers to character virtues as relations. On the contrary, Aristotle lists virtues as examples of qualities in *EN* I.6, 1096a25, and although he discusses the status of virtues as dispositions at length in *EN* II.5, he does not distinguish there between dispositions as first actualities of natural capacities and dispositions as relations among certain elements of the soul. However, perhaps the fact that Aristotle is explicit in the ethical treatises that virtues are not straightforwardly realizations of the natural capacities for character that humans have from birth is relevant here: Although *becoming virtuous* involves frequent realizations of these natural qualities in a good direction, such that they become altered for the better, character virtue itself cannot be

38 Practical wisdom is the proper virtue of practical reason that is acquired via moral education and *indirectly* via habituation, as habituation prepares the ground for moral education: For instance, Aristotle associates habituation with a form of "fine-tuning" and inundation of souls (*EN* X.9, 1179b16–18 and 1179b24–26; cf. *Pr.* XXII.11, 931a6–15) that makes men receptive to moral arguments (cf. *EN* I.13, 1102b26–1103a1 and X.9, 1180a7–11). Cf. Burnyeat 1980: 86 and Moss 2011: 239.

39 These latter two qualities are essential for what constitutes full virtue (*EN* II.4, 1105b3–4) and are also acquired *through habituation* (1105b4–5: "[they] arise from often performing just and temperate [actions]"). See Moss 2011: 210 for arguments against intellectualist readings of *EN* II.4, 1105a31–33.

identified with the resulting, stabilized dispositions, for these amount at most to *multiple, individual* natural virtues. Additionally, even though the ethical treatises never mention the idea that virtues are *themselves* relations between (internal) elements of the soul, they do mention how *having virtues* puts one in a good relation with regard to affections (see e.g. *EN* II.5, 1105b25–26 and *EE* II.1, 1220a33–34), thereby satisfying the second criterion for relations as specified in the *Physics* (VII.3, 246b9 and 247a3–4).

Furthermore, although the ethical treatises never define habituation as a process of perfection that always takes place with the alteration of the perceptive part of the soul, we saw above that they regularly associate habituation with alterations of (natural) character[40] and, more generally, with changes or movements of the soul.[41] And it is clear that not the whole soul is being changed: for, in his ethics, Aristotle works with a tripartite division of the soul, according to which the nutritive part is always set aside as having no share in human virtue (*EN* I.13, 1102a32–b12 and *EE* II.1, 1219b20–21; b26–39), while the two remaining parts each receive their own, exclusive virtue. The part that has reason receives the intellectual virtues, while the part that is receptive to reason receives virtue of character (*EN* I.13, 1103a3–10 and *EE* II.1, 1220a8–11; cf. *Pol.* VII.14, 1333a16–19). However, with one exception, Aristotle never calls this part the perceptive one; instead he calls it "that which is capable of appetite and, in general, of desire" (*EN* I.13, 1102b30; cf. *EE* II.4, 1221b27–32, *Pol.* VII.15, 1334b10–29, and *MM* I.5, 1185b1–12) or "the part that concerns character" (as opposed to the one concerning belief: *EN* VI.13, 1144b14–16). Only once does Aristotle contrast the perceptive and appetitive capacities of the soul *together* with nutrition (*EE* II.1, 1219b21–23), thereby suggesting that both are involved in character virtue. But this fits with the depiction of habituation in the *Politics*, according to which habituation affects and alters the natural character traits as well as the appetitive and perceptive capacities of boys, such that they are capable of seeing the actual good as good for them and to take pleasure in it.

In fact, according to Aristotle's exposition in *On the Soul*, the appetitive and perceptive capacities are *both* components of the perceptive part of the soul. For in this treatise (see *DA* II.2–4, especially at 413b11–16, and *DA* III.10, 433b1–4), Aristotle divides the human soul into a nutritive part, a perceptive part, and a thinking part, while characterizing all the remaining

[40] See e.g. *Pol.* VII.13, 1332a40–b3 and VIII.5, 1339a23–24.

[41] See e.g. *Pol.* VIII.5, 1340a12–14 and 1340a22–23; *EE* II.1, 1220a29–30 and 2, 1220a39–b3.

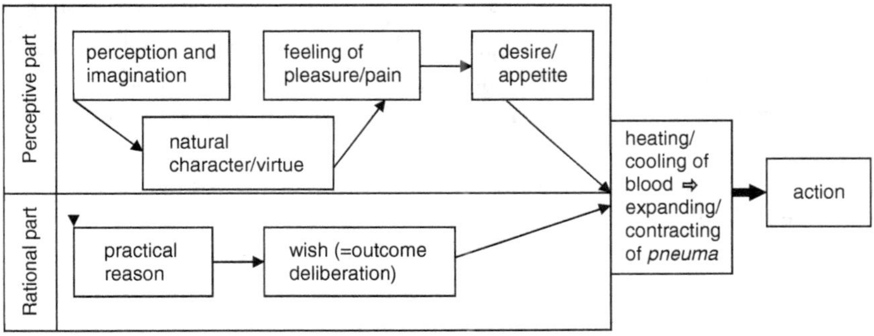

Figure 12.1 Schematic representation of the rational and perceptive part of the soul as involved in the production of human action.

capacities of the soul as belonging to one or more of these parts.[42] As for the perceptive part of the soul, it includes (a) the five perceptive capacities (touch, taste, vision, hearing, and smell), (b) imagination, (c) desire and appetite, and (d) pleasure and pain.[43] Each of these sets of capacities is causally dependent on the presence of the others, and together they form one functionally complex part of the soul (see my schematic representation in Figure 12.1).

Now, if we take seriously Aristotle's claim in the *Physics* that virtue-acquisition involves the alteration of the perceptive part of the soul, this must mean that all or most of the capacities constitutive of this part are altered via habituation. And, as we saw above, the ethical treatises provide evidence that this is indeed the case for at least three of these capacities (there is no evidence for habituation *directly* changing imagination, but since imagination in humans is tightly connected to the operation of the perceptive capacity,[44] these capacities are likely altered together).

First, habituation of the *perceptive capacities* involves a cognitive training through which one acquires the ability to recognize and judge correctly what is good and bad regarding moral phenomena,[45] but it also involves an alteration of the perceptive capacities themselves such that they are in a perfect mean[46] and are able to see correctly and actualize the appropriate feelings of pleasure and pain. For, according to Aristotle, every perception

[42] Corcilius and Gregoric 2010. These three soul-parts also figure in the function-argument in *EN* I.7, 1097b33–1098a7, which supports my suggestion that the political and natural sciences work with the same triparte division of the soul.

[43] See *DA* II.2, 413b22–24 and 3, 414b1–16; cf. *PA* II.17, 661a3–8.

[44] See *DA* III.3, 428b10–16, 429a1–2; and III.11, 433b31–434a9.

[45] See Moss 2012: *passim* on what she calls practical induction.

[46] Polansky 2007: 333.

is accompanied by a feeling of pleasure or pain and, vice versa, every feeling of pleasure or pain depends causally on the activity of perceptual discrimination (objects of perception do not need to be perceived *as* good or bad to be followed by feelings of pleasure or pain; instead, perceiving an object simply is also to feel pleasure or pain with regard to that object),[47] such that an *appropriate* feeling of pleasure and pain depends on the correct activity of perceptual discrimination.

Second, one's *capacity to feel pleasure and pain* is altered by the frequency and quality of the feelings experienced in response to one's perceptive activity: For feeling pleasure and pain in the right way, that is, in a way that tracks the actual good or bad of what one perceives, is a sign of character virtue (*EN* II.3, 1104b3–9; cf. *EE* VII.2, 1237a1–7 and *EN* III.12, 1119a11–17). Since the perceptional mean is naturally biased towards pleasure (*EN* II.9, 1109b7–9), habituation is needed to "make one feel pleasure and pain by the things one should" (*EN* II.3, 1104b12–13; cf. *EN* X.9, 1179b24–26 and *Pol.* VIII.5, 1339a20–25 and 1339a41–b4).

Third, since the kind of perception that involves feelings of pleasure or pain is causally followed by an appetite or desire to pursue or avoid the object of perception (*DA* III.7, 431a8–14), and since this desire – in humans – is a co-cause of action, the *capacities for appetites or desires* need to be appropriately calibrated as well, so as to respond correctly to feelings of pleasure or pain. Physiologically speaking, desires to avoid or pursue are heatings or coolings of the blood: That is, desires involve alterations of the blood that are themselves the result of some kind of alteration involved in perception or imagination (*MA* 6, 701a4–6 and 7, 701b13–23). These heatings or coolings of blood result in the contraction or expansion of the *pneuma* that is present in the blood, and this, then, results in locomotion or action.[48] Regarding this capacity, too, Aristotle emphasizes that the virtuous person must have the appropriate appetites and desires (see e.g. *EN* III.1, 1111a30–31 and III.11, 1118b15–21), and that their training is one of the objectives of habituation (*NE* I.3, 1095a2–13). In the virtuous person, then, regardless of whether his virtue is habituated or natural (*EN* VII.8, 1151a15–19), the actual good will appear as what is good for himself, and it will be pleasant and will produce a desire to pursue it.[49]

47 See Corcilius 2011 on the relation between perception, non-rational desire, and pleasure and pain.

48 See *MA* 7, 701b1–32 and 8, 701b33–702a21; and *EN* VII.3, 1147a34–35: "for appetite leads [him]: for it can move each of the [bodily] parts." Note that wish, the rational desire produced by practical reason as the outcome of deliberation, also causes such heatings and coolings of the blood, and that wish is by nature the more authoritative of the two desires (*DA* III.11, 434a12–15).

49 In men with *full* virtue, this desire or appetite will be in harmony with wish.

In addition to these three sets of capacities belonging to the perceptive part of the soul, the ethical treatises mention a fourth capacity – not mentioned in *On the Soul* – that is altered by habituation: namely, the natural capacities for character. These, too, I submit, belong to the perceptive part of the soul, and their role in Aristotle's moral psychology is that of a *mediator* between the activity of perception and the feelings of pleasure and pain: That is, unless one's natural character traits are themselves in a mean and "good," they either increase or decrease the amount of pleasure or pain that is experienced concomitant to the perceptual activity, regardless of the latter's accuracy (e.g. cowardice increases the amount of pain, and hence of chilling, experienced at the perception of real or apparent danger, beyond what is appropriate, and results in excessive flight or inaction). In this way, character is also responsible for how we perceive things (cf. *EN* III.5, 1114a31–b3 and *Insomn.* 2, 460a32–b11), and that is why it should be changed towards the better, and why Aristotle claims that having "good" natural character traits is advantageous in becoming virtuous.

We can now see what Aristotle means in *Physics* VII.3 when he claims that character virtue comes to be when the perceptive part of the soul is altered through perceptibles. For, as we saw, boys should be habituated by being exposed to morally good examples, by imitating those examples in play, and by listening to character-shaping music, all of which affects and thereby alters the perceptive capacities and influences the direction in which the natural character traits become stabilized. It is also clear what Aristotle has in mind when he says that pleasures and pains are alterations of the perceptive part of the soul: Perceptions – whether immediate perceptions or memories or expectations (the latter of which operate through "the likenesses" to affections: *MA* 8, 702a5–7) – are pleasant or painful, and since feeling pleasure and pain involves a heating or cooling of the blood that produces a desire to pursue or avoid, this is a kind of affection of the perceptive part of the soul (and, of course, of the sense organs with which the perceptive soul forms a unity). And finally, we can explain why Aristotle identifies the coming to be of pleasure and pain with the coming to be of virtues and vices of character, for it is through the pains and pleasures experienced via perceptions that the capacities of the perceptive part of the soul are altered.

In conclusion, assuming that Aristotle's account of virtue-acquisition in *Physics* VII.3 can indeed be used to (re-)construct a physiological theory of habituation, then we may infer that when these capacities of the perceptive part of the soul *are all in their best or worst condition, both internally and externally, in relation to each other*, that that is when virtue or vice is present. For, if character virtue is the proportionate *relation* between the capacities that are constitutive of the perceptive part of the soul, then it will be present when those capacities have been altered so as to be stably in the best condition possible and are related to each other in the appropriate way. Their coming to be is not itself a generation or an alteration, but the perfection that arises when all the constitutive elements of the perceptive part of the soul are in the best condition and are proportionately organized regarding each other, just as health[50] is the perfection that arises from such an organization of bodily elements.[51] If this account is correct, then perhaps it also helps explain what Aristotle means by the unity-thesis of virtue (*EN* VI.13, 1144b35–1145a2). For character virtue is only possible when *all* of the constitutive capacities of the perceptive soul are in their best condition and are proportionally organized, and the resulting virtuous disposition will constitute one, unified, but distinct psychological state.

[50] Cf. *Cael.* I.3, 270a27–29.

[51] Cf. Simplicius (*In Ar. Phys.* 1071.18): "It is clear that also the virtues of the soul belong to the relations, since they too are due proportions of those things in which they arise – emotions, and appetites, and such – just as the bodily [virtues are due proportions] of hot and cold [bodily elements] and such."

CHAPTER 13

Self-motion as other-motion in Aristotle's Physics

Ursula Coope

In *Physics* VIII.5, Aristotle sets out to provide a positive account of self-motion. He characterises self-motion as a kind of other-motion: a self-mover must be composed of two parts, one of which moves the other. The part that produces the movement is a kind of unmoved mover, in that it produces motion in a way that does not depend on it itself being moved. The contrast here is with a moved mover, something that produces motion just because it itself is moved by something else, as one carriage of a train might move another just because it in its turn is moved by the engine.

> τῆς ὅλης ἄρα τὸ μὲν κινήσει ἀκίνητον ὂν τὸ δὲ κινηθήσεται· μόνως γὰρ οὕτως οἷόν τέ τι αὐτοκίνητον εἶναι.
>
> In the whole thing, there is [a] that which imparts motion without itself being moved and [b] that which is moved, for only in this way is it possible for a thing to be self-moved. (258a2–4)

In this chapter, I shall discuss three questions raised by this account of self-motion. The first relates to the role of this account in *Physics* VIII. Aristotle's overarching aim here is to argue that there must be some first moved thing that is continuously in motion, and that its motion must be caused by an ultimate primary mover that is absolutely unmoved. Why is an account of self-motion given such a prominent place in this argument? After all, in *Metaphysics* L Aristotle argues for the existence of an unmoved mover without appealing in the same way to an account of self-motion.[1]

[1] In *On the Movement of Animals* I–IV, Aristotle argues that a self-mover must have, both within and outside itself, a kind of unmoved springboard that it pushes against when it moves itself. He goes on to ask whether the universe also moves itself in this way, by pushing back against something that is unmoved, and he argues at length that it does not: there is nothing that could serve as such a springboard for the whole universe. Note, though, that the possibility considered in these chapters of *On the Movement of Animals* is rather different from that discussed in *Physics* VIII. In *Physics* VIII, Aristotle argues that within a self-mover there must be an unmoved *mover*. In *On the Movement of Animals*, neither the internal nor the external unmoved thing is described as a mover. It is, rather, a

My second question concerns Aristotle's central argument that a self-mover must contain within it a kind of unmoved mover. One obvious opponent here is Plato, who, like Aristotle, thought that a chain of moved movers must have its origin in a self-mover, but who described self-movement as a kind of self-moving motion. Does Aristotle have any argument against this Platonic view that the ultimate origin of a chain of moved movers must be something that is itself in motion?

In the third section of my chapter, I consider an obvious objection that might be made to Aristotle's account of self-motion: can such an account really distinguish between a genuine self-mover (a unified thing, one part of which moves another) and two distinct things, one of which moves the other? For instance, can Aristotle avoid the consequence that when an animal moves a stick, the *animal plus stick* counts as a self-mover? Aristotle needs to have something to say in answer to this question if his account is really to be an account of self-motion, rather than an argument that there is, in the end, no such thing as self-motion.

As we shall see, the first of my questions bears on the other two. Once we understand the purpose for which Aristotle is giving this account of self-motion, we can see what constraints he is under in giving it, and which of the possible objections to it he most needs to answer.

I The role of the *Physics* VIII.5 account of self-movement in Aristotle's discussion

At this point in *Physics* VIII, Aristotle has already made the following claims:

(i) There always has been and always will be movement (VIII.1–2).
(ii) Every moving thing must be moved by something (VIII.4).
(iii) A chain of movers moved by other things cannot go on to infinity: it must have an origin (VIII.5, 256a4–b3).

From (ii) and (iii), he concludes:

(iv) The origin of a chain of moved movers must either be a self-mover or an unmoved mover (VIII.5, 257a26–27).

He then gives his account of self-motion, arguing that:

kind of springboard: something that the animal pushes against in moving itself. (For discussion, see the chapters by Coope and Morison, in Primavesi and Rapp (in press).)

(v) A self-mover must be a composite, one part of which moves without itself being moved, the other part of which is moved (VIII.5, 257a31–258b9).

He goes on to argue that:

(vi) There must be some first, continuous, eternal, and uniform circular locomotion. (The continued coming to be of animal self-movers depends on this.) (VIII.6, 258b16–259a6, 259a13–20)
(vii) There must be some single cause of this primary motion. (VIII.6, 259a6–20)
(viii) The unmoved mover within an animal self-mover is accidentally moved when the animal moves. An unmoved mover that is accidentally moved by itself in this way could not produce infinite continuous motion, and hence could not be the unmoved mover that explains this primary motion. (VIII.6, 259b20–260a10)
(ix) In fact, the cause of the primary motion must be something that is itself absolutely unmoved. (VIII.6, 259b22–31)[2]

My question is about the role, in this argument, of (v), the claim that a self-mover is composed of an unmoved mover and the thing it moves. An interesting feature of the argument I have set out above is that this claim seems somewhat tangential. In (iv), Aristotle sets out two alternatives: the mover that originates a chain of moved movers must be either a self-mover or an unmoved mover. However, what he goes on to say (in (viii) and (ix)) seems to rule out the possibility that what causes the primary continuous locomotion could be a self-mover, for he claims that the cause of *this* motion must be a mover that is absolutely unmoved. A mover that is absolutely unmoved cannot in any sense 'belong to' the thing it moves, since, according to Aristotle, a mover counts as (at least) accidentally moved if it belongs to a thing that is moved (or contains such a thing as a part) (254b9–10). This seems to imply that an absolutely unmoved mover cannot, together with what it moves, constitute a unified whole that counts as a self-mover. Hence, the origin of the ultimate primary continuous motion must be an unmoved mover *rather than* a self-mover.[3] Given this,

[2] Aristotle allows that certain heavenly bodies can be accidentally moved by something and yet moved continuously (as he must, if he is to maintain, as he does elsewhere, that the spheres on which the planets are set are accidentally moved by outer spheres), but he claims that something that accidentally moves itself cannot cause continuous motion (259b29–31).

[3] Further grounds for this conclusion can be found in the features Aristotle ascribes to self-movers in *Physics* VIII.4, when he is arguing that the simple bodies are not self-movers. He says there that self-movement is peculiar to living things and that what can move itself can stop itself (255a6–9).

Aristotle's attempt to establish the precise nature of *self-movers* can seem something of a digression.

Why, then, does Aristotle devote the second half of *Physics* VIII.5 to giving this account of self-motion? There are, I think, two main reasons why this account is important to Aristotle's overall argument. Both of them have to do with Aristotle's response to Plato.

In *Laws* X, Plato argued that the ultimate cause of motion was a self-motion (understood as a self-causing movement). Aristotle could have responded to this simply by claiming that such a cause would not be able to produce eternal continuous motion, but in fact he has a more radical objection. He argues that no origin of movement could be a self-causing movement, and hence that Plato's account is mistaken *even as an account of self-motion*. This argument is important not simply as a way of responding to Plato, but also because it provides crucial support for one of Aristotle's more positive claims. In particular, I shall argue, it is by understanding what is wrong with Plato's account of self-motion that we come to see why it is that every thing in motion must be moved by something (see section II below). This, then, is the first reason why Aristotle's positive account of self-motion is more than a digression: the argument he gives here against Plato provides further justification for his claim (ii): the claim that every moving thing must be moved by something.

However, I suspect that Aristotle also has a second reason for discussing self-motion more fully here. He wants to say that there is, after all, something right in Plato's claim about the primacy of self-motion (even though Plato had the wrong account of self-motion). An animal self-mover is a *first* mover: it (or more precisely, the mover within it) is the origin of a chain of moved movers. Aristotle wants to emphasise what animal self-movers have in common with the first mover of the primary eternal motion.[4] As we have

Moreover, the fact that each of the simple bodies only moves naturally in one direction shows that such movement cannot be self-movement (255a9–11). For an alternative interpretation, on which the first eternal movement of *Physics* VIII is a kind of self-movement, see Kosman 1994. Although I think Kosman is right to stress that the first eternal movement together with its mover constitutes something that is in many respects *like* a self-mover, the possibility that it *is* strictly speaking a self-mover is, I think, ruled out by Aristotle's claim that the first mover is not even accidentally moved, together with his account (in *Physics* VIII.4) of what it is to be accidentally moved.

[4] This perhaps explains why, even though he does not think that the eternal first motion is, strictly speaking, self-motion, he is nevertheless prepared to say that what moves itself is 'the principle (*archê*) of moved movers and the primary thing among things that are moved' (261a25–26, cf. similar remarks at *Physics* VIII.6, 259a33–b1 and at *MA* I, 698a7–8). What moves itself is primary among *sublunary* things that are moved, and the account he has given of the way in which a self-mover moves itself applies also to the way in which the eternal first movement is caused by its mover. That is why the fact that self-motion is a kind of locomotion is a reason for thinking that the eternal first motion will also be locomotion (261a23–26).

seen, this primary eternal motion cannot be a kind of self-motion, since its mover is absolutely unmoved. Nevertheless, I shall argue, Aristotle's account of self-motion in *Physics* VIII.5 is framed in such a way that it applies not only to genuine self-movers, but also to the conjunction of the first eternal moved thing and its unmoved mover. As such, it aims to give necessary *but not sufficient* conditions for being a self-mover. In this account, Aristotle includes only those features of self-movers that are also features of a particular conjunction of things that does *not* constitute a self-mover: the first eternal moved thing together with its mover.

This raises an obvious question. If Aristotle does not in fact think that the combination of the first eternal moved thing and its mover constitutes a self-mover, then why does he go to such pains to formulate an account in *Physics* VIII.5 that applies not only to genuine self-movers but also to this combination of the first eternal moved thing and its mover?

One possible answer is that providing such an account helps to establish something about the unity of the science of physics. In *Physics* II.7, Aristotle says that there are three different realms of study: one concerned with that which cannot undergo movement, a second concerned with eternal moving things, and a third concerned with perishable moving things (198a30–31). The study of things that cannot undergo movement is, he says, no part of physics (198a28). What, though, of the other two realms of study? Elsewhere, he implies that physics is a science concerned with both eternal and perishable moving things.[5] This raises a question: what justifies the assumption that there is one science of both eternal and perishable moving things, and that the same basic principles apply to both eternal and time-limited movement? In *Physics* VIII, this question arises with particular urgency, since Aristotle will argue that the mover that produces the primary eternal motion is itself absolutely unmoved (and hence, according to his own earlier remark (198a28), falls outside the scope of physics). Can a single science concern itself both with movement that is produced in this way and also with the kind of movement that has its origin in animal self-movers?

To answer this, it is not enough to show that (as Aristotle argues in *Physics* VIII.6) animal self-movers are causally dependent on things that are in eternal motion. Aristotle clearly thinks that X can be causally dependent on Y, even though X and Y do not fall under the same science.[6] What is

[5] See, in particular, *Metaph.* L.1, 1069a36–b2. For an illuminating discussion of Aristotle's views on the unity of physics, see Falcon 2005.

[6] As we have just seen, eternal motion is causally dependent on an absolutely unmoved mover, although such a mover does not fall within the scope of physics.

needed is an abstract account that will bring out the similarities between self-movers, on the one hand, and the first eternally moved thing together with its mover, on the other. It is precisely such an account, I argue, that Aristotle provides in the second half of *Physics* VIII.5. Because it applies both to animal self-movers and also to the primary eternally moved thing together with its mover, this account helps to justify the assumption that physics is a unified science.

I have argued that Aristotle provides the account of self-motion in *Physics* VIII.5 partly as a reply to Plato and partly as a way of emphasising what genuine self-movers have in common with the primary eternal moved thing together with its mover. In what follows, I first ask how Aristotle's argument fares as a response to Plato, and then discuss how the content of his account is affected by his wish to capture just those features of self-movers that they share with the first moved thing together with its mover.

II The argument that a self-mover must be composed of two parts: An unmoved mover and what it moves

I have claimed that Aristotle's account of self-movers here is written partly in opposition to Plato's view in the *Laws*, according to which the ultimate origin of a chain of moved movers is a self-causing movement. In this section, I shall ask whether Aristotle has any good argument against this Platonic view.

The core of Aristotle's argument comes in a single paragraph at 257a31–b13. First (257b2–6), he says that it is impossible that what moves itself should as a whole (and while remaining one and indivisible in form) both cause and undergo the same movement. This impossibility is illustrated with two examples: the same thing cannot at the same time and in the same respect be teaching and being taught; the same thing cannot at the same time and in the same respect be becoming healthy and producing health. Second (257b6–13), he says that what undergoes movement must be in potentiality whereas what causes movement must already be in activity (ἤδη ἐνεργείᾳ). This again is illustrated with an example: what produces heat must itself be hot. From this, it follows that what produces heat must be distinct from what is becoming hot (since what is becoming hot must not yet be hot).[7] Aristotle then generalises from this example: 'so too in

[7] 'It is that which is hot that produces heat, and in general that which produces the form possesses it. Consequently [if the mover is not distinct from the moved] the same thing in respect of the same thing will be at the same time both hot and not hot' (257b9–11: θερμαίνει τὸ θερμὸν καὶ ὅλως γεννᾷ τὸ ἔχον τὸ εἶδος. ὥσθ' ἅμα τὸ αὐτὸ κατὰ τὸ αὐτὸ θερμὸν ἔσται καὶ οὐ θερμόν).

every case where the mover must have a synonymous property'. From this he concludes, quite generally, that what first produces a movement must be distinct from what undergoes that movement.

Much of this passage simply relies on examples. As interpreters, we need to ask how Aristotle means to use these examples to defend his general conclusion about self-motion. After all, his aim is not merely to show that *in certain cases* the primary origin of a movement is something unchanging. Although this would establish (against Plato) that it is possible for the ultimate origin of a movement to be something that is not undergoing movement, it would not establish Aristotle's conclusion that within a self-mover there must be a part that is moved and a distinct, unmoved, part that is the mover.

Aristotle's final example (that the cause of becoming hot must be something that is already hot) suggests the following general argument:

(i) The primary cause of something's changing towards being F must be something that is itself already F.
(ii) What is undergoing a change towards being F cannot yet be F.
(iii) Nothing can at the same time and in the same respect both be F and not be F.

So: (iv) The primary cause of something's changing towards being F must be distinct from what undergoes the change towards being F.
So: (v) Since the mover within a self-mover is the primary cause (not merely an instrumental cause) of the self-mover's movement, the part of the self-mover that produces movement must be distinct from the part that undergoes that movement.

However, if this is Aristotle's argument, it seems to face two insuperable problems, both related to premise (i), the claim that a change towards being F must be caused by something that is already F. The first problem is that the scope of premise (i) is too restricted for it to do the work needed of it in this argument. Aristotle himself recognizes that (i) only holds for a certain range of cases, and, because of this, the conclusion he draws is carefully qualified: 'So too in every other case in which the mover must have the synonymous property' (257b11–12). This simply invites the question: what about those cases in which the mover *doesn't* have to have the synonymous property? How could Aristotle go from this carefully restricted conclusion to the general claim that, within a self-mover, what is moved must be distinct from the mover? These questions are especially pressing, since it is hard to see how premise (i) could apply to a particularly central kind of movement: locomotion. Locomotion is not naturally thought of as the acquisition of a form, and there is no obvious sense in which the agent of a locomotion must already be in the state that the moving thing is progressing towards. After all, it is clearly not true that what moves something to place P must itself already be in place P.[8] But Aristotle's argument needs to apply to locomotion, since he himself goes on to say that locomotion is the primary motion and that, in the strictest sense, the movement with which a thing moves itself is locomotion (253a14–15, 261a23–25).

Of course, there may be *a priori* grounds for accepting something like premise (i), in spite of the apparent counterexamples. However, when we consider what such *a priori* grounds might be we encounter a second problem. A plausible suggestion is that premise (i) is based on what might be called the 'transmission' principle of causation: the principle that what ultimately causes a thing to be a certain way must itself be that way. If causation is thought of as a kind of transmission of a property from one thing to another, then it is easy to see how this principle might be justified: how could an agent transmit a property to something else if it didn't itself have that property?[9] The problem with this is that one could accept this general principle without accepting Aristotle's application of it, and hence without accepting premise (i). In fact, one could use this same principle to argue for the very view Aristotle is attacking: the Platonic view that what causes motion must itself be in motion. The argument would go as follows. We are trying to explain X's motion. We have already agreed that if X is in motion, it must be being moved by something. But how can the mover transmit movement to X unless it is itself in movement?

Indeed, something like this view seems to lie behind Plato's reasoning in the *Laws*. Plato agrees that every movement has a cause and that chains of moved movers must have their origin in some primary cause. But he concludes that the primary cause is a self-moving movement. Plausibly, the basis for this view is this alternative application of the transmission principle: he assumes that what causes movement must itself be in movement. So, if we accept the transmission principle (the idea that what

[8] Of course, there are various ways in which one could try to fit locomotion into the model. For instance, one might say that the agent of locomotion has in its mind the goal towards which the locomotion is proceeding (as the craftsman has in his mind the form of the thing he is producing). But this would not be enough to save the apparent argument, since it is, at least *prima facie*, possible for one and the same thing (at the same time) both to have the goal F in mind and also to be (at) F.

[9] Waterlow 1982: 244–246 calls this the model of 'giving and receiving'.

ultimately causes a thing to be a certain way must itself be that way), there are two rival conclusions we might draw. We might conclude (with Plato) that what causes a thing to be in movement must itself be in movement (only then can it transmit movement to the thing it acts upon), or we might conclude (with Aristotle) that what causes a thing to become F must itself be F (only then can it transmit F-ness to the thing it acts upon). The transmission principle by itself cannot justify the Aristotelian, as opposed to the Platonic, conclusion.

The fact that this argument is vulnerable to these objections should make one doubt whether it is, in fact, the argument Aristotle had in mind. I shall argue that, in fact, Aristotle is defending here a more abstract principle: that what first causes motion must do so in virtue of being a certain way, not in virtue of itself undergoing motion. When Aristotle says that, in certain cases, the cause of *becoming F* is something that already *is F* he is providing an illustration of this more abstract principle, but he is not meaning to imply that all first causes operate by transmitting a form to the thing upon which they act.[10] His argument, thus, does not rely on the principle of contradiction, in the way that his remarks about the hot thing might suggest: he is not saying that the first cause must already be F, whereas what becomes F cannot yet be F. Instead, his point is that the first cause is a cause in virtue of stably being a certain way, not in virtue of changing or becoming. The sense in which a self-mover must be composite is, then, that it must have two aspects: *qua* mover it is unchanging (since it produces movement in virtue of stably being a certain way), whereas *qua* moved it is undergoing change. If this is Aristotle's argument, then his objection to Plato is that, on Plato's view, the ultimate explanation of a movement is itself a movement, rather than a state of being.

If this is right, then we need to ask how Aristotle would justify the claim that the ultimate explanation of movement must be a thing's stably being a certain way. I shall argue that Aristotle provides such a justification in a few difficult lines in our passage. In these lines, he draws a connection between movement, incompleteness, and potentiality:

> ἔτι διώρισται ὅτι κινεῖται τὸ κινητόν· τοῦτο δ' ἔστὶν δυνάμει κινούμενον, οὐκ ἐντελεχείᾳ, τὸ δὲ δυνάμει εἰς ἐντελέχειαν βαδίζει, ἔστιν δ' ἡ κίνησις ἐντελέχεια κινητοῦ ἀτελής. τὸ δὲ κινοῦν ἤδη ἐνεργείᾳ ἔστιν
>
> Further, it has been established that the moveable is what is moved. But this is moved in virtue of potentiality, not fulfilment. But what is in virtue of

[10] Or at least, the argument he gives does not rest on the truth of this universal claim.

> potentiality progresses towards fulfilment, and movement is the incomplete fulfilment of the movable. But the mover is already in activity. (257b6–9)

The difficulty of these lines stems partly from a problem of translation. The most obvious translation of τοῦτο δ' ἔστὶν δυνάμει κινούμενον, οὐκ ἐντελεχείᾳ would be 'this is moved in potentiality, not in fulfillment', but this translation seems to get the wrong sense. When the movable thing is being moved, it is not merely *potentially* being moved. I have, instead, given the translation: 'it is moved in virtue of potentiality, not fulfilment'.[11] To understand what might be meant by saying this, we shall need to look at what Aristotle says elsewhere about the relation between movement and potentiality.

The argument in these lines rests, I think, on two claims: first, the claim that what is in potentiality is posterior to, and causally dependent on, what is in fulfilment ('what is in virtue of potentiality progresses towards fulfilment'); and second, the claim that a moving thing is, as such, not completely in fulfilment ('movement is the incomplete fulfilment of the movable'). Before we consider how Aristotle is putting these two claims together in this argument, it will be helpful to look at each of them in turn.

The first claim is not directly about movement at all. It is, rather, about the relation between what is in fulfilment and what is in potentiality. Aristotle says that what is in fulfilment is prior, in a certain way, to what is in potentiality. In our passage, he expresses this priority relation by saying that what is in potentiality 'progresses towards' (*badizei eis*) fulfilment. With this notion of 'progressing towards' he means, I think, to capture a kind of teleological relation. *Being potentially* is a way of *being for the sake of some fulfilment.* Aristotle spells this point out much more fully in a discussion of potentiality and fulfilment in *Metaphysics Theta* 8. He says there that what comes to be progresses towards (*badizei epi*) a principle and end (*telos*) (1050a7–8). The end is an activity (*energeia*), and the thing that comes to be acquires its potentiality for the sake of such an activity. Thus, animals have sight (a potential) for the sake of seeing (an activity), and humans have the art of housebuilding (a potential) for the sake of building houses (an activity) (1050a10–12).

Aristotle takes this fact about teleological directedness to ground a further claim about causation: the primary cause of something that is in potentiality must be something that is in fulfilment. His thought seems to be that anything that is, by its very nature, directed towards (or for the

[11] In this, I follow Waterlow 1982: 244n.27.

sake of) some further thing must be a dependent kind of entity. It must depend on, and in particular derive its end-directedness from, some distinct cause. Ultimately, the source of such end-directedness must be something that is not similarly end-directed. The primary cause must be something that is just what it is, without reference to anything further. The primary cause must thus be something that is in fulfilment, not in potentiality. He illustrates this with examples of cases in which *being potentially F* is explained by *being F in fulfilment*. To be an acorn is to be potentially an oak. An acorn must, then, have as its primary cause something that is, not merely potentially but in fulfilment, an oak. To be a fluteplayer is to have the potential to play the flute; the primary cause of being a fluteplayer must be the activity of playing the flute (one becomes a fluteplayer by practising).[12]

This tells us that *being F in fulfilment* is causally prior to *being F in potentiality*, but it does not, by itself, tell us anything about the relation between causal priority and movement. For this, we need to turn to the second claim Aristotle makes in this passage: the claim that a moving thing is, as such, not completely in fulfilment. In making this claim, Aristotle is clearly referring back to his definition of *kinêsis* in *Physics* III.

He defines movement/change (*kinêsis*) there as 'the fulfilment (*entelecheia*) of what is potentially, *qua* such' (ἡ τοῦ δυνάμει ὄντος ἐντελέχεια, ᾗ τοιοῦτον) (201a9). What exactly this means is controversial. Interpreters have disagreed over what is meant by *entelecheia*, over what is meant by '*qua* such', and over how to understand the phrase 'what is potentially' (is this what is potentially *in some state* or what is potentially *changing*?).[13] For now, I want to leave these disputes to one side and instead look directly at something Aristotle presents as an advantage of this definition. He says that this definition allows us to capture what was right about earlier, confused, accounts of movement: accounts which represented change/movement as inequality, difference, or non-being. Such accounts were confused, in that they failed to provide necessary and sufficient conditions for movement (something can, for instance, be different without moving). But these accounts nevertheless correctly recognised something about the nature of movement: namely, that movement is essentially privative, or as Aristotle puts it, incomplete.

[12] See *Metaphy. Theta* 8, 1049b24–32. Aristotle draws the conclusion that the mover is already in activity at 1049b27, using almost exactly the same wording as at *Phys.* VIII.5, 257b9.

[13] For some contrasting answers to these questions, see Kosman 1969, Heinaman 1994, Kostman 1987, and A. Agnastopoulos 2010. I discuss the first three of these interpretations in Coope 2009.

We can learn something about what Aristotle means by this if we look at the reason he gives for thinking that movement is incomplete:

> ἥ τε κίνησις ἐνέργεια μὲν εἶναί τις δοκεῖ, ἀτελὴς δέ· αἴτιον δ' ὅτι ἀτελὲς τὸ δυνατόν, οὗ ἐστιν ἐνέργεια.
>
> Movement is thought to be activity of a kind, but incomplete. The cause is that the potential thing of which it is an activity is incomplete. (201b31–33)[14]

Movement is incomplete, Aristotle says, *because the corresponding* dunaton *(or potential thing) is incomplete.* This explanation is, at first sight, puzzling. Given Aristotle's general views about the priority of fulfilment (or actuality) to potentiality, one would expect features of what is capable of phi-ing to be explained by features of phi-ing, rather than vice versa. How, then, can Aristotle maintain that the incompleteness of movement is explained by the incompleteness of the corresponding potential thing?

Aristotle is, I think, claiming that movement is related in a special way to the corresponding potentiality. This follows from his definition of movement as 'the fulfilment of what is potentially, *qua* such'. Movement is the proper activity of what is potentially, *insofar as it is merely in potentiality.* Because of this, movement inherits certain features of the potential thing, in a way that other activities do not. Thus, Aristotle is *not* saying that the potentiality exercised in movement is incomplete and that there is, in contrast, some complete kind of potentiality exercised in the kind of activity that is not movement. Rather, he thinks that all potential things are, as such, incomplete, but that movement is related to potentiality in such a way that it inherits the incompleteness of the corresponding potential thing.[15] Aristotle is alluding to this special relation of movement to potentiality when he says, in our passage from *Physics* VIII, that the changeable thing is δυνάμει κινούμενον, οὐκ ἐντελεχείᾳ (what I translated as 'moved in virtue of potentiality, not fulfilment').

To be incomplete (*ateles*) is to be without an end. As we saw, a potential thing is directed at ('progresses towards') some fulfilment. The potential thing is thus *ateles*, because it is directed at some end (*telos*) that is other than it. Aristotle's claim is that movement, because of its connection to potentiality, is similarly end-directed.

[14] Here the claim is that movement or change is an incomplete activity (*energeia*), rather than (as in VIII) an incomplete fulfilment (*entelecheia*), but Aristotle seems to be using the two words interchangeably in these passages, so I think we can take him to be saying the same thing.

[15] Here I am in agreement with Kosman 2013: 66, who says, commenting on 201b31–33, 'This cannot mean that there is some special kind of incomplete ability, the realization of which is incomplete and is motion. For all mere ability is incomplete or unperfected – *atelês*.'

We are now in a position to arrive at an understanding of our passage by putting together these claims about incompleteness, posteriority, movement, and potentiality. As we saw earlier, Aristotle thinks that because of the way in which potentiality is directed towards fulfilment, a thing that is in potentiality cannot be causally primary. Being potentially F is an essentially other-directed way of being. Aristotle's point here is that movement also is an other-directed way of being. In this, movement is much like potential being. Just as the potential to be a musician is for the sake of actively being a musician, so also the movement *becoming a musician* is for the sake of actively being a musician. Just as the seed is for the sake of actively being the mature animal, so also growing (a change) is for the sake of actively being a mature animal. There is reason, then, to think that movement will be posterior to activity, just as we earlier argued that potential being is posterior to active being.

Aristotle holds that as movement inherits its incompleteness (and hence posteriority) from potentiality, so also it inherits its lack of causal primacy. A movement must have as its first cause something that explains its directedness towards an end, something that determines what would count as interference with, or completion of, that movement. What provides this explanation cannot be something that itself stands in need of a similar explanation. Thus, a movement must have its origin in something that is not, similarly, incomplete or end-directed. Thus, Aristotle can conclude that insofar as a thing is a first cause of movement, it must not only be active (rather than merely potential), but active in a *complete* way (and, hence, not undergoing movement).

Aristotle takes this argument to justify his claim that the first cause of a movement must be completely active, and hence that a movement cannot, ultimately, be explained by a movement. The challenge for such a view is to explain how Aristotle can reply to a Platonist, who holds that what produces a movement must do so in virtue of itself undergoing movement. The interpretation sketched above shows how Aristotle might respond. He argues that movement is a dependent kind of entity: it is an incomplete fulfilment. Because of this, every movement must have an agent that produces the movement in virtue of itself being completely fulfilled. Of course, Plato might disagree with the claim that movement is a dependent entity, a kind of incomplete fulfilment. Alternatively, he might say, with some plausibility, that the kind of movements he had in mind in the *Laws* were not what Aristotle is calling 'movements'. Plato lists as examples of such movements 'wish, reflection, diligence, counsel, opinion true and false, joy and grief, cheerfulness and fear, love and hate' (897a).

However, Aristotle does, I think, have an answer to this. Plato himself agrees with Aristotle that every movement requires an agent. That is why Plato thinks that the first movement must be a self-movement (rather than simply being an uncaused movement). What is the justification for this assumption that every movement must have an agent? Aristotle's answer, I want to suggest, is that it is because movement is an incomplete fulfilment that every movement must have an agent. As we have seen, in saying that movement is an incomplete activity, he is saying that it is a dependent entity that cannot itself be causally primary. In other words, it is because movement is in this sense dependent, that it requires an agent.

Aristotle does, then, have something to say to Plato. If Plato disagrees with Aristotle's characterisation of movement, he needs to find a different justification for his assumption that every movement has an agent. So Aristotle's challenge would be this: the very thing that justifies Plato's claim that every movement has an agent (and hence that leads him to say that the first movement must be a self-movement) also justifies the further claim that the first cause of a chain of movements must act as a cause in virtue of being completely in fulfilment, not in virtue of itself undergoing movement.

Moreover, we can now see a further way in which this discussion of self-movers contributes to Aristotle's overall argument. In the earlier chapters of *Physics* VIII, he argues that every movement has a mover, and claims that a chain of moved movers cannot go on to infinity. The second of these claims is (in these earlier chapters of *Physics* VIII) simply an assertion,[16] and the first is based merely on an examination of different cases: he tries to show that something has a mover whether it is moved unnaturally or naturally (and, if naturally, whether it is self-moved or not).[17] His account of self-motion in *Physics* VIII.5 provides some theoretical justification for both these claims, for it provides an argument that movement, as an incomplete (and hence dependent) being, must ultimately depend on something that is not similarly incomplete. This implies both that every movement must have a cause, and also that a movement's *first* cause must be something that is not itself a movement.

[16] As Wardy notes (1990: 89), Aristotle does defend this claim in the second half of *Physics* VII.1. My interpretation implies that Wardy is wrong to think that Aristotle needs this *Physics* VII.1 argument to justify what he says in *Physics* VIII.5.

[17] *Physics* VIII.4.

III Primary self-movers

I have claimed that the account of self-movers in *Physics* VIII.5 is framed in such a way as to apply not only to animal self-movers, but also to the first eternally moved thing together with its mover. In support of this claim, I turn now to Aristotle's discussion of what it is for something to be a primary self-mover. I shall argue that Aristotle is here responding to a natural objection to his account of self-motion. The manner in which Aristotle responds to this objection shows, I think, the care he takes in this account of self-movers to confine himself to features that genuine self-movers share with the first eternally moved thing together with its mover.[18]

The objection to which Aristotle is responding arises because of his view that self-motion is a kind of other-motion: for a thing to move itself is for one part of it to move another part.[19] The objection is that such an account cannot distinguish between what we would ordinarily call a self-mover and a conjunction of two things, one of which moves the other. Aristotle needs to have something to say about this, in part because animal self-motion is a kind of natural motion. When an animal moves itself from one place to another, the source of this movement is in the animal itself, and hence the movement is natural to the animal and the animal has a nature. Aristotle's account of nature would be undermined if he were forced to concede also that when, say, an animal moves a stick from one place to another, the source of this movement is within the animal+stick, and hence the animal+stick also has a nature (since it has an internal source of change and rest). Aristotle needs some account of how a genuine self-mover constitutes a unity, but the conjunction of such a self-mover with something else that it moves does not.

There are resources within Aristotle's philosophy for answering this. An obvious answer he might give is that the mover, within a self-mover, must be the form of the self-mover. The animal is a self-mover because its mover

[18] Another place where Aristotle is clearly tailoring this account to fit the eternal first motion and its mover is 257b20–25. Here he is considering the possibility that the two parts of a self-mover might move each other. One of his objections to this is that then it would not be necessary that each part move the other. (The point seems to be that this suggestion only explains why A moves B on the supposition that B moves A, and only explains why B moves A on the supposition that A moves B. It does not explain why A and B are in motion rather than not.) But if we are simply concerned with animal self-motion, it is not obvious why it should be desirable to have an account that presents such motion as necessary. The need to explain the *necessity* of movement arises when the movement we are trying to account for is eternal celestial movement.

[19] It is important to recognise that a 'part' here need not imply a physically separable part. The mover, for instance, could be the form of the thing.

(its soul) is its form; the animal+stick is not a self-mover, because its unmoved mover (the animal's soul) is not the form of the animal+stick. This answer would enable him to capture the idea that a self-mover must be a kind of unified whole (since it must have a form, which is its mover), and also to capture the idea that it is really a *self*-mover, since it implies that what moves the composite is itself accidentally moved.[20] Given that this is a good Aristotelian reply to our question, it is noteworthy that Aristotle does not give it here. In fact, in this section of *Physics* VIII.5 he says nothing about the mover being the soul or being a kind of form. The reason Aristotle does not give this answer here is, I think, because it would prevent his account of self-movers applying to the combination of the prime mover and what it first moves. The prime mover cannot be the form of the thing it first moves, since then it would itself be moved accidentally, and – as we have seen – Aristotle claims that the ultimate prime mover is not, even accidentally, moved.

Aristotle thus faces a difficult task in attempting to answer our question. On the one hand, he needs to distinguish between a self-moving composite (such as the soul+body) and a composite of one thing's moving another (such as the soul+body+stick). Such a distinction is needed if Aristotle is to avoid attributing an internal source of change and rest, and hence a nature, to all sorts of accidental conjunctions of self-movers with the things they move. On the other hand, Aristotle's account needs to apply not only to a self-moving composite, but also to the conjunction of the first eternally moved thing with its mover. The account of primary self-movers that Aristotle gives in the last part of VIII.5 (258a9–b9) is, I shall argue, designed to meet both these constraints.

The account is as follows. If A (the unmoved mover in the self–mover) moves B, and B in its turn moves C, then the whole, ABC, can be regarded as a self-mover of sorts, but primarily or strictly speaking the self-mover will just be AB. For instance, if the dog's soul moves its body, and its body in turn moves a stick, the primary self-mover will be the combination of the dog's body and soul, in other words, the dog. The primary self-mover, then, will be the unmoved mover together with the *first* thing that it moves. An animal is a self-mover because it is composed of an unmoved mover (presumably, the animal's soul) together with the first thing moved by that mover. This account allows for the conditions on being a primary self-mover to be satisfied also by the absolutely first mover together with the

[20] Aristotle himself makes this point about animal self-movers at *Phys.* VIII.6, 259b16–20. For the claim that the soul is the origin of an animal's self-movement, see *On the Soul* III.9–11.

thing it first moves, even though (as we have seen) the absolutely first mover cannot be the form or soul of the thing it first moves, since it is not accidentally moved by the movement it causes.

For this account to avoid obvious counterexamples, it is important that the kind of self-movement Aristotle has in mind here is self-locomotion. He says that this is, properly speaking, the only self-movement (253a14–15, 261a23–25). So if a hot thing, without itself undergoing change, causes something else to become hot, the hot thing and the thing it heats do not, together, satisfy these conditions on being a self-mover.[21] If one assumes (in accordance with what Aristotle himself asserts) that, at least in the sublunary realm, every origin of motion is a self-motion and the origin of the self-motion is an unmoved mover, then we can always trace a chain of moved movers back to an unmoved mover. The first of these moved movers, together with the unmoved mover will constitute a self-mover. Aristotle holds that, in the sublunary realm, this will be an animal.

Of course, one obvious question about this account is what it can say about the unity of B and of C (the first thing that is moved by the unmoved mover, and whatever is in its turn moved by that first thing). What makes B and C two distinct things, thus justifying the conclusion that B, rather than B plus C, is the first thing moved? Suppose B is itself composed of parts (as Aristotle has argued it must be, given that it is moved). What justifies the claim that B, as opposed to one of its parts, is the first thing moved?

As Aristotle points out, these questions can be pressed in such a way as to suggest that there is no *first* thing moved (and hence no primary self-mover). According to Aristotle, any thing that is moved is infinitely divisible: it has parts, which in turn have other parts, and so on ad infinitum.[22] Moreover, its movement depends on the movement of its parts (at least, in the sense that the movement of the parts is a necessary condition of the movement of the whole). This raises a problem for Aristotle, since it suggests that there may be no first thing moved: any putative first thing moved by the unmoved mover will in fact only be

[21] It is less obvious how Aristotle can avoid allowing that this *Physics* VIII.5 account applies to the conjunction of a magnet and the iron it attracts.

[22] He argues for this in *Physics* VI.4, and he reminds us of this claim at *Phys.* VIII.5, 257a33–b1, just before giving his argument that a self-mover does not move itself as a whole. The relevance of lines 257a33–b1 in their immediate context is not at all obvious. However, the point Aristotle makes here about infinite divisibility *is* important for understanding the *aporia* (at the end of *Physics* VIII.5) about the primary thing moved.

moved because certain other things (its parts) are themselves moved by the unmoved mover. Thus, any self-mover, AB, will have a more primary self-mover that is its part: 'AB will not be what is primarily moved by itself, since when something is taken away from AB, the remainder will continue to move itself' (258a30–32).

In raising this worry, Aristotle is taking up, and responding to, an argument that he himself gave in *Physics* VII.1. It is an argument for the conclusion that every moved thing is moved by something. Although of course he still agrees with this conclusion in *Physics* VIII, he realises that he cannot endorse the argument of *Physics* VII.1.[23] That argument starts out from the claim that anything that is moved is infinitely divisible. This is taken to imply that its movement depends on its parts, and hence on something that is distinct from it (since a thing is distinct from its parts). Since the parts are divisible, their movement will depend on the movement of their parts, and so on ad infinitum. In *Physics* VII.1 Aristotle concludes from this that a moved thing must be moved by something that is distinct from it.

In *Physics* VIII.5, Aristotle sees that he cannot endorse this argument, since, if it were successful, it would imply that there was no first thing moved. Every moved thing would depend for its movement on the other, prior, moved things that were its parts. This is a conclusion that Aristotle cannot accept, since it would undermine his argument that a chain of moved movers originates in a first mover together with a first thing moved by that mover.

His response, in *Physics* VIII.5, is to claim that the moved thing is potentially, but not actually, divided: 'so that if it is divided it will not continue in possession of the same nature' (258b2–3). Although anything that is moved can be divided into parts, this does not imply that a thing's movement is causally dependent on the movement of its parts. In fact, its parts may only move in virtue of being parts of the whole. A sign that this is so is that many self-movers no longer move themselves when divided. Hence, the fact that the moved thing is infinitely divisible does not undermine the claim that there are primary self-movers. A primary self-mover will be divisible into parts, but these parts will not be things that would, if separated from the primary self-mover, be self-movers in their own right.

[23] This, of course, raises interesting questions about the role of *Physics* VII in the *Physics* as a whole. It suggests that Aristotle did have *Physics* VII in mind when he wrote *Physics* VIII, but that he did not intend the two books to be part of a single work (since he took some of his arguments in VIII to supersede those he had given in VII).

As an account of animal self-movers, this response is not wholly successful. As Aristotle himself remarks elsewhere, some animals can be cut in two in such a way that their parts continue to move themselves. Does Aristotle have to deny that these are primary self-movers?[24] Moreover, even in the case of animals that cannot be cut in half and continue to move, there are usually parts one can remove that would still allow what remains of the animal to move itself. Is Aristotle committed to the claim that it is not strictly speaking the animal that is the self-mover, but the animal minus its hair and teeth and . . . ? Finally, it is not clear that even this subtracting procedure will always arrive at a definite first mover. It might turn out that the animal would remain a self-mover without part A or without part B, but not without both. In such a case *the animal minus part A* and *the animal minus part B* would both seem to have equal claim to be the primary self-mover.

Aristotle does not consider any of these problems here. I suspect part of the reason for this is that in this context he is not, primarily, interested in giving a full account of animal self-motion. Instead, he defends the claim that there can be a primary thing moved, because he will need to assume this claim in the account he goes on to give of the relation between the first eternal motion and its mover. In giving that account, he emphasises that there is one unified, continuous first motion. It is because of the unity and continuity of this first motion that it requires both a single subject and also a single absolutely first mover: 'if motion is continuous, it is one, and it is one only if its mover is one and the thing that undergoes the motion is one' (259a17–19). Aristotle's argument thus relies on there being not only a single continuous motion, but also a single unified subject for such a motion. If such a subject is to have the right kind of unity, its movement cannot be causally dependent on the movement of its parts. This is why it is so important for Aristotle, in *Physics* VIII.5, to explain what was wrong with his earlier argument in *Physics* VII.1.

I have claimed that Aristotle's account of self-movers in *Physics* VIII.5 is best understood when we consider its role in his more general argument. He is not attempting here to give a full account of what is involved in an animal's moving itself. Instead, he gives a highly abstract argument that appeals to his earlier discussion of the ontological status of motion. This account serves several purposes. It presents a challenge to Plato's alternative

[24] Perhaps he would not be worried by this conclusion. At *Progression of Animals* 707a24–b4, he implies that an animal that is divisible in this way is like a single body that contains within it many living beings (see also *Juv.* 468b9–15).

view of the origin of movement. More positively, it provides an argument that movement must, ultimately, be caused by something that is not itself a movement. Finally, by giving this account of self-motion, Aristotle shows that it is possible, at least at a certain level of abstraction, for there to be a single account that applies both to animal self-movers and also to the first eternal movement together with its mover. Insofar as the account is successful, it thus provides some basis for the view that physics, the study of motion, is a single unified science.

CHAPTER 14

The argument of Physics VIII

Andrea Falcon

I Introduction

Physics VIII is often regarded as an argument in two stages from the eternity of motion to the existence of a first mover which is eternal and absolutely unmoved. This twofold reading of *Physics* VIII raises an obvious concern about the boundaries of the science of nature. The first unmoved mover does not seem to be a proper object of study for the student of nature. Since this mover does not possess an internal principle of motion and rest, its study would seem to pertain to first rather than second philosophy. And yet the traditional reading does not only make the first unmoved mover an object of study for the student of nature; it also makes this mover the ultimate aim of the investigation conducted in *Physics* VIII. It is tempting to dismiss this concern by suggesting that in *Physics* VIII Aristotle does not enforce a division of labor between first and second philosophy, but that instead he builds a bridge between these two sciences. I will not accept this suggestion. I will argue that the investigation conducted in *Physics* VIII is not blind to the boundaries of natural philosophy. By my lights, Aristotle remains a disciplined investigator throughout the whole of *Physics* VIII. Even when Aristotle is concerned with the first unmoved mover, he does not lose sight of his stated goal, which is the explanation of the eternity of motion. As I will argue, his treatment of the unmoved mover does not, and need not, go beyond the boundaries of natural philosophy.

II The explanatory structure of *Physics* VIII in outline

My reading takes the lead from the opening lines of *On the Movement of Animals* (*MA*). Those lines allude to an investigation that clearly

Many thanks to Mariska Leunissen, Zoli Filotas, Emily Perry, and Charles Ives for their comments on a draft of this chapter.

corresponds to *Physics* VIII. We are told that this investigation develops in two stages, and that both stages are concerned with motion, or more precisely with eternal motion:

> Now, it was determined earlier that the source of all other motions is that which moves itself, and [that the source] of this [is] that which is not subject to motion, and that the first mover must be unmoved, *when it was determined concerning eternal motion, whether or not there is such a thing, and if there is, what it is.* (*MA* 1, 698a8–11)

In this passage, Aristotle seems to have in mind the two stages of scientific investigation outlined at the beginning of the second book of the *Posterior Analytics*. These stages are introduced by two questions: the *ei esti* (whether it is) and the *ti esti* (what it is) question. By answering an *ei esti* question we establish whether something is the case (or not). A fact thus established calls for a causal explanation. It is quite possible that a thing's cause becomes available at the same time as the fact is established, but this is the exception rather than the rule. Typically, we *first* establish a fact by answering an *ei esti* question, and *then* we look for its causal explanation. When we ask a *ti esti* question, we ask for a definition that specifies the relevant cause. Consider how Aristotle combines the search for a definition and the search for a causal explanation in the *Posterior Analytics*: It is the same, he says, to search what a thing is (*ti esti*) and why it is (*dia ti*) (*Apo.* II.2, 90a15). Aristotle illustrates this point with the help of the following example: The definition of a lunar eclipse is privation of light coming from the moon due to the interposition of the earth. This definition tells us what a lunar eclipse is by singling out the cause of the lunar eclipse, namely the interposition of the earth between the moon and the sun.

With this in mind we can turn to the opening lines of *Physics* VIII. The *ei esti* question is introduced right at the start of the book with the following words:

> Did motion ever come into existence not having existed before, and will it go out of existence again, so that nothing is in motion? Or is it the case that motion neither comes into existence nor goes out of existence but instead always existed and always will exist, belonging to things as something immortal and unceasing, as if it were a kind of life of all beings constituted by nature? (*Phys.* VIII.1, 250b11–15)

Following what Aristotle himself tells us at the outset of *On the Movement of Animals*, I read this passage as introducing an investigation into the existence of eternal motion. By the end of *Physics* VIII.1, Aristotle is confident that he has established that motion is eternal. He tells us that there was never a time when there was no motion, and there will never be a time when there will be no motion (252b5–6). Thus, there exists eternal motion. If this reading is accepted, the end of the first chapter marks the end of the *ei esti* investigation. I will discuss the role of *Physics* VIII.2 in due course.[1] For the time being, I am content to say that Aristotle turns to the *ti esti* question only at the beginning of *Physics* VIII.3. We have seen that answering a *ti esti* question entails looking for a causal explanation, so it is not really surprising that the chapter begins with a *dia ti* question: Why in the world (*dia ti pote*) are some things at one time in motion and at another time at rest (253a23–24)? It is in the context of answering this question that Aristotle introduces, in *Physics* VIII.6, a first principle that is eternal and causes motion without being itself subject to motion, namely the first unmoved mover. However, Aristotle cannot be content with the mere introduction of such a principle but is required, *by his own conception of the scientific enterprise*, to try to get to the most causally precise description of that principle. This description is reached only at the end of *Physics* VIII.10. It crucially depends on some of the results reached in *Physics* VIII.7–9. But to understand how these chapters contribute to answering the *ti esti* question we have to go back to *Physics* VIII.6, and in particular to the claim that there must be a first principle that is eternal and causes motion without being itself subject to motion. This principle can only move eternally in one and the same invariable way. By reflecting on this apparently innocuous fact, Aristotle is able to infer that there must exist a *single* eternal motion. Note that the existence of a single eternal motion cannot be logically inferred from the claim that there exists eternal motion (*Physics* VIII.1). It can be secured only by reflecting on the nature of the first principle (*Physics* VIII.6). But as soon as this result is in place, we need to know what kind of motion this is and why. This is the task that Aristotle sets for himself in *Physics* VIII.7–9.

I have offered an outline of *Physics* VIII. On the proposed reading, there is a single plan running from the beginning to the end of *Physics* VIII. My goal in the pages to follow is to shed further light on this plan. I will argue that Aristotle is following a deliberate argumentative strategy in which eternal motion is progressively illuminated, and eventually explained, by arriving at the most precise description of its cause. In other words, Aristotle adopts a strategy of progressive revelation. As a result of this strategy, it is only at the very end of *Physics* VIII that we (following

[1] See below, section IV.

Aristotle) are able to answer the *ti esti* question, and hence to say what eternal motion is.

On this reading, eternal motion remains the focus of study throughout *Physics* VIII. One might object that this kind of focus does not sit well with Aristotle's programmatic remarks in *Physics* VIII.1. There, Aristotle tells us that the investigation he is about to launch contributes "not only to the study of nature but also to the investigation about the first principle" (251a5–8), which suggests that the book is also about the first principle. To block this objection, it is sufficient to introduce the distinction between primary and secondary objects of study. Since proper knowledge depends on grasping the relevant cause(s), it is not merely an accident that Aristotle is concerned with the first unmoved mover as the cause that explains why motion is eternal. On the contrary, as we have already seen, Aristotle is required to search not simply for the relevant cause but also for the most precise description of that cause. But this fact alone does not make the unmoved mover the primary object of study in *Physics* VIII (thereby turning its investigation into a metaphysical one). At most, it makes it a secondary object of study. The primary investigation, which remains firmly within the realms of natural philosophy, is about motion, or more precisely about eternal motion.

III The *ei esti* stage of investigation (*Physics* VIII.1)

Physics VIII.1 is an attempt to answer the question whether motion is eternal, or so I have argued. As soon as the *ei esti* question is introduced, Aristotle outlines three possible answers to the question. He links them to the names of Democritus, Anaxagoras, and Empedocles. Aristotle is not interested in a full review of what has been said on the topic of the eternity of motion by his predecessors. Rather, he is content to highlight what his (our) theoretical options are. They are the following: (1) motion is eternal (Democritus); (2) motion is not eternal (Anaxagoras); (3) motion and rest alternate eternally (Empedocles).

The *ei esti* question is answered by invoking the definition of motion given at the beginning of *Physics* III.[2] It is also possible to detect a less obvious reference to the treatment of time offered in *Physics* IV.[3] Evidently, Aristotle uses the results established in *Physics* III and IV as his *scientific* starting-points (*archai*). He is able to arrive at the conclusions that there was never a time when there was no motion and that there will never be a

[2] *Phys.* VIII.1, 251a8–10. [3] *Phys.* VIII.1, 251b10–13.

time when there will be no motion, while starting from principles developed within his own science of nature, and while doing so solely with the help of conceptual tools he developed earlier in the *Physics*. What his predecessors have said on the topic of motion and time does not seem to play any significant role in how the conclusion stated *in calce* at the end of *Physics* VIII.1 is reached.

If this reading of *Physics* VIII.1 is correct, then dialectic understood as the discussion of the so-called reputable opinions (*endoxa*) does not contribute much to answering the *ei esti* question. Its role is a different one, and one that does not become fully apparent until in the second part of the chapter. By then, Aristotle has already established that there is always a motion prior to any given motion, and that there is always a motion subsequent to any given motion. At that point, Aristotle returns to the positions defended by Anaxagoras, Empedocles, and Democritus in order to review how they failed in their attempt to offer an adequate causal explanation for the positions they have defended. His discussion of their views (and their failures) helps us see what is expected from an attempt to answer the *ti esti* question, if it is to be successful. The ultimate criteria of success that can be extracted from the discussion of Anaxagoras, Empedocles, and Democritus are two: Our answer (1) must do justice to the existence of order and regularity in the natural world (*contra* Anaxagoras), and (2) must explain eternal motion in such a way that it leaves nothing unexplained (*contra* Empedocles and Democritus).

So let us see, briefly, what Aristotle thinks we can learn from the errors of his predecessors. *Pace* Anaxagoras, there is no disorder in nature and in the things that happen in accordance with nature (252a11–12). By Aristotle's lights, this is an empirical fact and a fact that the cosmological account advanced by Anaxagoras, which posits *nous* as an agent initiating motion, contradicts. Unlike Anaxagoras, Empedocles does not violate the empirical truth that there is order (*taxis*) in nature. In fact, his cosmological account may be taken as a vindication of the claim that there is order in nature. However, Empedocles fails to offer an adequate explanation of the particular order he envisions. More specifically, it is not sufficient to say that change and rest alternate, and to say that there are equally long periods of change and rest. It is necessary to find a cause of that particular arrangement. In particular, Aristotle says that we should never posit anything without an argument, and that our argument must be either inductive or deductive (252a23–25). When Aristotle turns to Democritus, he draws attention to the following principle that the latter invoked in support of his claim that motion is eternal: "things happened in the past as they happen now" (252a35). Democritus may have used this principle in his explanation of the collision of atoms. His explanation may have combined two claims: (1) that there is always a collision prior to any given collision (and presumably there will always be a collision subsequent to any given collision), and (2) that no further cause for this fact needs to be given. In short, things happened in the past as they happen now. Aristotle objects to this explanatory strategy with the help of an example which is familiar to the reader of the *Posterior Analytics*: It is always true that a triangle's internal angles have the property of adding up to 180 degrees, but this fact nevertheless has a cause. Put differently, it is not sufficient to say that things happened in the past as they happen now: A true causal explanation of why things happen always in the same way must be supplied.[4]

IV From the *ei esti* to the *ti esti* question (*Physics* VIII.2)

My starting-point has been what Aristotle himself says in the opening lines of *On the Movement of Animals*. Those lines invite the application of the two-stage theory of scientific inquiry outlined in the second book of the *Posterior Analytics* to *Physics* VIII. According to this theory, it is possible to launch into a search for the cause(s) explaining why something is the case only when that fact has been established. By the end of *Physics* VIII.1 Aristotle has established that there is eternal motion. The search for the cause of this fact is a long and complex journey. It is only at the end of *Physics* VIII.10 that Aristotle appears to be confident that he has reached an adequate description of the cause and is in the position to say what eternal motion is. In the following sections, I will outline some aspects of his argumentative strategy.

Before moving on, however, a few more words on the conclusion reached at the end of *Physics* VIII.1 are in order. That conclusion not only commits Aristotle to the thesis that there is no beginning and no end to motion; it also commits him to the claim that there are no gaps in the infinite chain of motion. This follows from the rejection of the position that Aristotle ascribes to Empedocles. Aristotle's view is that there is always

[4] Democritus seems to have employed the same explanatory strategy in his account of animal generation. In *Generation of Animals*, Aristotle tells us that Democritus invoked the fact that "this is how things always come to be" as sufficient explanation of the formation of bodily parts (*GA* II.6, 742b17–23). In this case, Aristotle gives some additional information concerning how Democritus may have reasoned: to ask the question why (*dia ti*) is to seek for a beginning (*archê*); however, what is always is indefinite (*apeiron*), and there is no beginning (*archê*) of the indefinite (*apeiron*). Aristotle objects to this explanatory strategy, as he does in *Physics* VIII, by recalling his favorite geometrical example of the triangle and its angles, as described above (*GA* II.6, 742b24–29).

motion without interruption.[5] At the end of *Physics* VIII.1, nothing precludes the existence of a *single* motion that is both continuous and eternal. And yet, as we have already said, the existence of such a motion cannot be derived from the existence of an eternal, unbroken chain of motion. In other words, the existence of a single motion that is both eternal and continuous is not a given in *Physics* VIII. It is a result to be secured in the process of answering the *ti esti* question.[6]

What I have said so far draws attention to the fact that the conclusion reached at the end of *Physics* VIII.1 is in need of further articulation and elaboration. Let us therefore return to the theory of scientific inquiry outlined in the second book of the *Posterior Analytics.* There, Aristotle is not just content to say that the *ei esti* question is to be answered before asking the *ti esti* question; he also tells us that by answering the *ei esti* question we gain some initial grasp of what the thing itself is. More precisely, we acquire *non-accidental* knowledge that the thing exists. For Aristotle, this is knowledge that is already directed towards the *ti esti* of the thing (93a28–29). What Aristotle may have in mind is a certain epistemic state that, among other things, gives us a fix on the thing to be explained and at the same time motivates us to go further in our inquiry.[7] I suggest that this is exactly the state we are in at the end of *Physics* VIII.1.

In *Physics* VIII.2, Aristotle introduces two objections to the conclusion that has just been established, namely that motion is eternal. It is important to understand that his discussion of these objections is not meant to be a confirmation of the conclusion reached in *Physics* VIII.1. This conclusion is not in need of further confirmation. At this stage of the inquiry, we do not know yet whether there is a *single* eternal motion. But we do know that there is always motion without interruption. Our knowledge of this fact is as firm as it can possibly be. Therefore, the question for the reader of *Physics* VIII is this: What role does *Physics* VIII.2 play in the argument, if it does not serve to confirm the results achieved in *Physics* VIII.1? I submit that the discussion of the two objections is not looking backward to the conclusion established in *Physics* VIII.1 but rather forward to the subsequent investigation. In dealing with these objections, Aristotle does not only give us a sense of the investigation to come but also orients it in a certain direction.

[5] Cf. how the conclusion reached at the end of *Physics* VIII.1 is recalled at the beginning of *Phys.* VIII.6, 258b10–12: "*since there is always to be motion without interruption*, there must be something eternal that first imparts motion (whether one or many), and the first mover must be unmoved."

[6] See below, section V.

[7] This aspect of Aristotle's theory of scientific inquiry has been studied by Rob Bolton. See, in particular, Bolton 1976: 514–544 and 1987: 120–166 (especially at 130–142).

The first objection is that motion is a self-terminating process, since all motion seems to be from a contrary to a contrary (252b9–12). Although this general observation does not really prove that motion cannot be eternal, it can be taken as initial evidence against the conclusion established in *Physics* VIII.1. If all motion is bounded by contraries, then it is more difficult to believe that motion is eternal. In his reply to this first objection, Aristotle anticipates that his argument will show that all motion does not take place between contraries, for there is one type of motion, namely circular locomotion, which is *not* bounded by contraries.

The second objection begins with an observation about the behavior of inanimate objects: These objects are in motion at one time and at rest at another; it would seem that they should be either always in motion or always at rest (252b12–16). Aristotle builds on this objection by adding an observation about the behavior of animate beings. It would seem that these beings can initiate their own motion without any prior motion. If this is the case, these beings are *prima facie* evidence for the thesis that there can be motion without a prior motion (259b17–28). Aristotle rejects the claim that animate beings are self-movers in the sense that they can initiate their motion *without any prior motion*. This qualification is important. Aristotle does not mean to deny that animate beings are self-movers; he only disputes how self-motion is typically understood. Clearly, he thinks that it is an empirical fact that some physical motions caused by the environment take place in the animal. Those physical motions are not only prior to animal motion, but also causally connected to it. In other words, there is a sense in which the animal is causally responsible for its bodily displacement. But that bodily displacement causally depends on other, prior motions taking place in the body of the animal. Aristotle ends this section by promising that self-motion will become clear at a later stage of his discussion (253a20–21). His remark confirms that the chapter as a whole is looking forward to the second stage of the argument offered in *Physics* VIII.

V The *ti esti* stage of investigation (*Physics* VIII.3–6)

Throughout *Physics* VIII.2, the focus remains on motion. Aristotle promises to tell us more about circular motion, the motion of inanimate objects, and the motion of animate beings. Evidently, these are the types of motion which are to be integrated into a single causal explanation of why motion is eternal. It is from the motion of the animate and inanimate things that the subsequent investigation gets started. These things are at one time in motion and at another time at rest. This is a fact that is clear and obvious to everyone.[8] It is also a fact that calls for an explanation, as it seems to be *prima facie* evidence against the conclusion reached in *Physics* VIII.1, namely that motion is eternal. This explains why *Physics* VIII.3 opens with the following question: Why in the world are some things at one time in motion and at another time at rest?

This question is presented as the starting-point (*archê*) of a new investigation. While the investigation up to this point was pre-causal (as it was prompted by an *ei esti* question), the new investigation is a causal one, as it is introduced by a why-question (a *dia ti* question). Aristotle is remarkably forthcoming as to where this causal investigation will take us. While the starting-point (*archê*) of the investigation is why some things partake of motion and rest, the anticipated conclusion is that there are three types of thing that correspond also to three modes of being with respect to motion: for while some things are not subject to motion, other things are always in motion, and still others partake of motion and rest. This is how Aristotle puts it:

> The starting-point (*archê*) of our investigation is the same as the starting-point of the difficulty mentioned above: Why in the world (*dia ti pote*) are some things at one time in motion and at another time at rest? Now, it is necessary *either* that all things are always at rest, *or* that all things are always in motion, *or* that some things are in motion and some are at rest. In this last case, again, [it is necessary] *either* that the things that are in motion are always moved and the things that are at rest are always at rest *or* that all the things are by nature equally at rest and in motion. *Or* there remains also a third possibility: for it is possible that some things are not moved, others are always moved, and still others partake of both [motion and rest]. This is what we must say, for this solves all the difficulties [introduced in *Physics* VIII.2] and is the end (*telos*) of our investigation. (*Phys.* VIII.3, 253a22–32)

In stating the *telos* of his investigation, Aristotle does not make use of causal language. And yet, it is clear that he will have to do more than just establish that there are three types of thing (or, if you prefer, three modes of being with respect to motion). If Aristotle does not want to be open to the criticism he has leveled against his predecessors, he will have to offer a causal explanation for the arrangement envisioned in the opening lines of *Physics* VIII.3. More precisely, he will have to explain this arrangement, and explain it in such a way that there is no need for us to go any further in the

[8] *Phys.* VIII.6, 259a27.

search for an explanation. Hence, he will have to find the *ultimate* cause of this order.[9]

This remarkable explanatory feat is achieved at the end of *Physics* VIII.6. By then, Aristotle has not just established that there are three types of thing (three ways of being with respect to motion); he has also shown *how* they are causally interconnected. At that stage of the argument, Aristotle seems to be confident that the question introduced in the opening lines of *Physics* VIII.3 can be answered by identifying the cause which explains why there is this threefold relation to motion:

> It has become clear from what has been said also what puzzled us at the start, namely why in the world (*dia ti pote*) is it not the case that all things are *either* in motion *or* at rest *or* some always in motion and some always at rest, but some things are [in motion] at one time but not at another. *The cause* (aition) *of this is now clear*: Some things are moved by an eternal thing that is unmoved, and for this reason they are always in motion, while others [are moved] by something that is moved and changed so that they too must change. What is unmoved, as was said, because it remains simple, identical, and in itself, will move with a motion that is one and simple. (*Phys.* VIII.6, 260a17–19)

The argument leading to the discovery of the eternal mover which moves by being absolutely unmoved is best understood as an inductive inference to the only possible causal explanation of a fact that is forced upon us by experience, namely that some things partake of both motion and rest.[10] I will not engage in a detailed study of the remarkably complex argument offered in *Physics* VIII.3–6. Here I am content to point out that Aristotle does not simply reason from effect (motion) to cause (the first eternal unmoved mover); rather, he reasons from a *remote* effect (the fact that some things partake of motion and rest) to its *ultimate* cause (the first eternal unmoved mover). The relevance of these qualifications will become clear in due course.[11] For the time being, I only add that the ultimate cause is introduced in *Physics* VIII.6 to explain why motion is eternal in the sense of being a continuous and inexhaustible chain of motion. Aristotle argues that a collection of perishable unmoved movers, no matter how many they are, does not suffice to explain why there is always motion without

[9] Recall what Aristotle says in connection with Democritus: To look for an explanatory principle is to look for the sort of thing that explains eternal motion in such a way that the investigation can stop when the principle is found because at that point nothing is left unexplained.

[10] Cf. *Phys.* VIII.3, 254a35–b1: "against all these [views] one piece of evidence (*pistis*) is sufficient: *We see* (*horômen*) that some things are in motion at one time and are at rest at another time."

[11] See below, section VII.

interruption. In addition to the existence of perishable unmoved movers, the existence of an eternal unmoved mover is necessary.[12]

I have already noted that the existence of a *single* eternal motion cannot be derived from the conclusion that motion is eternal (*Physics* VIII.1). The existence of such a motion must be secured in the process of looking for an answer to the question why motion is eternal, or so I have argued. Now, the existence of a single eternal motion is inferred towards the end of *Physics* VIII.6. It is inferred from a reflection on the very nature of the eternal mover which moves by being absolutely unmoved. Precisely because the first eternal mover cannot be moved even in an accidental way, it must move always in one and the same way. At first sight, this may look like a fairly innocuous point. And yet it is sufficient to establish that, in addition to the first unmoved mover, there must be some single motion that is eternal and continuous:

> But if indeed there exists something that is always of such a nature as to move something while being itself unmoved and eternal, then it is necessary that the first thing moved by this is eternal too. This is also clear from the fact that there is no other way for generation and corruption and change to exist in the other things unless there is something moved that moves. For the unmoved [mover] will move in the same way and with one and the same motion, since it will not change in relation to what is moved. But what is moved by something that, though it is in motion, is moved directly by what is unmoved, since it will be in a different relation to the things [that it moves], it will not be the cause of the same motion, but because it is in contrary places or forms it will cause each of the other things to be moved in contrary ways, and will cause them to be at one time at rest and at another time in motion. (*Phys.* VIII.6, 259b32–260a10)

At this stage of the argument, we know very little about this motion. We do not know, for instance, what kind of motion it is. But we do know that it enters as an indispensable causal factor into the explanation of the eternal cycle of generation and corruption. More specifically, the first unmoved mover cannot account directly for motion as experienced on earth. Hence, an intermediate mover is required. The eternal motion caused by the first unmoved mover functions as such a mover. Needless to say, there cannot be a single eternal motion without a thing that undergoes this motion. In

[12] In the opening lines of *Physics* VIII.6, the question of how many unmoved movers are needed is clearly flagged: "since there is always to be motion without interruption, there must be something eternal that first imparts motion (*whether one or many*), and the first mover must be unmoved" (*Phys.* VIII.6, 258b10–12). Note that in the course of *Physics* VIII.6, Aristotle argues for the view that one eternal motion is sufficient to account for the existence of a single eternal inexhaustible chain of motion.

Physics VIII, however, Aristotle says very little about this being. In fact, he is content to say that it performs the function of a *moved* mover in the explanation of the eternity of motion.

If this reading is correct, what Aristotle says in the above passage cannot be dismissed as an afterthought. On the contrary, it does some important explanatory work. It helps us articulate our initial grasp (and confused idea) of the eternity of motion. The eternal chain of motion which does not admit of a beginning and an end, and does not admit of gaps either, turns out to be a fairly complex state of affairs entailing the existence of a single thing that is eternally moved in addition to the existence of an eternal cycle of generation and corruption. Perhaps this result can be recast as follows. By the end of *Physics* VIII.6, we have discovered not only the *cause* of eternal motion, which is to say the first mover that is absolutely unmoved, but also the proper *subject* that undergoes eternal motion, namely the first moved thing.

It may be helpful to recall how Aristotle himself refers to the results he has achieved in *Physics* VIII.6. In the first book of *On Generation and Corruption*, Aristotle refers to *Physics* VIII as one of the investigations concerned with motion. Interestingly enough, he seems to think that his argument has established two things: (1) that there is a first unmoved mover that is not subject to motion; and (2), that there is something else that is eternally moved by it. Note that Aristotle is quite forthcoming in describing the first unmoved mover as an *efficient* cause of motion:

> We must deal with these [questions] to the extent that it is possible, and give the cause of the fact that there is always generation – both unqualified and qualified generation. Since one cause is that from which we say that motion originates, and another is matter, it is the latter that we have to discuss here. *For we have dealt with the other cause earlier in our writings on motion when we said that there is something that is not subject to motion through all time, and something else that is always moved.* (*GC* I.3, 317a1–5)

VI The *ti esti* stage of the investigation continued (*Physics* VIII.7–9)

The conclusion of *Physics* VIII.1 is that there must be a chain of motion with no beginning and no end and no gaps in between. The argument advanced at the end of *Physics* VIII.6 helps us see that such a chain requires the existence of a thing which is always moved in the same invariable way. Admittedly, we know very little about the nature of its motion. By reflecting on the nature of the first principle of motion, we can safely conclude that this motion must not only be single, but also simple. More specifically, if the mover must be absolutely unmoved, then it can only impart a simple motion. In this case, simplicity entails uniformity. Still, we are not yet in the position to say what kind of motion this is. Hence, our investigation cannot stop here. Let us put it differently: Even if we have found the proper *subject* and the proximate efficient *cause* of motion, we still do not know the *form* of this motion. The task Aristotle sets for himself in *Physics* VIII.7–9 is to establish that the first moved mover is moved in a circle. Circular motion is the only form of locomotion that can be both primary and eternal.

Aristotle is quite forthcoming about his goal in *Physics* VIII.7–9. His goal is to establish whether there is a motion that is continuous, and, if there is one, what it is.[13] It turns out to be the motion undergone by the first mover that is eternally moved, and it enters into the causal system as the primary, i.e. first, motion:

> But once we have made another beginning (*archê*), these matters will become clearer. It is to be investigated whether it is possible that there exist one motion that is continuous or not, and if it is possible, what it is, and what is the first motion: For it is clear that if it is necessary that there is always motion, this motion is first and continuous because the first mover moves this motion, which must be one and the same, continuous, and first. (*Phys.* VIII.7, 260a20–26)

In the stretch of text that begins with chapter 7 and ends with chapter 9, Aristotle proves that:

1. locomotion is prior to generation, alteration, growth, and corruption
2. circular motion is prior to rectilinear motion
3. hence, circular locomotion is the primary motion.

Part of the argument for the second claim consists in showing that:

4. no motion other than circular locomotion can be eternal and continuous.

For Aristotle, there can only be finite rectilinear motion. But no finite rectilinear motion can be eternal and continuous. Two reasons are given for this claim. First, every motion takes place between contraries. Finite rectilinear motion is no exception to the rule. Let us treat A and B as contraries. Since the thing that undergoes rectilinear motion from A to B

[13] Note that Aristotle adopts the *ei esti* and the *ti esti* questions in this stretch of text. In other words, even the search for the primary, i.e. first, motion, offered in *Physics* VIII.7–9 unfolds according to the two-stage theory of scientific investigation outlined in the *Posterior Analytics*.

or from B to A must turn back as soon as it has reached either A or B, we are confronted with two motions rather than a single motion, namely motion from A to B and motion from B to A. Second, turning back necessarily involves coming to a stop, as there must be an interval of time between the motion from A to B and the motion from B to A. What we obtain is an eternal back and forth between A and B, rather than a single continuous eternal motion. Circular locomotion is the only motion that can be continuous and eternal. Aristotle treats circular locomotion as motion from one point to the same point, namely motion from A to A. Contrast this motion with finite rectilinear motion. In the case of finite rectilinear motion, it is possible for one thing to move from A to A, but only after having stopped at B. Moreover, motion from A and motion to A are contrary motions.[14] It is possible to have one thing moving with a single motion from A and to A provided that the motion in question is in a circle. This is the only motion that can be continuous, as the thing that undergoes motion need not stop when it has reached A.

It might be helpful to draw attention to a prominent feature of Aristotle's argumentative strategy. This feature is especially (but by no means exclusively) evident in *Physics* VIII.7–9. There, Aristotle is content to speak of circular locomotion in very general (i.e. abstract) terms. It is fairly clear that the circular locomotion that enters into the explanation of eternal motion is the eternal, continuous, and indeed uniform, circular locomotion of the outermost celestial sphere. This is the sphere of the fixed stars, which is moved always in the same way by the first unmoved mover. Still, it is remarkable that Aristotle never describes it in this way. Nor does he distinguish the motion of the heaven of the fixed stars from the motions of the so-called planets. To understand why Aristotle never engages in what we might want to call (for lack of anything better) celestial physics, we need to recall that the ultimate goal of *Physics* VIII is to explain why there always was and always will be motion. It turns out that this explanation requires the introduction of a causal intermediary between the first unmoved mover and the eternal cycle of generation of corruption. This causal intermediary is celestial motion. Aristotle is interested in celestial motion insofar as it enters as an *indispensable* moved mover into the eternal chain of motion.[15] A full

[14] Recall that the identity of motion is established by the contraries involved in the process. Motion from A is shorthand for motion from A to B, and motion to A for motion from B to A.

[15] Contrast this eternal chain of motion with the following chain of motion: man–stick–stone. In this second chain, the stick is a *dispensable* moved mover. While the stone is moved by the stick, it is clear that the man could move the stone without employing a stick. The eternal circular motion envisioned as an intermediary between the first unmoved mover and the eternal cycle of generation

study of the nature of celestial motion goes emphatically beyond the scope of *Physics* VIII. Such a study is the task of celestial physics. It is in the context of celestial physics that Aristotle offers a study of the heaven of the fixed stars, including an explanation of its distinctive orientation. Furthermore, it is in the context of celestial physics that the circular motion of the heaven of the fixed stars is distinguished from the circular motion of the so-called planets. Finally, it is in this context that Aristotle introduces the highly controversial claim that the heavens are made of a special simple body, unique to them.

VII Aristotle's second treatment of the cause (*Physics* VIII.10)

The last words of *Physics* VIII.9 are a careful recapitulation of the argument of *Physics* VIII up to this point:

> We have argued that there was always motion and always will be motion throughout all time [*Physics* VIII.1], and we have explained what the principle of this eternal motion is [*Physics* VIII.3–6]. We have also explained which is the primary motion and which is the only motion that can be eternal [*Physics* VIII.7–9], and we have pronounced the first mover to be unmoved [*Physics* VIII.6]. (*Phys.* VIII.9, 266a7–9)

One might be tempted to think that the argument has reached its natural conclusion at the end of *Physics* VIII.9. Interestingly enough, Aristotle does not stop there. He immediately goes on to announce a second treatment of the cause, namely the first unmoved mover:

> Now, we have to say that this [i.e. the first unmoved mover] must be without parts and must have no magnitude. (*Phys.* VIII.10, 266a10–11)

At least at first sight, this return to the cause of eternal motion is surprising. Recall that Aristotle has already offered an inference to the ultimate cause of eternal motion. We have seen that the argument that started at *Physics* VIII.3 resulted not only in the discovery of such a cause, but also in the discovery of the genuine subject of eternal motion. Arguably, the greatest triumph of *Physics* VIII.6 is the conclusion that, in addition to the first mover that moves by being absolutely unmoved, there must be something else that is eternally moved by the first mover. This other thing enters into the explanation of the eternity of motion as an indispensable moved mover. At that point, it was quite natural to wonder what kind of motion this second mover undergoes and why. Hence, it was not really surprising to

and corruption is not this sort of moved mover. Without its presence, there would be no eternal cycle of generation and corruption.

see Aristotle turn his attention to this important question in *Physics* VIII.7–9. But once it is established that there is a genuine subject of motion, and that this subject performs circular locomotion, which is the only motion that can be single and continuous, we might be inclined to think that the explanatory work has been completed. And yet Aristotle does not seem to think so, because he returns to the cause of eternal motion. Why?

The interpreters who take *Physics* VIII to be an argument in two stages from the eternity of motion to the first unmoved mover have no trouble explaining the final chapter of *Physics* VIII. By their lights, *Physics* VIII.10 is the culmination of the investigation conducted in *Physics* VIII. At first sight, it would seem to be more difficult to find a satisfactory reading of this chapter for someone who adopts the view I have been advocating, namely that *Physics* VIII as a whole is concerned with eternal motion. Recall, however, that we do not have full knowledge of something unless we have knowledge of its proper and adequate cause. The qualification "proper and adequate" is the key to understanding the role that our chapter plays in the argument of *Physics* VIII. For Aristotle, it is never sufficient to point to the relevant cause; it is also imperative to seek out the most causally precise description of that cause. And it is only at the end of the chapter that such a description is available to us.[16]

It cannot be just a coincidence that the second treatment of the cause of eternal motion uses *the proximate effect* as its starting-point. This effect was not available when Aristotle offered the first treatment of the cause. We have seen that, in *Physics* VIII.3–6, Aristotle reasoned from *the remote effect*, which is to say from the fact that some things are at one time in motion and at another time are at rest, to its ultimate cause, namely the first eternal unmoved mover.[17] It is only at the end of *Physics* VIII.9 that *the proximate effect*, namely the primary circular locomotion, is available. In *Physics* VIII.7–9, Aristotle is able to establish that the primary circular locomotion is a single, continuous, eternal, and indeed infinite, motion. What seems to be especially important is that this motion is *infinite*. At least twice Aristotle returns to this feature.[18] Also, in light of this fact, it cannot be just a coincidence that at the beginning of *Physics* VIII.10, in gathering premises for his second treatment of the cause of eternal motion, Aristotle establishes that nothing finite can impart an infinite motion, and that an infinite power cannot reside in a finite magnitude.

[16] This is how I understand the methodological recommendation to look for the most precise cause. See, for instance, *Phys.* II.3, 195b21–23.

[17] See above, section V. [18] *Phys.* VIII.8, 261b27 and 265a10.

There is at least one other aspect of *Physics* VIII.10 which is worth mentioning: Aristotle's focus on projectile motion, which presents a *prima facie* difficulty for the thesis that everything that is moved is moved by something: How can something be in motion when the mover is no longer in contact with the thing that is moved? And yet there is a more specific, and indeed pressing, reason to deal with this *aporia* in *Physics* VIII.10. But in order to appreciate why Aristotle engages in this excursus on projectile motion, we need to recall, briefly, how he solves the *aporia*. His solution is that the original mover gives the power of being a mover either to air or to water: Both air and water are suited to being moved and to moving themselves. In other words, projectile motion is explained as a case in which the object is moved by a sequence of consecutive movers in contact with one another. Now, one might try to apply this solution to the case of eternal circular locomotion. In particular, one might try to explain this motion as the result of an infinite succession of consecutive finite powers transmitting motion from one to the other. Aristotle blocks this explanatory route by making it clear that projectile motion merely *appears* to be a case of continuous motion. A requirement for continuity is that motion is caused by a single mover. Hence, positing an infinite sequence of finite powers which jointly cause an infinite motion will not do. The resulting motion will be infinite but will not be continuous. Yet continuity is established as a key feature of eternal circular locomotion in *Physics* VIII.7–9.

We are finally in a position to turn to the second treatment of the cause of eternal motion. This treatment is offered in the final stretch of text (*Physics* VIII.10, 267a21–b26). This treatment is not just a recapitulation of what was said before. It has a new starting-point and does not end in the same way. It goes as follows:

1. There must be a single continuous motion.
2. There must be a single subject of motion.
3. There must be a single mover.
4. This mover must be either moved or unmoved.
5. The chain of motion must end with a single unmoved mover.
6. This is the only mover which can always move because it is not affected and its motion is effortless.
7. It moves with a uniform motion.
8. Because action at distance is not possible, this mover must be located either at the periphery or in the middle of the universe.
9. It is located at the periphery.

The treatment is not complete. A final step is missing. Aristotle invokes two claims he has proven in *Physics* VIII.10, namely that nothing finite can impart an infinite motion, and that an infinite power cannot reside in a finite magnitude. Since there cannot be an infinite magnitude, he is left with the following conclusion:

10. The unmoved mover must be without magnitude, without parts, and must be indivisible.

At this point we have reached the most precise possible description of the (efficient) cause of eternal motion. This description is secured from within natural philosophy and solely with the help of the conceptual resources available to Aristotle in *Physics* VIII. Note that the description of the ultimate principle of motion is largely negative, which is exactly what is expected, given that this principle is not a proper object of study for natural philosophy. More directly, if one looks at what Aristotle says on the topic of the first principle, one quickly realizes that there is not much of a positive description beyond the fact that this principle is a mover. Sure, the first principle is an unmoved mover. But this is a feature shared with a number of other (lesser) principles. The analysis of self-motion offered in Physics VIII.5 suggests that souls (both human and animal souls) are unmoved movers. One might reply that the first unmoved mover is eternal, whereas these other unmoved movers are perishable. One might also add that the first principle is absolutely unmoved, whereas the souls are incidentally moved. Still, it is not clear that these qualifications get us any closer to the essential nature of the first principle. For a positive treatment of this principle we have to look elsewhere. In *Lambda*, Aristotle has a great deal to say on the first principle. He describes it as a substance whose essence is pure actuality. He also identifies it with *nous* and the good. Clearly, these positive descriptions are not available in *Physics* VIII. *Physics* VIII does not even refer to this principle as a divine principle (a god). By contrast, in *Lambda*, Aristotle does not only refer to the first principle as a god; he also identifies it with the good (*Lambda* 7, 1072b25, 28–29, 30).

VIII The scientific definition of eternal motion

I have argued that *Physics* VIII is a single extended argument concerned with eternal motion. The unity and integrity of this argument becomes fully apparent as soon as the theory of science outlined in the *Posterior Analytics* is applied to *Physics* VIII. This theory requires Aristotle *first* to establish that there is eternal motion and *then* to look for a definition that singles out the cause of the eternity of motion. To understand, however, what gets defined, and thereby explained, we need to recall that the argument advanced in *Physics* VIII establishes not only the existence of eternal motion, but also the existence of a *single* eternal motion. The transition from the existence of eternal motion – in the sense of the existence of one unbroken chain of motion with no beginning and no end – to the existence of a *single* eternal motion is made at the end of *Physics* VIII.6. What gets defined, and thereby explained, is the latter, the single eternal motion that is identical with the circular motion caused by the first principle.

Admittedly, Aristotle does not give us such a definition. But at this point it is not difficult to offer a definition that, in addition to giving us the most precise causal description of the (efficient) cause of eternal motion, also specifies the type of motion as well as the relevant subject of that motion: Eternal motion is the primary circular locomotion of that which is moved by the first eternal unmoved mover which is indivisible, without parts, and without magnitude.

Bibliography

A. Ancient, medieval, and Renaissance tradition

Aquinas, Thomas 1884. *In* VIII *Libros Physicorum Aristotelis* (*Opera Omnia*, ed. Leonina). Rome

Averroes 1562. *Aristotelis De Physico Audito Libri Octo cum Averrois Commentariis* (vol. IV *In Aristotelis opera quae extant . . . Averrois Cordubensis in ea opera omnes qui a nos pervenere commmentairi*). Venice (facsimile reprint Frankfurt am Main: Minerva 1962)

Blackwell, R. J., R. J. Spath, and W. E. Thirlkel 1884. *Commentary on Aristotle's* Physics *by St Thomas Aquinas*. New Haven: Yale University Press

Bodnár, I., M. Chase, and M. Share 2012. *Simplicius, On Aristotle's* Physics *8.1–5*. London, Ithaca, NY: Duckworth and Cornell University Press

Bruns, I. (ed.) 1887. *Alexandri Aphrodisiensis praetor commentaria scripta minora:* De Anima *liber cum mantissa*. Commentaria in Aristotelem Graeca 2.1. Berlin.

Coninbricenses (Coimbra Commentators) 1594. *In Octos Libros Physicorum Aristotelis*. Lyons (facsimile reprint Hildesheim: G. Olms 1984)

Diels, H. (ed.) 1882. *Simplicius, Aristotelis Physicorum Libros Quattuor Priores Commentaria*, in his *Commentaria in Aristotelem Graeca*, vol. IX. Berlin: G. Reimer

(ed.) 1895. *Simplicii in Aristotelis Physicorum Libri Quattuor Posteriores Commentaria*, in his *Commentaria in Aristotelem Graeca*, vol. X. Berlin: G. Reimer

Hayduck, M. (ed.) 1891. *Alexander of Aphrodisias, in Aristotelis Metaphysica Commentaria*. Berlin: G. Reimer

Hossfeld, P. (ed.) 1987–93. *Albertus Magnus. Physica* (Albertus Magnus, *Opera Omnia*, ed. B. Gayer, vol. IV 1–2). Ashendorf

Hussey, Edward (ed. and trans.) 1983. *Aristotle's* Physics *Books* III *and* IV. Oxford: Clarendon Press

Jowett, Benjamin (ed. and trans.) 1875. *Timaeus:* Plato. Translated with an introduction by Benjamin Jowett. Oxford

McKirahan, R. 2001. *Simplicius, On Aristotle's Physics 8.6–10*. London, Ithaca, NY: Duckworth and Cornell University Press

Pacius, Julius 1595. *Aristotelis Naturalis Auscultationis libri* VIII. Frankfurt am Main (facsimile repr. Frankfurt am Main: Minerva 1964)

1596. *Aristotelis Naturalis Auscultationis*. Frankfurt am Main (repr. Hildesheim: G. Olms 1964)

Rashed, M. 2005. *Aristote:* De la génération et la corruption. Edition, traduction et commentaire. Paris: Les Belles Lettres

(ed.) 2011. *Alexandre d'Aphrodise, Commentaire perdu à la Physique d'Aristote (Livres* IV–VIII). Berlin, New York: Walter de Gruyter

Ross, W. D. (ed.) 1928. *The Works of Aristotle*. Translated into English under the editorship of W. D. Ross. Oxford: Clarendon Press

Schenkl, H. (ed.) 1900. *Themistii in Aristotelis* Physica *Paraphrasis. Commentaria in Aristotelem Graeca*, vol. V. 2. Berlin: G. Reimer

Todd, R. B. 2008. *Themistius, Aristotle,* Physics *5–8*. London, Ithaca, NY: Duckworth and Cornell University Press

Vitelli, G. (ed.) 1887. *Ioannis Philoponi in Aristotelis Physicorum Libri Quinque Posteriores Commentaria. Commentaria in Aristotelem Graeca*, vol. XII. Berlin: G. Reimer

(ed.) 1887. *Philoponus, Commentary on Aristotle's 'Physics', Commentaria in Aristotelem Graeca*, vols. XVI–XVIII. Berlin: G. Reimer

Zabarella, J. 1602. *Opera Physica*. Frankfurt am Main (facsimile repr. Verona: Minerva 2009)

1607. *De Rebus Naturalibus Libri* XXX. Frankfurt am Main (facsimile repr. Frankfurt am Main: Minerva 1996)

1608. *Opera Logica*. Frankfurt am Main (facsimile repr. Frankfurt am Main: Minerva 1966)

B. Modern translations, monographs, and references

Anagnostopoulos, Andreas 2010. 'Change in Aristotle's *Physics* III', *Oxford Studies in Ancient Philosophy* 39: 33–79

Anagnostopoulos, Georgios (ed.) 2009. *A Companion to Aristotle*. Chichester: Wiley–Blackwell

Angioni, L. 2010. *Aristóteles:* Física I–II. Campinas: Editora Unicamp

Balme, D. M. 1987. 'Aristotle's Biology was not Essentialist', in Gotthelf and Lennox (eds.), 291–312

Barnes, J. (ed.) 1984. *Revised Oxford Translation: The Complete Works of Aristotle*. Princeton University Press

1993. *Aristotle, Posterior Analytics*. Oxford: Clarendon Press

Barnes, J., M. Schofield, and R. Sorabji (eds.) 1975. *Articles on Aristotle, Volume* I*: Science*. London: Duckworth

Beere, J. 2009. *Doing and Being – An Interpretation of Aristotle's* Metaphysics Theta. Oxford University Press

Berti, E. (ed.) 1981. *Aristotle on Science: The* Posterior Analytics. Padua: Editrice Antenore

Bodnár, I. 1997. 'Movers and Elemental Motions in Aristotle', *Oxford Studies in Ancient Philosophy* 15: 81–117

2012. 'Commentary on 246b3–246b20', in Maso, Natali, and Seel (eds.), 73–84

Bolton, R. 1976. 'Essentialism and Semantic Theory in Aristotle: *Posterior Analytics* II 7–10', *Philosophical Review* 85: 514–544

1987. 'Definition and Scientific Method in Aristotle's *Posterior Analytics* and *Generation of Animals*', in Gotthelf and Lennox (eds.), 120–166

1991. 'Aristotle's Method in Natural Science: *Physics* I', in Judson (ed.), 1–29

1997. 'The Material Cause: Matter and Explanation in Aristotle's Natural Science', in Kullmann and Föllinger (eds.), 97–126

2011. 'Why Does Aristotle Need Four Causes?', in Couloubaritsis and Delcomminette (eds.), 27–46

Bonitz, H. 1961. *Index aristotelicum*. Hildesheim: G. Olms

1969. *Aristotelische Studien: Fünf Teile in einem Band*. First edition, Berlin 1862. Berlin: Walter de Gruyter

Bostock, D. 2006. *Space, Time, Matter, and Form: Essays on Aristotle's* Physics. Oxford University Press

Bowen, A. C. and C. Wildberg (eds.) 2009. *New Perspectives on Aristotle's* De Caelo. Louvain: Brill Academic Publishers

Bowin, J. 2011. 'Aristotle on Various Types of Alteration in *De Anima* II.5', *Phronesis* 56: 138–161

Bradie, M. and F. D. Miller Jr. 1999. 'Teleology and Natural Necessity', in L. D. P. Gerson (ed.), *Aristotle, Critical Assessments*. London: Routledge, 75–89; reprint from (1984) *History of Philosophy Quarterly* 1: 133–146

Broadie [Waterlow], S. 1982. *Nature, Change, and Agency in Aristotle's* Physics. Oxford: Clarendon Press

Broadie, S. 1990. 'Nature and Craft in Aristotelian Teleology', in Devereux and Pellegrin (eds.), 389–403

2004. '*On Generation and Corruption* 1.4: Distinguishing Alteration – Substantial Change, Elemental Change, and First Matter in *GC*', in De Haas and Mansfeld (eds.), 123–150

2009. 'Heavenly Bodies and First Causes', in G. Anagnostopoulos (ed.), 230–241

Brunschwig, J. 1991. 'Qu'est-ce que "La Physique" d'Aristote?', in F. de Gandt and S. Souffrin (eds.), *La Physique d'Aristote et les Conditions d'une Science de la Nature*. Paris: Vrin, 11–40

Burnyeat, M. F. 1980. 'Aristotle on Learning to be Good', in Rorty (ed.), 69–92

Cambiano, G. 2012. 'The Desire to Know: *Metaphysics* A1', in Steel (ed.), 1–42

Carteron, H. (ed.) 1926–31. *Aristote:* Physique. 2 vols. Paris: Les Belles Lettres

Castelli, L. M. 2012. *Aristotele:* Fisica, *Libro* IV. Rome: Carocci Editore

Cavagnaro, E. 2002. *Aristotele e il Tempo*. Bologna: Il Mulino

Charles, D. 1984. *Aristotle's Philosophy of Action*. London: Duckworth

1991. 'Teleological Causation in the Physics', in Judson (ed.), 101–128

2000. *Aristotle on Meaning and Essence*. Oxford University Press

(ed.) 2010a. *Definition in Greek Philosophy*. Oxford University Press

2010b. 'Definition and Explanation in the *Posterior Analytics* and *Metaphysics*', in Charles (ed.), 286–328

2010c. '*Metaphysics Theta* 7 and 8: Some Issues Concerning Actuality and Potentiality', in Lennox and Bolton (eds.), 168–197

2012. 'Teleological Causation', in Shields (ed.), 227–266

Charlton, W. 1970. repr. 1992. *Aristotle's* Physics *Books 1 & 2: Translated from the Greek with Introduction and Notes*. Oxford: Clarendon Press

Cherniss, H. F. 1944. *Aristotle's Criticism of Plato and the Academy*. Baltimore: Johns Hopkins University Press

Code, A. 1976. 'The Persistence of Aristotelian Matter', *Philosophical Studies* 29.6: 357–367

1987. 'Soul as Efficient Cause in Aristotle's Embryology', *Philosophical Topics* 15.2: 51–59

1988. 'Reply to Michael Frede's "Being and Becoming in Plato"', *Oxford Studies in Ancient Philosophy suppl. vol.*, 53–60

1995. 'Potentiality in Aristotle's Science and Metaphysics', *Pacific Philosophical Quarterly* 76.3–4: 405–418

1997. 'The Priority of Final Causes over Efficient Causes in Aristotle's *PA*', in Kullmann and Föllinger (eds.), 127–143

Coope, Ursula 2004. 'Aristotle's Account of Agency in *Physics* III.3', *Proceedings of the Boston Area Colloquium in Ancient Philosophy* 20: 201–221

2005. *Time for Aristotle:* Physics IV.*10–14*. Oxford University Press

2007. 'Aristotle on Action', *Aristotelian Society suppl. vol.*, 81.1: 109–138

2009. 'Change and its Relation to Potentiality and Actuality', in G. Anagnostopoulos (ed.), 277–291

2012. 'Commentary on 246a10–246b3', in Maso, Natali, and Seel (eds.), 57–72

in press. 'Aristotle, *De Motu Animalium*, Chapters 2 and 3', in Primavesi and Rapp (eds.)

Cooper, J. M. 1982. 'Aristotle on Natural Teleology', in Schofield and Nussbaum (eds.), 197–222

1987. 'Hypothetical Necessity and Natural Teleology', in A. Gotthelf and J. G. Lennox (eds.) *Philosophical Issues in Aristotle's Biology*. Cambridge University Press, 243–274

1996. 'Reason, Moral Virtue, and Moral Value', in Frede and Striker (eds.), 81–114

Corcilius, K. 2011. 'Aristotle's Definition of Non-Rational Pleasure and Pain and Desire', in I. Miller (ed.), *Aristotle's* Nicomachean Ethics*: A Critical Guide*. Cambridge University Press, 117–143

Corcilius, K. and P. Gregoric 2010. 'Separability vs. Difference: Parts and Capacities of the Soul in Aristotle', *Oxford Studies in Ancient Philosophy* 39: 81–119

Couloubaritsis, L. and S. Delcomminette (eds.) 2011. *La Causalité chez Aristote*. Paris, Brussels: Vrin/Ousia

Crisp, R. (trans.) 2000. *Aristotle:* Nicomachean Ethics. Cambridge University Press

Crivelli, P. 2012. *Plato's Account of Falsehood: A Study of the Sophist*. Cambridge, New York: Cambridge University Press

Davidson, Donald 1970. 'Events as Particulars', *Noûs* 4: 25–32

De Haas, F. A. J. 2012. 'Commentary on 247b1–247b13', in Maso, Natali, and Seel (eds.), 99–108

De Haas, Frans and Jaap Mansfeld (eds.) 2004. *Aristotle:* On Generation and Corruption, *Book* 1. *Symposium Aristotelicum*. Oxford: Clarendon Press

Denniston, D. 1934. *The Greek Particles*. Oxford University Press

Depew, D. 1997. 'Etiological Approaches to Biological Aptness in Aristotle and Darwin', in Kullmann and Föllinger (eds.), 209–227

Devereux, D. 2011. 'Aristotle on the Form and Definition of a Human Being: Definitions and their Parts in *Metaphysics* Z.10 & 11', *Proceedings of the Boston Area Colloquium in Ancient Philosophy* 26: 167–196

Devereux, D. and P. Pellegrin (eds.) 1990. *Biologie, Logique et Métaphysique chez Aristote*. Paris: Editions du Centre National de la Recherche Scientifique

Dirlmeier, F. (trans.) 1956. *Aristoteles* Nikomachische Ethik. Berlin: Akademie Verlag

Dretske, F. I. 1980. 'The Intentionality of Cognitive States', *Midwest Studies in Philosophy* 5: 281–294

Düring, I. (ed.) 1969. *Naturphilosophie bei Aristoteles und Theophrast*. Heidelberg: Lothar Stiehm Verlag

Ebrey, D. unpublished 'Aristotle's Motivation for Matter'. UCLA dissertation

Falcon, Andrea 2005. *Aristotle and the Science of Nature: Unity without Uniformity*. Cambridge University Press

Fazzo, S. 2013. 'Heavenly Matter in Aristotle, *Metaphysics Lambda* 2', *Phronesis* 58: 160–175

Fine, K. 1999. 'Things and their Parts', *Midwest Studies in Philosophy* 23.1: 61–74

Föllinger, S. (ed.) 2010. 'Was ist "Leben"? Aristoteles' Anschauungen zur Entstehung und Funktionsweise von "Leben"', *Philosophie der Antike* 27.

Frede, M. 1980. 'The Original Notion of Cause', in Schofield, Burnyeat, and Barnes (eds.), 217–249

1990. 'An Empiricist View of Knowledge: Memorism', in Stephen Everson (ed.) *Companions to Ancient Thought, Volume* 1*: Epistemology*. Cambridge University Press, 225–250

1996. 'Aristotle's Rationalism', in Frede and Striker (eds.), 157–173

2000. 'Introduction', in Frede and Charles (eds.), 1–52

2004. 'Aristotle's Account of the Origin of Philosophy', *Rhizai* 1: 9–44

Frede, M. and D. Charles (eds.) 2000. *Aristotle's* Metaphysics Book Lambda. Oxford: Clarendon Press

Frede, M. and G. Striker (eds.) 1996. *Rationality in Greek Thought*. Oxford University Press

Freeland, C. 1987. 'Aristotle on Bodies, Matter and Potentiality', in James G. Lennox and Allan Gotthelf (eds.), *Philosophical Issues in Aristotle's Biology*. Cambridge University Press, 392–407

Freudenthal, G. 1995. *Aristotle's Theory of Material Substance, Heat and Pneuma, Form and Soul.* Oxford: Clarendon Press

Galton, A. and R. Mizoguchi 2009. 'The Water Falls but the Waterfall does not Fall: New Perspectives on Objects, Processes and Events', *Applied Ontology* 4.2: 71–107

Gauthier, R. E. and J. Y. Jolif (trans.) (1968) L' Ethique à Nicomaque*: Introduction, traduction et commentaire.* 2 vols. Louvain, Paris: Publications Universitaires

Gill, M. L. 1980. 'Aristotle's Theory of Causal Action in *Physics* III 3', *Phronesis* 25.2: 129–147

1989. *Aristotle on Substance: The Paradox of Unity.* Princeton University Press

2009. 'The Theory of the Elements in *De Caelo* 3 and 4', in Bowen and Wildberg (eds.), 139–162

Gill, M. L. and P. Pellegrin (eds.) 2006. *Blackwell Companions to Philosophy, A Companion to Ancient Philosophy.* Malden, MA: Blackwell Publishing Ltd

Gill, M. L. and J. G. Lennox (eds.) 1994. *Self-Motion: From Aristotle to Newton.* Princeton University Press

Godfrey-Smith, P. 1994. 'A Modern History Theory of Functions', *Nous* 28.3: 344–362

2014. *The Philosophy of Biology.* Princeton University Press

Gotthelf, A. (ed.) 1985. *Aristotle on Nature and Living Things.* Pittsburgh, PA, Bristol (UK): Methesis

2012. *Teleology, First Principles and Scientific Method in Aristotle's Biology.* Oxford University Press

Gotthelf, A. and J. Lennox (eds.) 1987. *Philosophical Issues in Aristotle's Biology.* Cambridge University Press

Graham, Daniel 1999. *Aristotle* Physics *Book* VIII. *Translated with a Commentary.* Oxford: Clarendon Press

Grant, A. (ed.) 1866. *The Ethics of Aristotle.* 2 vols. London: Longmans, Green, and Co.

Guthrie, W. K. C. 1946. *'Notes on Some Passages in the Second Book of Aristotle's* Physics*', Classical Quarterly* 40: 70–76

Guyer, P. (ed. and trans.) and E. Matthews (trans.) 2000. *Kant's Critique of the Power of Judgment (The Cambridge Edition of the Works of Immanuel Kant).* Cambridge University Press, 2000.

Halliwell, S. 2002. *The Aesthetics of Mimesis.* Princeton University Press

Hamelin, O. (trans.) 1931. *Aristote:* Physique II, *Traduction et Commentaire.* Paris: Vrin

Hardie, R. P. and R. K. Gaye 1930. 'Physics', in Ross (ed.), 315–446

Haslanger, S. 2003. 'Persistence through Time', in Loux and Zimmerman (eds.), 315–354

Heinaman, Robert 1994. 'Is Aristotle's Definition of Change Circular?', *Apeiron* 27: 25–37

Hursthouse, Rosalind 2006. 'The Central Doctrine of the Mean', in Kraut (ed.), 96–115

Irwin, T. 1988. *Aristotle's First Principles*. Oxford: Clarendon Press
1990. *Aristotle's First Principles*. Oxford University Press
2006. 'Aquinas, Natural Law, and Aristotelian Eudaimonism', in Kraut (ed.), 323–341
Irwin, T. and G. Fine 1995. *Aristotle: Selections*. Indianapolis, IN: Hackett
Joachim, H. H. 1922. *Aristotle.* On Coming-to-be and Passing-Away*: A Revised Text with Introduction and Commentary*. Oxford: Clarendon Press
Johansen, T. K. 2004. *Plato's Natural Philosophy*. Cambridge University Press
2012a. 'Capacity and Potentiality: Aristotle's *Metaphysics Theta* 6–7 from the Perspective of the *De Anima*', *Topoi* 31.2: 209–220
2012b. *The Powers of Aristotle's Soul.* Oxford University Press
Johnson, M. R. (2005). *Aristotle on Teleology*. Oxford University Press
Johnson, M. R. and D. S. Hutchinson 2005. 'Authenticating Aristotle's *Protrepticus*', *Oxford Studies in Ancient Philosophy* 29: 193–294
Johnson, W. E. 1921. *Logic. Part 1*. Cambridge University Press
Jones, B. 1974. 'Aristotle's Introduction of Matter', *Philosophical Review* 83.4: 474–500
Judson, L. (ed.) 1991a. *Aristotle's* Physics*: A Collection of Essays*. Oxford: Clarendon Press
1991b. 'Chance and "Always or For the Most Part" in Aristotle', in Judson (ed.), 73–100
2005. 'Aristotelian Teleology', *Oxford Studies in Ancient Philosophy* 29: 341–366
Kahn, C. 1985. 'The Place of the Prime Mover in Aristotle's Teleology', in Gotthelf (ed.), 183–205
Katayama, Errol G. 1999. *Aristotle on Artifacts*. Albany: State University of New York Press
2009. 'Necessity and Artifacts in Aristotle', unpublished manuscript
2011. 'Soul and Elemental Motion in Aristotle's *Physics* VIII.4', *Apeiron* 44.2: 163–190
Kelsey, S. 2003. 'Aristotle's Definition of Nature', *Oxford Studies in Ancient Philosophy* 25: 59–87
2008. 'The Place of I 7 in the Argument of *Physics* 1', *Phronesis* 53.2: 180–208
2010. 'Hylomorphism in Aristotle's *Physics*', *Ancient Philosophy* 30: 107–124
Kirwan, C. 1971. *Aristotle's Metaphysics, Books Γ, Δ, and E*. Oxford: Clarendon Press
Kosman, Aryeh 1969. 'Aristotle's Definition of Motion', *Phronesis* 14.1: 41–62
1994. 'Aristotle's Prime Mover', in Gill and Lennox (eds.), 135–154
2013. *The Activity of Being: An Essay on Aristotle's Ontology*. Cambridge, MA: Harvard University Press
Kostman, James 1987. 'Aristotle's Definition of Change', *History of Philosophy Quarterly* 4.1: 3–16
Kraut, R. (ed.) 2006. *The Blackwell Guide to Aristotle's* Nicomachean Ethics. Malden: Blackwell Publishing
Kühner, R. and B. Gerth 1892–98. *Ausführliche Grammatik der griechischen Sprache*. 2 parts, 4 vols. Hanover: Hahnsche

Kullmann, W. and S. Föllinger (eds.) 1997. *Aristotelische Biologie: Intentionen, Methoden, Ergebnisse*. Stuttgart: Steiner Verlag

Lacey, A. R. (trans.) 1997. *Simplicius: On Aristotle's* Physics 2. London, Ithaca, NY: Duckworth and Cornell University Press

Lawrence, G. 2011. 'Acquiring Character: Becoming Grown-Up', in Pakaluk and Pearson (eds.), 233–283

Lennox, J. G. 1982. 'Teleology, Chance and Aristotle's Theory of Spontaneous Generation', *Journal of the History of Philosophy* 20: 219–238

1984. 'Aristotle on Chance', *Archiv für Geschichte der Philosophie* 66: 52–60

2001a. *Aristotle's Philosophy of Biology*. Cambridge University Press

2001b. *Aristotle's* Parts of Animals *Books* I-IV. Oxford: Clarendon Series

2004. 'The Place of Zoology in Aristotle's Natural Philosophy', in R. W. Sharples (ed.), *Philosophy and the Sciences in Antiquity*. London: Ashgate, 58–70

2005. 'Getting a Science Going: Aristotle on Entry Level Kinds', in Wolters (ed.), 87–100

2006. 'Aristotle's Biology and Aristotle's Philosophy', in Gill and Pellegrin (eds.), 292–315

2008. 'As if we were Investigating Snubness: Aristotle on the Prospects for a Single Science of Nature', *Oxford Studies in Ancient Philosophy* 35: 149–186

2009a. '*De Caelo* 2.2 and its Debt to *De incessu animalium*', in Bowen and Wildberg (eds.), 187–214

2009b. 'Aristotle on Mind and the Science of Nature', in Rossetto, Tsianikas, Couvalis, and Palaktsoglou (eds.), 1–18

2010. 'Aristotle's Natural Science: The Many and the One', in James Lesher (ed.), *From Inquiry to Demonstrative Knowledge: New Essays in Aristotle's Posterior Analytics, Apeiron* 43.2–3 (June–September): 1–23

in press. *Seeking and Knowing: Aristotle on Norms of Inquiry*. Cambridge University Press

Lennox, J. G. and R. Bolton (eds.) 2010. *Being, Nature, and Life in Aristotle: Essays in Honor of Allan Gotthelf*. Cambridge University Press

Leunissen, M. 2009. 'Why Stars Have No Feet: Teleological Explanations in Aristotle's Cosmology', in Bowen and Wildberg (eds.), 245–271

2010. *Explanation and Teleology in Aristotle's Science of Nature*. Cambridge University Press

2012. 'Aristotle on Natural Character and its Implications for Moral Development', *Journal of the History of Philosophy* 50.4: 507–530

2013. 'Becoming Good Starts with Nature: Aristotle on the Moral Advantages and the Heritability of Good Natural Character', *Oxford Studies in Ancient Philosophy* 44: 99–127

Lewis, F. 1988. 'Teleology and Material/Efficient Causes in Aristotle', *Pacific Philosophical Quarterly* 69: 54–98

Loux, M. J. and D. W. Zimmerman (eds.) 2003. *The Oxford Handbook of Metaphysics*. Oxford University Press

Mansion, A. 1946. *Introduction à la* Physique *Aristotélicienne*. 2nd edn. Louvain and Paris: L'Institut Supérieur de Philosophie

Marmadoro, Anna 2007. 'The Union of Cause and Effect in Aristotle: *Physics* III 3', *Oxford Studies in Ancient Philosophy* 32: 205–232

Maso, S., C. Natali, and G. Seel (eds.) 2012. 'Reading Aristotle's *Physics* VII.3: "What is Alteration?"' *Proceedings of the International ESAP–HYELE Conference.* Las Vegas: Parmenides Publishing

Matthen, M. 2001. 'Holistic Presuppositions of Aristotle's Cosmology', *Oxford Studies in Ancient Philosophy* 20: 171–199

Meikle, Scott 1994. 'Aristotle on Money', *Phronesis* 39.1: 26–44

Menn, S. in press. '*Physics* II and its Relation to *Physics* I–IV'

Millikan, R. 1989. 'In Defense of Proper Functions', *Philosophy of Science* 56: 288–302

Moravcsik, J. 1994. 'Essences, Powers, and Generic Propositions', in Scaltsas, Charles, and Gill (eds.), 229–344

Morison, Benjamin in press. 'Aristotle, *De Motu Animalium*, Chapters 4 and 5', in Primavesi and Rapp (eds.)

Morrison, D. 1985. 'Separation in Aristotle's *Metaphysics*', *Oxford Studies in Ancient Philosophy* 3: 125–157

Moss, J. 2011. '"Virtue Makes the Goal Right": Virtue and *Phronesis* in Aristotle's *Ethics*', *Phronesis* 56.3: 204–261

2012. *Aristotle on the Apparent Good, Perception, Phantasia, Thought, and Desire.* Oxford University Press

Nagel, E. 1977. 'Teleology Revisited: Goal-Directed Processes in Biology', *Journal of Philosophy* 74: 261–301

2012. *Mind and Cosmos.* Oxford University Press

Natali, C. 2012. 'Commentary on 247b13–248a9', in Maso, Natali, and Seel (eds.), 109–117

Nussbaum, M. 1978. *Aristotle's* De Motu Animalium. Princeton University Press

Oehler, K. 1984. *Der Unbewegte Beweger des Aristoteles.* Frankfurt am Main: V. Klostermann

Pakaluk, M. and G. Pearson (eds.) 2011. *Moral Psychology and Human Action in Aristotle.* Oxford University Press

Pellegrin, P. 2000. *Aristote,* Physique. Paris: Flammarion

Penner, Terrence 1970. 'Verbs and the Identity of Actions – A Philosophical Exercise in the Interpretation of Aristotle', in O. P. Wood and G. Pitcher (eds.), *Ryle: A Collection of Critical Essays.* London: Macmillan, 393–453

Peramatzis, M. 2011. *Priority in Aristotle's* Metaphysics *(Oxford Aristotle Studies).* Oxford University Press

Polansky, R. 2007. *Aristotle's* De Anima. Cambridge University Press

Prantl, K. (trans.) [1854] 1978. Repr. *Aristoteles' Acht Bücher* Physik. Leipzig: Aalen

Primavesi, O. and C. Rapp (eds.) in press. 'Aristotle's *De Motu Animalium*', *Proceedings of the XIX Symposium Aristotelicum.* Oxford University Press

Quarantotto, D. 2005. *Causa Finale, Sostanza, Essenza in Aristotele.* Naples: Bibliopolis

2010. 'Aristotle on the Soul as a Principle of Biological Unity', in Follinger (ed.), 35–53

2014. 'Aristotle's Natural Teleology Seen from Above: A "Cosmogony" of the Means–Goal Relation', in M. Tuominen, S. Heinämaa, and V. Mäkinen (eds.), *Aristotelianism and the Critique of Modernity*. Leiden, Boston: Brill

Reeve, C. D. C. 1992. *Practices of Reason: Aristotle's* Nicomachean Ethics. New York: Oxford University Press

2012. *Action, Contemplation, and Happiness: An Essay on Aristotle*. Cambridge, MA: Harvard University Press

Roark, T. 2012. *Aristotle on Time*. Cambridge University Press

Rorty, A. O. (ed.) 1980. *Essays on Aristotle's* Ethics. Berkeley: California University Press

Ross, W. D. (ed.) 1924. *Aristotle's* Metaphysics: *A Revised Text with Introduction and Commentary*. Oxford: Clarendon Press

(ed.) 1936. *Aristotle's* Physics: *A Revised Text with Introduction and Commentary*. Oxford: Clarendon Press

(ed.) 1949. *Aristotle's* Prior *and* Posterior Analytics: *A Revised Text with Introduction and Commentary*. Oxford: Clarendon Press

1950. *Aristotelis* Physica, *recognovit brevique adnotatione critica instruxit W. D. Ross*. Oxford: Clarendon Press

Rossetto, M., M. Tsianikas, G. Couvalis, and M. Palaktsoglou (eds.) 2009. 'Greek Research in Australia', Proceedings of the Biennial International Conference of Greek Studies, June 2009. Flinders University Department of Languages

Rowe, C. and S. Broadie (trans. and comm.) 2002. *Aristotle:* Nicomachean Ethics. Oxford University Press

Sauvé Meyer, S. 1992. 'Aristotle, Teleology, and Reduction', *The Philosophical Review* 101: 791–825

Scaltsas, T., D. Charles, and M. L. Gill (eds.) 1994. *Unity, Identity, and Explanation in Aristotle's* Metaphysics. Oxford University Press.

Scharle, Margaret 2008. 'Elemental Teleology in Aristotle's *Physics* II.8', *Oxford Studies in Ancient Philosophy* 34: 147–183

Schofield, M. and M. C. Nussbaum (eds.) 1982. *Language and Logos*. Cambridge University Press

Schofield, M., M. F. Burnyeat, and J. Barnes (eds.) 1980. *Doubt and Dogmatism*. Oxford: Clarendon Press

Sedley, D. 1991. 'Is Aristotle's Teleology Anthropocentric?' *Phronesis* 36: 179–197

2000. '*Metaphysics* Λ 10', in Frede and Charles (eds.), 327–350

2007. *Creationism and its Critics in Antiquity*. Berkeley: University of California Press

2010. 'Teleology, Platonic and Aristotelian', in Lennox and Bolton (eds.), 5–29

Shields, C. 2008. 'Substance and Life in Aristotle', in J. Mouracade (ed.), *Aristotle on Life, (Apeiron,* special issue, 41.3), 129–152

2012a. *The Oxford Handbook to Aristotle*. Oxford University Press

2012b. 'Being *qua* Being', in Shields (ed.), *The Oxford Handbook to Aristotle*, 343–371

Solmsen, F. 1960. *Aristotle's System of the Physical World: A Comparison with his Predecessors*. Cornell University Press

Steel, C. (ed.) 2012. *Aristotle's Metaphysics Alpha: Symposium Aristotelicum*. Oxford University Press

Stewart, J. A. 1892. *Notes on the* Nicomachean Ethics *of Aristotle*. 2 vols. Oxford: Clarendon Press

Stout, R. 1996. *Things That Happen Because They Should*. Oxford University Press

Strohl, M. S. 2011. 'Pleasure as Perfection: *Nicomachean Ethics* 10.4–5', *Oxford Studies in Ancient Philosophy* 41: 257–287

Taylor, C. C. W. (trans.) 2006. *Aristotle's* Nicomachean Ethics *Books* II–IV. Oxford University Press

Torstrik, A. 1875. 'Peri Tuchês kai tou Automatou', *Hermes* 9: 425–470

Tracey, T. J. 1969. *Physiological Theory and the Doctrine of the Mean in Plato and Aristotle*. The Hague: Mouton

Ugaglia, M. 2012. *Aristotele:* Fisica, *Libro* III. Rome: Carocci Editore

Viano, C. 2012. 'Commentary on 246b20–247a19', in Maso, Natali, and Seel (eds.), 85–98

Wagner, H. (trans.) 1989. *Aristoteles: Physikvorlesung*. Berlin: Akademie Verlag

Wardy, R. 1990. *The Chain of Change: A Study of Aristotle's* Physics VII. Cambridge University Press

Waterfield, R. 1996. *Aristotle's* Physics *(translation; with introduction and notes by D. Bostock)*. Oxford University Press

Waterlow (Broadie), Sarah 1982. *Nature, Change and Agency in Aristotle's* Physics. Oxford: Clarendon Press

White, M. J. 2009. 'Aristotle on the Infinite, Space, and Time', in G. Anagnostopoulos (ed.), 260–276

Whiting, J. 1986. 'Human Nature and Intellectualism in Aristotle', *Archiv für Geschichte der Philosophie* 68.1: 70–95

2002. 'Locomotive Soul: The Parts of Soul in Aristotle's Scientific Works', *Oxford Studies in Ancient Philosophy* 22: 141–200

Wicksteed, Ph. H. and F. M. Cornford 1929. *Aristotle, The* Physics. 2 vols. *The Loeb Classical Library*. Cambridge, MA: Harvard University Press

Wieland, W. 1970. *Die aristotelische Physik*. Göttingen: Vandenhoeck & Ruprecht

1975. 'The Problem of Teleology', in Barnes, Schofield, and Sorabji (eds.), 141–160

Williams, C. J. F. 1982. *Aristotle's* De Generatione et Corruptione *(translation with notes)*. Oxford: Clarendon Press

Witt, Charlotte 1989. *Substance and Essence in Aristotle*. Ithaca, NY: Cornell University Press

2012. 'Aristotle on Deformed Animal Kinds', *Oxford Studies in Ancient Philosophy* 43 (winter): 83–106

2015. 'As if by Convention Alone: The Unstable Ontology of Aristotle's *Ethics*', in Devin Henry and Karen M. Nielsen, *Bridging the Gap between Aristotle's Science and Ethics*. Cambridge University Press.

Wolters, G. (ed.) 2002. *Homo Sapiens und Homo Faber. Festschrift Mittelstrass.* Berlin: Walter De Gruyter

Wright, L. 1973. 'Functions', *The Philosophical Review* 82.2: 139–68

1976. *Teleological Explanations: An Etiological Analysis of Goals and Functions.* Berkeley, CA: University of California Press

Zeller, E. 1879. *Die Philosophie der Griechen in ihrer geschichtlichen Entwicklung.* Part II, section 2. Leipzig: R. Reisland

Index

For EU product safety concerns, contact us at Calle de José Abascal, 56–1°, 28003 Madrid, Spain or eugpsr@cambridge.org.

www.ingramcontent.com/pod-product-compliance
Lightning Source LLC
LaVergne TN
LVHW011525200926
848484LV00001B/157

* 9 7 8 1 1 0 7 0 3 1 4 6 3 *